T0092968

Entropy and Information Optics

Connecting Information and Time, Second Edition

OPTICAL SCIENCE AND ENGINEERING

Founding Editor
Brian J. Thompson
University of Rochester
Rochester, New York

RECENTLY PUBLISHED

Entropy and Information Optics: Connecting Information and Time, Second Edition,
   Francis T. S. Yu
Laser Beam Shaping Applications, Second Edition, *Fred M. Dickey and Todd E. Lizotte*
Fundamentals and Basic Optical Instruments, *edited by Daniel Malacara Hernández*
Tunable Laser Applications, Third Edition, *edited by F. J. Duarte*
Laser Safety: Tools and Training, Second Edition, *edited by Ken Barat*
Optical Materials and Applications, *edited by Moriaki Wakaki*
Lightwave Engineering, *Yasuo Kokubun*
Handbook of Optical and Laser Scanning, Second Edition, *Gerald F. Marshall*
   and Glenn E. Stutz
Computational Methods for Electromagnetic and Optical Systems, Second Edition,
   John M. Jarem and Partha P. Banerjee
Optical Methods of Measurement: Wholefield Techniques, Second Edition, *Rajpal S. Sirohi*
Optoelectronics: Infrared-Visible-Ultraviolet Devices and Applications, Second Edition,
   edited by Dave Birtalan and William Nunley
Photoacoustic Imaging and Spectroscopy, *edited by Lihong V. Wang*
Polarimetric Radar Imaging: From Basics to Applications, *Jong-Sen Lee and Eric Pottier*
Near-Earth Laser Communications, *edited by Hamid Hemmati*
Slow Light: Science and Applications, *edited by Jacob B. Khurgin and Rodney S. Tucker*
Dynamic Laser Speckle and Applications, *edited by Hector J. Rabal and Roberto A. Braga Jr.*
Biochemical Applications of Nonlinear Optical Spectroscopy, *edited by Vladislav Yakovlev*
Optical and Photonic MEMS Devices: Design, Fabrication and Control, *edited by Ai-Qun Liu*
The Nature of Light: What Is a Photon?, *edited by Chandrasekhar Roychoudhuri,*
   A. F. Kracklauer, and Katherine Creath
Introduction to Nonimaging Optics, *Julio Chaves*
Introduction to Organic Electronic and Optoelectronic Materials and Devices,
   edited by Sam-Shajing Sun and Larry R. Dalton
Fiber Optic Sensors, Second Edition, *edited by Shizhuo Yin, Paul B. Ruffin, and Francis T. S. Yu*
Terahertz Spectroscopy: Principles and Applications, *edited by Susan L. Dexheimer*
Photonic Signal Processing: Techniques and Applications, *Le Nguyen Binh*

*Please visit our website **www.crcpress.com** for a full list of titles*

Entropy and Information Optics

Connecting Information and Time, Second Edition

Francis T. S. Yu

CRC Press
Taylor & Francis Group
Boca Raton London New York

CRC Press is an imprint of the
Taylor & Francis Group, an **informa** business

CRC Press
Taylor & Francis Group
6000 Broken Sound Parkway NW, Suite 300
Boca Raton, FL 33487-2742

First issued in paperback 2019

© 2017 by Taylor & Francis Group, LLC
CRC Press is an imprint of Taylor & Francis Group, an Informa business

No claim to original U.S. Government works

ISBN-13: 978-1-138-55549-5 (hbk)
ISBN-13: 978-0-367-89202-9 (pbk)

This book contains information obtained from authentic and highly regarded sources. Reasonable efforts have been made to publish reliable data and information, but the author and publisher cannot assume responsibility for the validity of all materials or the consequences of their use. The authors and publishers have attempted to trace the copyright holders of all material reproduced in this publication and apologize to copyright holders if permission to publish in this form has not been obtained. If any copyright material has not been acknowledged please write and let us know so we may rectify in any future reprint.

Except as permitted under U.S. Copyright Law, no part of this book may be reprinted, reproduced, transmitted, or utilized in any form by any electronic, mechanical, or other means, now known or hereafter invented, including photocopying, microfilming, and recording, or in any information storage or retrieval system, without written permission from the publishers.

For permission to photocopy or use material electronically from this work, please access www.copyright.com (http://www.copyright.com/) or contact the Copyright Clearance Center, Inc. (CCC), 222 Rosewood Drive, Danvers, MA 01923, 978-750-8400. CCC is a not-for-profit organization that provides licenses and registration for a variety of users. For organizations that have been granted a photocopy license by the CCC, a separate system of payment has been arranged.

Trademark Notice: Product or corporate names may be trademarks or registered trademarks, and are used only for identification and explanation without intent to infringe.

Library of Congress Cataloging-in-Publication Data

Names: Yu, Francis T. S., 1932- author.
Title: Entropy and information optics : connecting information and time / Francis T.S. Yu.
Description: Second edition. | Boca Raton : Taylor & Francis, CRC Press,2017. | Includes bibliographical references and index.
Identifiers: LCCN 2017032627| ISBN 9781138555495 (hardback : alk. paper) | ISBN 9781315149561 (ebook)
Subjects: LCSH: Optical communications. | Entropy (Information theory) | Optics. | Optical data processing.
Classification: LCC TK5103.59 .Y8 2017 | DDC 621.382/7--dc23
LC record available at https://lccn.loc.gov/2017032627

Visit the Taylor & Francis Web site at
http://www.taylorandfrancis.com

and the CRC Press Web site at
http://www.crcpress.com

Contents

From the Series Editor

Fundamentally optical beams and optical systems transmit and analyze information. The information can be analog or digital. It can be three-dimensional, two-dimensional, or one-dimensional. It can be in the traditional form of an image or information that is coded and/or compressed. The light beam carrying the information can be incoherent, coherent, or even partially coherent.

In the early days of this important field, the concepts of communication theory had a major impact on our understanding and our descriptions of optical systems. The initial impetus was to deal with images and image quality. Concepts of impulse responses and transfer functions caused considerable rethinking about the design and evaluation of optical systems. Resolution criteria were only the beginning; fidelity, fidelity defect, relative structural content, and correlation quantity were concepts introduced by E. H. Linfoot in 1964. Formal definitions of entropy and information content were to follow, and the field continues to expand, driven by the explosion of high-speed, high-data-rate, and high-capacity communication systems.

This volume discusses the fundamentals and the applications of entropy and information optics by means of a sampling of topics in this field, including image restoration, wavelet transforms, pattern recognition, computing, and fiber-optic communication.

Brian J. Thompson

Preface

Every substance is attached with a piece of information that includes all the building blocks in our universe; one cannot ignore information when dealing with science.

Time ignited the creation of our universe, but it is the created substances that presented to us the existence of time.

The discovery of the laser in the 1960s prompted the building of new optical communication and information processing systems and equipment. The impact of fiber-optic communication and optical signal processing provided unique evidence of the relationship between light and information science. As we are aware, light not only is the main source of energy that supports life but also is a very important information carrier. It was my first objective to describe the profound relationship between light and information science: my earlier book, *Optics and Information Theory* (Wiley, 1976), had been used as a text by numerous universities and by engineers in the United States and abroad. The book had been revised and incorporated with additional materials with a new title, *Entropy and Information Optics* (Marcel Dekker, 2000). This revised edition had been used in part as a course text in my classes while I was at Pennsylvania State University, before I retired in 2004. Research scientists, as well as my students, have found the book very stimulating and enlightening. And this was one of the major impetuses that prompted me to update the book with new additional materials, connecting information to modern science and creation, as well as to time and the temporal universe.

Nonetheless, the relationship between information and science has provided the basic driving force for research on and the development of high-speed, high-data-rate, and high-capacity communication systems. This trend started more than 70 years ago due to the development of Norbert Wiener's optimum communication and Claude Shannon's information theory in the 1940s. And the trend is still vigorously continued in an even wider application due to the newly developed optical and photonic technology. The reason for this success may be deduced from the imaginative relationship between information and science, as described in this book.

In short, I would like to express my appreciation to my colleagues and friends for their valuable and enthusiastic encouragement; without their support, this new edition would not have been written.

Francis T. S. Yu

Author

Francis T. S. Yu received his BSEE degree from Mapua Institute of Technology, Manila, Philippines, and his MS and PhD degrees in electrical engineering from the University of Michigan, Ann Arbor. He has been a consultant to several industrial and government laboratories. He is an active researcher in the fields of optical signal processing, holography, information optics, optical computing, neural networks, photorefractive optics, fiber sensors, and photonic devices. He has published more than 500 papers, in which more than 300 are refereed. He is a recipient of the 1983 Faculty Scholar Medal for Outstanding Achievement in Physical Sciences and Engineering, the 1984 Outstanding Researcher in the College of Engineering, the 1993 Premier Research Award from the Penn State Engineering Society, and the SPIE 2004 Dennis Garbor Award; is the corecipient of the 1998 Institute of Electrical and Electronics Engineers Donald G. Fink Prize Paper Award; and was named Evan Pugh Professor of Electrical Engineering in 1985 at Penn State, honorary professor of Nankai University in 1995, and honorary professor of the National Chiao-Tung University in Taiwan. He has served as associate editor, editorial board member, and guest editor for various international journals. He is the author and coauthor of nine books. Dr. Yu is a life fellow of the Institute of Electrical and Electronics Engineers and a fellow of The Optical Society, SPIE, and Photonics Society of Chinese-American.

1 Introduction to Information Transmission

In the physical world, light not only is part of the mainstream of energy that supports life, but also provides us with important sources of information. One can easily imagine that without light present, civilization would never have emerged. Furthermore, humans are equipped with a pair of exceptionally good, although not perfect, eyes. With the combination of an intelligent brain and remarkable eyes, humans were able to advance themselves above the rest of the animals in the world. It is undoubtedly true that if humans had not been equipped with eyes, they would not have evolved into their present form. In the presence of light, humans are able to search for the food they need and the art they enjoy, and to explore the unknown. Thus, light, or rather *optics*, has provided us with a very useful source of information whose application can range from very abstract artistic to very sophisticated scientific uses.

The purpose of this text is to discuss the relationship between optics and information transmission. However, it is emphasized that it is not our intention to consider the whole field of optics and information, but rather to center on an area that is important and interesting to our readers.

Prior to going into a detailed discussion of optics and information, we devote this first chapter to the fundamentals of information transmission. However, it is noted that *entropy information* was not originated by optical physicists, but rather by a group of mathematically oriented electrical engineers whose original interest was centered on electrical communication. Nevertheless, from the very beginning of the discovery of entropy information, interest in the application has never totally been absent from the optical standpoint. As a result of the recent advances in modern information optics and optical communication, the relationship between optics and entropy information has grown more rapidly than ever.

Although everyone seems to know the word *information*, a fundamental theoristic concept may not be the case. Let us now define the meaning of information. Actually, information may be defined in relation to several different disciplines. In fact, information may be defined according to its applications, but with identical mathematical formalism, as developed in the next few sections. From the viewpoint of pure mathematics, information theory is basically a *probabilistic concept*. We see in Section 1.1 that without probability, there would be no information theory. But, from a physicist's point of view, information theory is essentially an *entropy theory*. In Chapter 4, we see that without the fundamental relationship between physical entropy and information entropy, information theory would have no useful application in physical science. From a communication engineer's standpoint, information theory can be considered an *uncertainty theory*. For example, the more uncertainty there is about a message we have received, the greater the amount of information the message contained.

Since it is not our intention to define information for all fields of interest, we quickly summarize: the beauty and greatness of entropy of information is its application to all fields of science. Application can range from the very abstract (e.g., economy, music, biology, and psychology) to very sophisticated hard-core scientific research. However, in our present introductory version, we consider the concept of information from a practical communication standpoint. For example, from the information theory viewpoint, a perfect liar is as good an informant as a perfectly honest person, provided of course that we have the *a priori* knowledge that the person is a perfect liar or perfectly honest. One should be cautious not to conclude that if one cannot be an honest person, one should be a liar. For, as we may all agree, the most successful crook is the one that does not look like one. Thus, we see that information theory is a guessing game, and is in fact a *game theory*.

In general, an information transmission system can be represented by a block diagram, as shown in Fig. 1.1. For example, in simple optical communication, we have a message (an information source) shown by means of written characters, for example, Chinese, English, French, or German. Then we select suitable written characters (a code) appropriate to our communication. After the characters are selected and written on a piece of paper, the information still cannot be transmitted until the paper is illuminated by visible light (the transmitter), which obviously acts as an information carrier. When light reflected from the written characters arrives at your eyes (the receiver), a proper decoding (translating) process takes place, that is, character recognition (decoding) by the user (your mind). Thus, from this simple example, we can see that a suitable encoding process may not be adequate unless a suitable decoding process also takes place. For instance, if I show you a Chinese newspaper, you might not be able to decode the language, even if the optical channel is assumed to be perfect (i.e., noiseless). This is because a suitable decoding process requires *a priori* knowledge of the encoding scheme (i.e., appropriate information storage), for example, *a priori* knowledge of the Chinese characters. Thus, the decoding process can also be called a *recognition process*.

Information theory is a broad subject that cannot be fully discussed in a few sections. Although we only investigate the theory in an introductory manner, our discussion in the next few sections provides a very useful application of entropy information to optics. Readers who are interested in a rigorous treatment of information theory are referred to the classic papers by Shannon [1–3] and the text by Fano [4].

Information theory has two general orientations: one developed by Wiener [5,6], and the other by Shannon [1–3]. Although both Wiener and Shannon share a common probabilistic basis, there is a basic distinction between them.

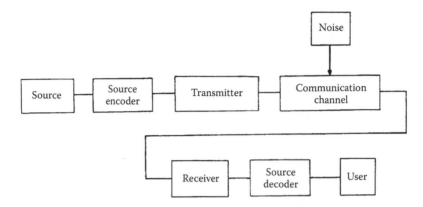

FIG. 1.1 Block diagram of a communication system.

The significance of Wiener's work is that if a signal (information) is corrupted by some physical means (e.g., noise or nonlinear distortion), it may be possible to recover the signal from the corrupted one. It is for this purpose that Wiener develops the theories of correlation detection, optimum prediction, matched filtering, and so on. However, Shannon's work is carried a step further. He shows that the signal can be optimally transferred provided that it is properly encoded. That is, the signal to be transferred can be processed before and after transmission. In the encoding process, he shows that it is possible to combat the disturbances in the communication channel to a certain extent. Then, by a proper decoding process, the signal can be recovered optimally. To do this, Shannon develops the theories of information measure, channel capacity, coding processes, and so on. The major interest in Shannon's theory is efficient utilization of the communication channel.

A fundamental theorem proposed by Shannon can be considered the most surprising result of his work. The theorem can be stated approximately as follows. Given a stationary finite memory information channel having a channel capacity C, if the binary information transmission rate R of the message is smaller than C, there exists channel encoding and decoding processes for which the probability of error in information transmission per digit can be made arbitrarily small. Conversely, if the information transmission rate R is larger than C, there exists no encoding and decoding processes with this property; that is, the probability of error in information transmission cannot be made arbitrarily small. In other words, the presence of random disturbances in a communication channel does not, by itself, limit transmission accuracy. Rather, it limits the transmission rate for which arbitrarily high transmission accuracy can be accomplished.

In summarizing this brief introduction to information transmission, we point out again the distinction between the viewpoints of Wiener and Shannon. Wiener assumes, in effect, that the signal in question can be processed after it has been corrupted by noise. Shannon suggests that the signal can be processed both before and after its transmission through the communication channel. However, the main objectives of these two branches of information transmission

are basically the same, namely, faithful reproduction of the original signal.

1.1 INFORMATION MEASURE

We have in the preceding discussed a general concept of information transmission. In this section, we discuss this subject in more detail. Our first objective is to define a measure of information, which is vitally important in the development of modern information theory. We first consider discrete input and discrete output message ensembles as applied to a communication channel, as shown in Fig. 1.2. We denote the sets of input and output ensembles $A = \{a_i\}$ and $B = \{b_j\}$, respectively, $i = 1, 2, \ldots, M$, and $j = 1, 2, \ldots, N$. It is noted that AB forms a *discrete product space*.

Let us assume that a_i is an input event to the information channel, and b_j is the corresponding output event. Now we would like to define a measure of information in which the received event b_j specifies a_i. In other words, we would like to define a measure of the amount of information provided by the output event b_j *about* the corresponding input event a_i. We see that the transmission of a_i through the communication channel causes a change in the probability of a_i, from an *a priori* $P(a_i)$ to an *a posteriori* $P(a_i/b_j)$. In measuring this change, we take the logarithmic ratio of these probabilities. It turns out to be appropriate for the definition of information measure. Thus, the amount of information provided by the output event b_j about the input event a_i can be defined as

$$I(a_i; b_j) \triangleq \log_2 \frac{P(a_i/b_j)}{P(a_i)} \quad \text{bits} \qquad (1.1)$$

It is noted that the base of the logarithm can be a value other than 2. However, the base 2 is the most commonly used in

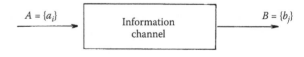

FIG. 1.2 Input–output communication channel.

information theory. Therefore, we adopt this base value of 2 for use in this text. Other base values are also frequently used, for example, $\log_{10}$ and $\ln = \log_e$. The corresponding units of information measure of these different bases are hartleys and nats. The *hartley* is named for R. V. Hartley, who first suggested the use of a logarithmic measure of information [7], and *nat* is an abbreviation for *natural unit*. Bit, used in Eq. (1.1), is a contraction of *binary unit*.

We see that Eq. (1.1) possesses a symmetric property with respect to input event a_i and output event b_j:

$$I(a_i; b_j) = I(b_j; a_i) \tag{1.2}$$

This symmetric property of information measure can be easily shown:

$$\log_2 \frac{P(a_i, b_j)}{P(b_j)P(a_i)} = \log_2 \frac{P(b_j/a_i)}{P(b_j)}$$

According to Eq. (1.2), the amount of information provided by event b_j about event a_i is the same as that provided by a_i about b_j. Thus, Eq. (1.1) is a measure defined by Shannon as *mutual information* or the amount of information transferred between event a_i and event b_j.

It is clear that if the input and output events are *statistically independent*, that is, if $P(a_i, b_j) = P(a_i)P(b_j)$, then $I(a_i; b_j) = 0$.

Furthermore, if $I(a_i; b_j) > 0$, then $P(a_i, b_j) > P(a_i)P(b_j)$; that is, there is a higher joint probability of a_i and b_j. However, if $I(a_i; b_j) < 0$, then $P(a_i, b_j) < P(a_i)P(b_j)$; that is, there is a lower joint probability of a_i and b_j.

As a result of the conditional probabilities $P(a_i/b_j) \leq 1$ and $P(b_j/a_i) \leq 1$, we see that

$$I(a_i; b_j) \leq I(a_i) \tag{1.3}$$

and

$$I(a_i; b_j) \leq I(b_i) \tag{1.4}$$

where

$$I(a_i) \triangleq -\log_2 P(a_i) \tag{1.5}$$

$$I(b_j) \triangleq -\log_2 P(b_j) \tag{1.6}$$

$I(a_i)$ and $I(b_j)$ are defined as the respective *input* and *output self-information* of event a_i and event b_j. In other words, $I(a_i)$ and $I(b_j)$ represent the amount of information provided at the input and output of the information channel of event a_i and event b_j, respectively. It follows that the mutual information of event a_i and event b_j is equal to the self-information of event a_i if and only if $P(a_i/b_j) = 1$; that is,

$$I(a_i; b_j) = I(a_i) \tag{1.7}$$

It is noted that if Eq. (1.7) is true for all i, that is, the input ensemble, then the communication channel is *noiseless*. However, if $P(b_j/a_i) = 1$, then

$$I(a_i; b_j) = I(b_j) \tag{1.8}$$

If Eq. (1.8) is true for all the output ensemble, then the information channel is *deterministic*.

It is emphasized that the definition of measure of information can be extended to higher product spaces. For example, we can define the mutual information for a product ensemble ABC:

$$I(a_i; b_j/c_k) \triangleq \log_2 \frac{P(a_i/b_j c_k)}{P(a_i/c_k)} \tag{1.9}$$

Similarly, one can show that

$$I(a_i; b_j/c_k) = I(b_j; a_i/c_k) \tag{1.10}$$

$$I(a_i; b_j/c_k) \leq I(a_i/c_k) \tag{1.11}$$

and

$$I(a_i; b_j/c_k) \leq I(b_j/c_k) \tag{1.12}$$

where

$$I(a_i/c_k) \triangleq -\log_2 P(a_i/c_k) \tag{1.13}$$

and

$$I(b_j/c_k) \triangleq -\log_2 P(b_j/c_k) \tag{1.14}$$

represent the *conditional self-information*.

Furthermore, from Eq. (1.1) we see that

$$I(a_i; b_j) = I(a_i) - I(a_i/b_j) \tag{1.15}$$

and

$$I(a_i; b_j) = I(b_j) - I(b_j/a_i) \tag{1.16}$$

From the definition of

$$I(a_i b_j) \triangleq -\log_2 P(a_i, b_j) \tag{1.17}$$

the self-information of the point (a_i, b_j) of the product ensemble AB, one can show that

$$I(a_i; b_j) = I(a_i) + I(b_j) - I(a_i b_j) \tag{1.18}$$

Conversely,

$$I(a_i b_j) = I(a_i) + I(b_j) - I(a_i; b_j) \qquad (1.19)$$

In concluding this section, we point out that for the mutual information $I(a_i; b_j)$ (i.e., the amount of information transferred through the channel), there exists an upper bound, $I(a_i)$ or $I(b_j)$, whichever comes first. If the information channel is noiseless, then the mutual information $I(a_i; b_j)$ is equal to $I(a_i)$, the input self-information of a_i. However, if the information channel is deterministic, then the mutual information is equal to $I(b_j)$, the output self-information of b_j. Moreover, if the input–output of the information channel is statistically independent, then no information can be transferred. It is also noted that when the joint probability $P(a_i; b_j) < P(a_i)P(b_j)$, then $I(a_i; b_j)$ is negative; that is, the information provided by event b_j about event a_i further deteriorates, as compared with the statistically independent case. Finally, it is clear that the definition of the measure of information can also be applied to a higher product ensemble; namely, $ABC \dots$ produce space.

1.2   ENTROPY INFORMATION

In Section 1.1, we defined a measure of information. We saw that information theory is indeed a branch of probability theory.

In this section, we consider the measure of information as a random variable, that is, information measure as a random event. Thus, the measure of information can be described by a probability distribution $P(I)$, where I is the self-, conditional, or mutual information.

Since the measure of information is usually characterized by an ensemble average, the average amount of information provided can be obtained by the ensemble average

$$E[I] = \sum_I IP(I) \qquad (1.20)$$

where E denotes the ensemble average, and the summation is over all I.

If the self-information a_i in Eq. (1.5) is used in Eq. (1.20), then the average amount of self-information provided by the input ensemble A is

$$E[I(a_i)] = \sum_I IP(I) = \sum_{i=1}^{M} P(a_i)I(a_i) \qquad (1.21)$$

where $I(a_i) = -\log_2 P(a_i)$.

For convenience in notation, we drop the subscript i; thus, Eq. (1.21) can be written as

$$I(A) \triangleq -\sum_A P(a) \log_2 P(a) \triangleq H(A) \qquad (1.22)$$

where the summation is over the input ensemble A.

Similarly, the average amount of self-information provided at the output end of the information channel can be written as

$$I(B) \triangleq -\sum_B P(b) \log_2 P(b) \triangleq H(B) \qquad (1.23)$$

In fact, Eqs. (1.22) and (1.23) are the starting points of Shannon's [1–3] information theory. These two equations are in essentially the same form as the *entropy equation* in statistical thermodynamics. Because of the identical form of the entropy expression, $H(A)$ and $H(B)$ are frequently used to describe *information entropy*. Moreover, we see in the next few chapters that Eqs. (1.22) and (1.23) are not just mathematically similar to the entropy equation, but they represent a profound relationship between science and information theory [8–10], as well as between optics and information theory [11,12].

It is noted that entropy H, from the communication theory point of view, is mainly a measure of *uncertainty*. However, from the statistical thermodynamic point of view, entropy H is a measure of *disorder*.

In addition, from Eqs. (1.22) and (1.23), we see that

$$H(A) \geq 0 \qquad (1.24)$$

where $P(a)$ is always a positive quantity. The equality of Eq. (1.24) holds if $P(a) = 1$ or $P(a) = 0$. Thus, we can conclude that

$$H(A) \leq \log_2 M \qquad (1.25)$$

where M is the number of different events in the set of input events A, that is, $A = \{a_i\}$, $i = 1, 2, \dots, M$. We see that the equality of Eq. (1.25) holds if and only if $P(a) = 1/M$, that is, if there is *equiprobability* of all the input events.

In order to prove the inequality of Eq. (1.25), we use the well-known inequality

$$\ln u \leq u - 1 \qquad (1.26)$$

Let us now consider that

$$H(A) - \log_2 M = -\sum_A p(a) \log_2 p(a) - \sum_A p(a) \log_2 M$$
$$= \sum_A p(a) \log_2 \frac{1}{Mp(a)}$$

$$(1.27)$$

By the use of Eq. (1.26), one can show that

$$H(A) - \log_2 M \leq \sum_A \left[\frac{1}{M} - p(a)\right] \log_2 e = 0 \qquad (1.28)$$

Thus, we have proved that the equality of Eq. (1.25) holds if and only if the input ensemble is equiprobable, $p(a) = 1/M$.

We see that the entropy $H(A)$ is maximum when the probability distribution a is equiprobable. Under the maximization condition of $H(A)$, the amount of information provided is the *information capacity* of A.

To show the behavior of $H(A)$, we describe a simple example for the case of $M = 2$, that is, for a *binary source*. Then the entropy equation (1.22) can be written as

$$H(p) = -p \log_2 p - (1-p) \log_2 (1-p) \qquad (1.29)$$

where p is the probability of one of the events.

From Eq. (1.29), we see that $H(p)$ is maximum if and only if $p = \dfrac{1}{2}$. Moreover, the variation in entropy as a function of p is plotted in Fig. 1.3, in which we see that $H(p)$ is a symmetric function, having a maximum value of 1 bit at $p = \dfrac{1}{2}$.

Similarly, one can extend this concept of ensemble average to the conditional self-information:

$$I(B/A) \triangleq -\sum_B \sum_A p(a,b) \log_2 p(b/a) \triangleq H(B/A) \quad (1.30)$$

We define $H(B/A)$ as the conditional entropy of B given A. Thus, the entropy of the product ensemble AB can also be written as

$$H(AB) = -\sum_A \sum_b p(a,b) \log_2 p(a,b) \qquad (1.31)$$

where $p(a, b)$ is the joint probability of events a and b.

From the entropy equations (1.22) and (1.30), we have the relation

$$H(AB) = H(A) + H(B/A) \qquad (1.32)$$

FIG. 1.3 Variation of $H(p)$ as a function of p.

Similarly, we have

$$H(AB) = H(B) + H(A/B) \qquad (1.33)$$

where

$$H(A/B) = -\sum_A \sum_B p(a,b) \log_2 p(a/b) \qquad (1.34)$$

From the relationships of Eq. (1.26), we also show that

$$H(B/A) \le H(B) \qquad (1.35)$$

and

$$H(A/B) \le H(A) \qquad (1.36)$$

where the equalities hold if and only if a and b are statistically independent.

Furthermore, Eqs. (1.35) and (1.36) can be extended to a higher product ensemble space. For example, with a triple product space ABC, we have the conditional entropy relation

$$H(C/AB) \le H(C/B) \qquad (1.37)$$

in which the equality holds if and only if c is statistically independent of a for any given b, that is, if $p(c/ab) = p(c/b)$.

It is noted that extension of the conditional entropy relationship to a higher product ensemble, for example, source encoding, is of considerable importance. Since the conditional entropy is the average amount of information provided by the successive events, it cannot be increased by making the successive events dependent on the preceding ones. Thus, we see that the information capacity of an encoding alphabet cannot be made maximum if the successive events are interdependent. Therefore, the entropy of a message ensemble places a lower limit on the average number of coding digits per code word:

$$\bar{n} \ge \frac{H(A)}{\log_2 D} \qquad (1.38)$$

where $\bar{n}$ is the average number of coded digits and D is the number of the *coding alphabet*; for example, for binary coding, the number of the coding alphabet is 2. It is emphasized that the lower limit of Eq. (1.38) can be approached as closely as we desire for encoding sufficiently long sequences of independent messages. However, long sequences of messages also involve a more complex coding procedure.

We now turn our attention to defining the *average mutual information*. We consider first the conditional average mutual information:

$$I(A;b) \triangleq \sum_A p(a/b) I(a;b) \qquad (1.39)$$

where

$$I(a;b) = \log_2 \frac{p(a/b)}{p(a)}$$

Although the mutual information of an event a and an event b can be negative, $I(a; b) < 0$, the average conditional mutual information can never be negative:

$$I(A;b) \geq 0 \tag{1.40}$$

with the equality holding if and only if events A are statistically independent of b, that is, $p(a/b) = p(a)$, for all a.

By taking the ensemble average of Eq. (1.39), the average mutual information can be defined

$$I(A;B) \triangleq \sum_B p(b)I(A;b) \tag{1.41}$$

Equation (1.41) can be written as

$$\begin{aligned} I(A;B) &\triangleq \sum_B \sum_A p(a,b) \log_2 \frac{p(a/b)}{p(a)} \\ &= \sum_A \sum_B p(a,b)I(a,b) \end{aligned} \tag{1.42}$$

Again, one can show that

$$I(A;B) \geq 0 \tag{1.43}$$

The equality holds for Eq. (1.43) if and only if a and b are statistically independent. Moreover, from the symmetric property of $I(a, b)$ (Eq. (1.2)), it can be easily shown that

$$I(A;B) = I(B;A) \tag{1.44}$$

where

$$I(B;A) \triangleq \sum_A \sum_B p(a,b) \log_2 \frac{p(b/a)}{p(b)} \tag{1.45}$$

Furthermore, from Eqs. (1.3) and (1.4), one can show that

$$I(A;B) \leq H(A) = I(A) \tag{1.46}$$

and

$$I(A;B) \leq H(B) = I(B) \tag{1.47}$$

This says that the mutual information (the amount of information transfer) cannot be greater than the entropy (the amount of information provided) at the input or the output ends of the

information channel, whichever comes first. We see that if the equality holds for Eq. (1.46), then the channel is noiseless; however, if the equality holds for Eq. (1.47), then the channel is deterministic.

From the entropy equation (1.31), we can show that

$$H(AB) = H(A) + H(B) - I(A;B) \tag{1.48}$$

From the relationship of Eq. (1.48) and the conditional entropy of Eqs. (1.32) and (1.33), we have

$$I(A;B) = H(A) - H(A/B) \tag{1.49}$$

and

$$I(A;B) = H(B) - H(B/A) \tag{1.50}$$

Equations (1.49) and (1.50) are of interest to us in determination of the mutual information (the amount of information transfer). For example, if $H(A)$ is considered the average amount of information provided at the input end of the channel, then $H(A/B)$ is the average amount of *information loss* (e.g., due to noise) in the channel. It is noted that the conditional entropy $H(A/B)$ is usually regarded as the *equivocation* of the channel. However, if $H(B)$ can be considered the average amount of information received at the output end of the channel, then $H(B/A)$ is the average amount of information needed to specify the noise disturbance in the channel. Thus, $H(B/A)$ may be referred to as the *noise entropy* of the channel. Since the concept of mutual information can be extended to a higher product ensemble, we can show that [13]

$$I(A;BC) = I(A;B) + I(A;C/B) \tag{1.51}$$

and

$$I(BC;A) = I(B;A) + I(C;A/B) \tag{1.52}$$

By the symmetric property of the mutual information, we define *triple mutual information*:

$$\begin{aligned} I(a;b;c) &\triangleq I(a;b) - I(a;b/c) = I(a;c) - I(a;c/b) \\ &= I(b;c) - I(b;c/a) \end{aligned} \tag{1.53}$$

Thus, we have

$$\begin{aligned} I(A;B;C) &\triangleq \sum_A \sum_B \sum_C p(a,b,c) I(a;b;c) \\ &= I(A;B) - I(A;B/C) \\ &= I(A;C) - I(A;C/B) \\ &= I(B;C) - I(B;C/A) \end{aligned} \tag{1.54}$$

In view of Eq. (1.54), it is noted that $I(A; B; C)$ can be positive or negative in value, in contrast to $I(A; B)$, which is never negative.

Furthermore, the concept of mutual information can be extended to an A^n product ensemble:

$$I(a_1; a_2; \ldots; a_n) \triangleq \log_2 \frac{[\pi p(a_i, a_j)][\pi p(a_i, a_j, a_k, a_l)] \cdots}{[\pi p(a_i)][\pi p(a_i, a_j, a_k)] \cdots} \quad (1.55)$$

where π denotes the products over all possible combinations. Furthermore, Eq. (1.55) can be written as

$$I(a_1; a_2; \ldots; a_n) = I(a_1; a_2; \ldots; a_{n-1}) - I(a_1; a_2; \ldots; a_{n-1}/a_n) \quad (1.56)$$

The average mutual information is therefore

$$\begin{aligned} I(A_1; A_2; \ldots; A_n) &= \sum_{A^n} p(a_1, a_2, \ldots, a_n) I(a_1; a_2; \ldots; a_n) \\ &= \sum H(A_i) - \sum H(A_i A_j) + \sum H(A_i A_j A_k) \\ &\quad - \cdots (-1)^{n-1} H(A_1 A_2 \cdots A_n) \end{aligned} \quad (1.57)$$

where the summations are evaluated over all possible combinations.

In concluding this section, we remark that generalized mutual information may have interesting applications for communication channels with multiple inputs and outputs. We see in the next few sections that the definition of mutual information $I(A; B)$ eventually leads to a definition of *information channel capacity*. Finally, the information measures we have defined can be easily extended from a discrete space to a continuous space:

$$H(A) \triangleq - \int_{-\infty}^{\infty} p(a) \log_2 p(a) \, da \quad (1.58)$$

$$H(B) \triangleq - \int_{-\infty}^{\infty} p(b) \log_2 p(b) \, db \quad (1.59)$$

$$H(B/A) \triangleq - \int_{-\infty}^{\infty} \int_{-\infty}^{\infty} p(a, b) \log_2 p(b/a) \, da \, db \quad (1.60)$$

$$H(A/B) \triangleq - \int_{-\infty}^{\infty} \int_{-\infty}^{\infty} p(a, b) \log_2 p(a/b) \, da \, db \quad (1.61)$$

and

$$H(AB) \triangleq - \int_{-\infty}^{\infty} \int_{-\infty}^{\infty} p(a, b) \log_2 p(a/b) \, da \, db \quad (1.62)$$

where the p's are the probability density distributions.

1.3 COMMUNICATION CHANNELS

In the preceding sections, we discussed the measure of information and we noted that the logarithmic measure of information was the basic starting point used by Shannon in the development of information theory. We pointed out that the main objective of the Shannon information theory is efficient utilization of a communication channel. Therefore, in this section we turn our attention to the problem of transmission of information through a prescribed communication channel with certain noise disturbances.

As noted in regard to Fig. 1.2, a communication channel can be represented by an input–output block diagram. Each of the input events a can be transformed into a corresponding output event b. This transformation of an input event to an output event may be described by a transitional (conditional) probability $p(b/a)$. Thus, we see that the input–output ensemble description of the transitional probability distribution $P(B/A)$ characterizes the channel behavior. In short, the conditional probability $P(B/A)$ describes the random noise disturbances in the channel.

Communication channels are usually described according to the type of input–output ensemble and are considered *discrete* or *continuous*. If both the input and output of the channel are discrete events (discrete spaces), then the channel is called a discrete channel. But if both the input and output of the channel are represented by continuous events (continuous spaces), then the channel is called a continuous channel. However, a channel can have a discrete input and a continuous output, or vice versa. Then, accordingly, the channel is called a discrete-continuous or continuous-discrete channel.

The terminology of the concept of discrete and continuous communication channels can also be extended to spatial and temporal domains. This concept is of particular importance in an optical spatial channel, which is discussed in Chapter 3. An input–output optical channel can be described by input and output spatial domains, which can also be functions of time.

As noted in Section 1.2, a communication channel can have multiple inputs and multiple outputs. Thus, if the channel possesses only a single input terminal and a single output terminal, it is a *one-way channel*. However, if the channel possesses two input terminals and two output terminals, it is a *two-way channel*. In addition, one can have a channel with n input and m output terminals.

Since a communication channel is characterized by the input–output transitional probability distribution $P(B/A)$, if the transitional probability distribution remains the same for all successive input and output events, then the channel is a *memoryless channel*. However, if the transitional probability distribution changes for the preceding events, whether in the input or the output, then the channel is a *memory channel*. Thus, if the memory is finite, that is, if the transitional probability depends on a finite number of preceding events, then the channel is a *finite memory* channel. Furthermore, if the transitional probability distribution depends on stochastic processes and the stochastic processes are assumed to be nonstationary, then the channel is a *nonstationary channel*. Similarly, if the stochastic processes the transitional probability depends on are *stationary*, then the channel is a *stationary channel*. In short, a communication channel can be fully described by the characteristics of its transitional probability distribution, for example, a discrete nonstationary memory channel.

Since a detailed discussion of various communication channels is beyond the scope of this book, we evaluate two of the simplest, yet important channels, namely, memoryless discrete channel and continuous channel.

1.4 MEMORYLESS DISCRETE CHANNELS

One of the simplest communication channels is the memoryless one-way discrete channel. Again, we denote the input ensemble by A and the output ensemble by B. To characterize the behavior of the channel, we give the corresponding transitional probability distribution $P(b/a)$. We denote any one of the ith input events of A by α_i, and the corresponding output event of B by β_i. Let the input to the channel be a sequence of n arbitrary events of A:

$$\alpha^n = \alpha_1 \alpha_2 \cdots \cdots \alpha_n \qquad (1.63)$$

and the corresponding output sequence be

$$\beta^n = \beta_1 \beta_2 \cdots \cdots \beta_n \qquad (1.64)$$

where α_i and β_j are any one of the input and output events of A and B, respectively.

Since the transitional probabilities for a memoryless channel do not depend on the preceding events, the composite transitional probability is

$$P(\beta^n / \alpha^n) = P(\beta_1 / \alpha_1) P(\beta_2 / \alpha_2) \cdots P(\beta_n / \alpha_n) \qquad (1.65)$$

The joint probability of the output sequence β^n is

$$P(\beta^n) = \sum_{A^n} P(\alpha^n) P(\beta^n / \alpha^n) \qquad (1.66)$$

where the summation is over the A^n product space.

From the preceding sections, the average mutual information between the input and output sequences of a^n and β^n can be written as

$$I(A^n; B^n) = H(B^n) - H(B^n / A^n) \qquad (1.67)$$

where B^n is the output product space.

We also see that from Eq. (1.32), the entropy of B^n can be written as

$$
\begin{aligned}
H(B^n) &= -\sum_{B^n} P(B^n) \log_2 P(B^n) \\
&= H(B_1) + H(B_2/B_1) + H(B_3/B_2 B_1) + \cdots \qquad (1.68) \\
&\quad + H(B_n/B_{n-1} \cdots B_1)
\end{aligned}
$$

where

$$H(B_i / B_{i-1} \ldots B_1) \triangleq -\sum_{B^i} P(\beta_i) \log_2 P(\beta_i / \beta_{i-1} \cdots \beta_1) \qquad (1.69)$$

The conditional entropy of B^n given A^n can be written as

$$H(B^n / A^n) = -\sum_{A^n} \sum_{B^n} P(\alpha^n) P(\beta^n / \alpha^n) \log_2 P(\beta^n / \alpha^n) \qquad (1.70)$$

From Equation (1.70), it can be shown that

$$H(B^n / A^n) = \sum_{i=1}^{n} H(B_i / A_i) \qquad (1.71)$$

where

$$H(B_i / A_i) \triangleq -\sum_{A_i} \sum_{B_i} P(\alpha_i) P(\beta_i / \alpha_i) \log_2 P(\beta_i / \alpha_i) \qquad (1.72)$$

$H(B_i/A_i)$ is the conditional entropy of the ith output event β_i given the input event α_i.

By substitution of Eqs. (1.68) and (1.71) into Eq. (1.67), we have

$$I(A^n; B^n) = \sum_{i=1}^{n} [H(B_i / B_{i-1} \cdots B_1) - H(B_i / A_i)] \qquad (1.73)$$

From the definition of the average mutual information in the preceding section, we see that $I(A^n; B^n)$ measures the amount of information, on average, provided by the n output events about the given n input events. Therefore, $I(A^n; B^n)/n$ is the amount of mutual information, on average, per event. Moreover, the channel is assumed to be memoryless; that is, $P(b/a)$ are independent of the preceding event. Thus, $I(A^n; B^n)/n$ is a function of $P(\alpha^n)$ and n. Therefore, the capacity of the channel is the maximum value of $I(A^n; B^n)/n$ for a possible probability distribution of the input sequences $P(\alpha^n)$ and length n; that is, the capacity of the channel can be defined as

$$C \triangleq \max_{P(\alpha_n), n} \frac{I(A^n; B^n)}{n} \quad \text{bits/event} \qquad (1.74)$$

It is also noted that if the input events are statistically independent (i.e., from a memoryless information source), then the channel capacity of Eq. (1.74) can be written as

$$C \triangleq \max_{P(a)} I(A; B) \quad \text{bits/event} \qquad (1.75)$$

That is, the channel capacity is the maximization of $I(A;B)$ over the input probability distribution of $P(a)$.

It is emphasized that evaluation of the channel capacity of Eq. (1.74) or (1.75) is by no means simple; it can be quite involved. However, we illustrate a few examples in which the maximization is particularly simple. In this case, we restrict ourselves to *discrete uniform channels*. It is generally convenient to characterize a channel by means of a *transition probability matrix*:

$$[P] = \begin{bmatrix} P(b_1/a_1) & P(b_2/a_1) & \cdots & P(b_m/a_1) \\ P(b_1/a_2) & P(b_2/a_2) & \cdots & P(b_m/a_2) \\ \cdot & \cdot & \cdots & \cdot \\ \cdot & \cdot & \cdots & \cdot \\ \cdot & \cdot & \cdots & \cdot \\ P(b_1/a_n) & P(b_2/a_n) & \cdots & P(b_m/a_n) \end{bmatrix} \quad (1.76)$$

With this channel matrix, we can now define a uniform channel. If the rows of the transition probability matrix are permutations of identical sets of probabilities, for example, P_1, P_2, ..., P_m, then the channel is said to be *uniform from input*. However, if the columns of the transition probability matrix are permutations of the same set of probabilities, then the channel is said to be *uniform from output*.

If a channel is uniform from input, then the conditional entropy $H(B/a_i)$ can be shown to be

$$H(B/a_i) = -\sum_B P(b/a_i) \log_2 P(b/a_i)$$
$$= -\sum_{j=1}^{m} P_j \log_2 P_j = H(B/a) \quad (1.77)$$

which is identical for all a_i of A. The significant interpretation of Eq. (1.77) is that if the channel is uniform from input, then the transmission of any input event a of A is disturbed in the same manner by the noise in the channel. It is also noted that if the channel is uniform from output, then a given equiprobable (i.e., uniform in probability) input ensemble of a, $[P(a) = 1/n]$, will give rise to an *equiprobable* output ensemble $[P(b) = 1/m]$.

Now if a channel is uniform from both input and output, then it is said to be a *doubly uniform channel* or simply a *uniform channel*. In the following, we evaluate the capacity for a special type of uniform channel, namely, an *n-ary symmetric channel*.

Let the transition probability matrix of an *n*-ary symmetric channel be

$$[P] = \begin{bmatrix} 1-p & \dfrac{p}{n-1} & \dfrac{p}{n-1} & \cdots & \dfrac{p}{n-1} \\ \dfrac{p}{n-1} & 1-p & \dfrac{p}{n-1} & \cdots & \dfrac{p}{n-1} \\ \vdots & \vdots & \vdots & \cdots & \vdots \\ \dfrac{p}{n-1} & \dfrac{p}{n-1} & \dfrac{p}{n-1} & \cdots & 1-p \end{bmatrix} \quad (1.78)$$

To evaluate the channel capacity, we first evaluate the average mutual information $I(A; B)$. To do so, we calculate the conditional entropy $H(B/A)$. But from Eq. (1.77), we see that for a channel uniform from input, the ensemble over B is independent of every a:

$$H(B/A) = -\sum_A P(a) \sum_B P(b/a) \log_2 P(b/a)$$
$$= -\sum_B P(b/a) \log_2 P(b/a) = H(B/a) \quad (1.79)$$

Now we seek to maximize the average mutual information $I(A; B)$. Since $I(A; B) = H(B) - H(B/A)$, from Eq. (1.75) we see that $I(A; B)$ is maximum when $H(B)$ is maximum, and that the maximum value of $H(B)$ occurs only when every output event of b is equiprobable. However, it is noted that in general it is not true that there exists an input probability distribution of $P(a)$ such that every output event is equiprobable. But it is true for a doubly uniform channel that equiprobability of the input events produces equiprobability of the output events. Therefore, the capacity of an *n*-ary uniform channel is

$$C = \log_2 n + \sum_B P(b/a) \log_2 P(b/a) \quad (1.80)$$

By substituting transitional probabilities of Eq. (1.78) into Eq. (1.80), we have

$$C = \log_2 n + (1-p) \log_2(1-p) + p \log_2 \frac{p}{n-1} \quad (1.81)$$

which can also be written as

$$C = \log_2 n - p \log_2(n-1) - H(p) \quad (1.82)$$

where

$$H(p) \triangleq -\left[p \log_2 p + (1-p) \log_2(1-p) \right] \quad (1.83)$$

It is interesting to note that if $n = 2$, then the *n*-ary channel matrix of Eq. (1.78) reduces to

$$[P] = \begin{bmatrix} 1-p & p \\ p & 1-p \end{bmatrix} \quad (1.84)$$

which is a *binary symmetric channel*, as shown in Fig. 1.4. Thus, from Eq. (1.82) the capacity of a binary symmetric channel is

$$C = 1 + p \log_2 p + (1-p) \log_2(1-p) = 1 - H(p) \quad (1.85)$$

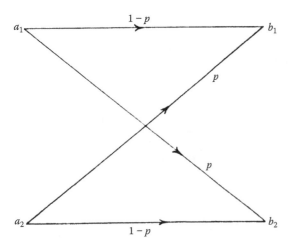

FIG. 1.4   Binary symmetric channel.

In concluding this section, we evaluate a simple channel that is uniform from input but not from output, namely, a *binary symmetric erasure channel*, as shown in Fig. 1.5. The corresponding transition probability matrix is

$$[P] = \begin{bmatrix} 1-p-q & p & q \\ p & 1-p-q & q \end{bmatrix} \qquad (1.86)$$

Since the channel is uniform from input, it follows that equiprobability of the input events produces a maximum value of the output entropy $H(B)$. It can be readily checked that the maximum value of $H(B)$ occurs when the input events are equiprobable, that is, $P(a_1) = P(a_2) - \frac{1}{2}$. Thus, the output probability distribution is

$$P(b_1) = P(b_2) = \frac{1}{2}(1-q) \quad P(b_3) = q \qquad (1.87)$$

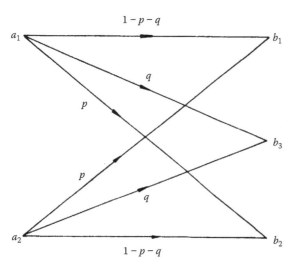

FIG. 1.5   Binary symmetric erasure channel.

We see that $H(B)$ can be evaluated:

$$
\begin{aligned}
H(B) &= -\sum_B P(b)\log_2 P(b) \\
&= (1-q)[1-\log_2(1-q)] - q\log_2 q]
\end{aligned}
\qquad (1.88)
$$

From Eq. (1.86), we have the conditional entropy $H(B/A)$:

$$H(B/A) = -\Big[(1-p-q)\log_2(1-p-q) + p\log_2 p + q\log_2 q\Big] \qquad (1.89)$$

Thus, the channel capacity is

$$
\begin{aligned}
C &= (1-q)\Big[1-\log_2(1-q)\Big] \\
&\quad + \Big[(1-p-q)\log_2(1-p-q) + p\log_2 p\Big]
\end{aligned}
\qquad (1.90)
$$

We see that if $p = 0$, the channel capacity can be written as

$$C = 1 - q \qquad (1.91)$$

The capacity is equal to that of a noiseless binary channel minus the erasure probability q.

1.5   CONTINUOUS CHANNELS WITH ADDITIVE NOISE

We now consider information transmission through continuous channels. We restrict our discussion mainly to the case of additive noise.

A channel is said to be continuous if and only if the input and output ensembles are represented by continuous Euclidean spaces. For simplicity, we restrict our discussion to only the one-dimensional case, although it can be easily generalized to a higher dimension.

Again, we denote by A and B the input and output ensembles, but this time, A and B are continuous random variables of a communication channel, as in Fig. 1.2. It is also noted that a continuous channel can be either *time-discrete* or *time-continuous*. We first discuss time-discrete channels and then consider time-continuous channels.

Like a discrete channel, a continuous channel is said to be memoryless if and only if its transitional probability density $p(b/a)$ remains the same for all successive pairs of input and output events. A memoryless continuous channel is said to be disturbed by an *additive noise* if and only if the transitional probability density $p(b/a)$ depends only on the difference between the output and input random variables, $b - a$:

$$p(b/a) = p(c) \qquad (1.92)$$

where $c = b - a$.

Thus, for additive channel noise the conditional entropy $H(B/a)$ can be shown:

$$H(B/a) = -\int_{-\infty}^{\infty} p(b/a)\log_2 p(b/a)\,db$$
$$= -\int_{-\infty}^{\infty} p(c)\log_2 p(c)\,dc \qquad (1.93)$$

We see that $H(B/a)$ is independent of a, which is similar to the fact of Eq. (1.79), where the channel is uniform from input. The average conditional entropy is

$$H(B/A) = \int_{-\infty}^{\infty} p(a)H(B/a)\,da$$
$$= -\int_{-\infty}^{\infty} p(c)\log_2 p(c)\,dc = H(B/a) \qquad (1.94)$$

It is noted that the definition of capacity for a discrete channel can also be applied to a continuous channel:

$$C \triangleq \max_{p(a)} I(A;B) \qquad (1.95)$$

Thus, in evaluation of the channel capacity, we first evaluate the average mutual information:

$$I(A;B) = H(B) - H(B/A) \qquad (1.96)$$

where

$$H(B) \triangleq -\int_{-\infty}^{\infty} p(b)\log_2 p(b)\,db \qquad (1.97)$$

and

$$H(B/A) \triangleq -\int_{-\infty}^{\infty}\int_{-\infty}^{\infty} p(a)p(b/a)\log_2 p(b/a)\,da\,db \qquad (1.98)$$

We see that from Eq. (1.94), $H(B/A)$ depends only on $P(b/a)$. Thus, if one maximizes $H(B)$, then $I(A;B)$ is maximized. But it is noted that $H(B)$ cannot be made infinitely large, since $H(B)$ is always restricted by certain physical constraints, namely, the available power. This power constraint corresponds to the mean square fluctuation of the input signal (e.g., a mean square current fluctuation):

$$\sigma_a^2 = \int_{-\infty}^{\infty} a^2 p(a)\,da \qquad (1.99)$$

Without loss of generality, we assume that the average value of the additive noise is zero:

$$\bar{c} = \int_{-\infty}^{\infty} cp(c)\,dc = 0 \qquad (1.100)$$

Then the mean square fluctuation of the output signal can be written as

$$\sigma_b^2 = \int_{-\infty}^{\infty} b^2 p(b)\,db \qquad (1.101)$$

Since $b = a + c$ (i.e., signal plus noise), one can show that

$$\sigma_b^2 = \int_{-\infty}^{\infty}\int_{-\infty}^{\infty} (a+c)^2 p(a)p(c)\,da\,dc = \sigma_a^2 + \sigma_c^2 \qquad (1.102)$$

where

$$\sigma_c^2 = \int_{-\infty}^{\infty} c^2 p(c)\,dc \qquad (1.103)$$

From Eq. (1.102), we see that setting an upper limit to the mean square fluctuation of the input signal is equivalent to setting an upper limit to the mean square fluctuation of the output signal. Thus, for a given mean square value of σ_b^2, one can show that for the corresponding entropy, derived from $p(b)$, there exists an upper bound:

$$H(B) \le \frac{1}{2}\log_2\left(2\pi e\sigma_b^2\right) \qquad (1.104)$$

where the equality holds if and only if the probability density $p(b)$ has a Gaussian distribution, with zero mean and the variance equal to σ_b^2.

Since from the additivity property of the channel noise $H(B/A)$ depends solely on $p(c)$ (see Eq. (1.94)), we see that

$$H(B/A) \le \frac{1}{2}\log_2\left(2\pi e\sigma_c^2\right) \qquad (1.105)$$

where the equality holds when $p(c)$ has a Gaussian distribution, with zero mean and the variance equal to σ_c^2.

Thus, if the additive noise in a memoryless continuous channel has a Gaussian distribution, with zero mean and the variance equal to N, where N is the average noise power, then the average mutual information satisfies the inequality

$$I(A;B) \le \frac{1}{2}\log_2\left(2\pi e\sigma_b^2\right) - \frac{1}{2}\log_2(2\pi eN) \le \frac{1}{2}\log_2\frac{\sigma_b^2}{N} \qquad (1.106)$$

Since the input signal and the channel noise are assumed to be statistically independent,

$$\sigma_b^2 = \sigma_a^2 + \sigma_c^2 = \sigma_a^2 + N \qquad (1.107)$$

Therefore, Eq. (1.106) can also be written as

$$I(A;B) \le \frac{1}{2}\log_2\frac{\sigma_a^2 + N}{N} \qquad (1.108)$$

In calculating the channel capacity, we maximize the average mutual information of Eq. (1.108). We see that the equality of Eq. (1.108) holds if and only if the input signal is *also* Gaussianly distributed, with zero mean and the variance equal to S:

$$C = \frac{1}{2} \log_2 \left(1 + \frac{S}{N} \right) \qquad (1.109)$$

It is noted that Eq. (1.109) is one of the formulas obtained by Shannon for an additive Gaussian channel.

However, it should be cautioned that if the additive noise does *not* have a Gaussian distribution, then in general there does not exist an input probability density distribution of $p(a)$ so that the corresponding output probability density of $p(b)$ has a Gaussian distribution.

To define the *entropy power*, we let the equality hold for Eq. (1.104) and replace σ_b^2 with $\bar{\sigma}_b^2$. Thus, for a given value of $H(B)$, the entropy power of B is

$$\bar{\sigma}_b^2 \triangleq \frac{1}{2\pi e} 2^{2H(B)} \qquad (1.110)$$

We see that $\bar{\sigma}_b^2$ is equivalent to the variance of a random variable having a Gaussian distribution, with the entropy $H(B)$. Thus, from Eq. (1.104) we see that

$$\sigma_b^2 \geq \bar{\sigma}_b^2 \qquad (1.111)$$

That is, the variance of the output random variable is greater than its entropy power.

Since $b = a + c$, the sum of two statistically independent random variables a and c, from Eq. (1.111) we have the inequality

$$\bar{\sigma}_a^2 + \bar{\sigma}_c^2 \leq \bar{\sigma}_b^2 \leq \sigma_a^2 + \sigma_c^2 \qquad (1.112)$$

where $\bar{\sigma}_a^2$ and $\bar{\sigma}_c^2$ are the entropy powers of the input ensemble A and of the additive noise ensemble c. The equality of Eq. (1.112) holds if and only if the input signal and additive noise are both Gaussianly distributed, with zero means and variances equal to σ_a^2 and σ_c^2, respectively.

We now consider a memoryless continuous channel disturbed by an additive but *non-Gaussian* noise, with zero mean, variance equal to N, and entropy power $\bar{\sigma}_c^2$. If the mean square fluctuation of the input signal cannot exceed a certain value S, then from Eq. (1.112), one can show that the channel capacity is bounded from below and from above:

$$\frac{1}{2} \log_2 \left(1 + \frac{S}{\bar{\sigma}_c^2} \right) \leq C \leq \frac{1}{2} \log_2 \frac{N+S}{\bar{\sigma}_c^2} \qquad (1.113)$$

The equality holds if and only if the additive channel noise is Gaussianly distributed, with zero mean and the variance equal to N, that is, $\bar{\sigma}_c^2 = N$.

Furthermore, in view of Eq. (1.105), we see that the noise entropy is maximum (for a given value of σ_c^2) if the random noise c is Gaussianly distributed, with zero mean. Thus, the noise disturbance is expected to be more severe for additive Gaussian noise. Therefore, from Eq. (1.113) and the fact that $\sigma_c^2 \geq \bar{\sigma}_c^2$, the capacity of a memoryless continuous channel disturbed by additive non-Gaussian noise with zero mean and the variance equal to N, where the mean square fluctuation of the input signal does not exceed a given value of S, is

$$C \geq \frac{1}{2} \log_2 \left(1 + \frac{S}{N} \right) \qquad (1.114)$$

which is larger than the capacity of an additive Gaussian channel (Eq. (1.109)). We also see that the equality holds for Eq. (1.114) if and only if the additive noise has a Gaussian distribution, with zero mean and the variance equal to N.

We can now evaluate the most well-known channel in information theory, namely, a memoryless, time-continuous, band-limited, continuous channel. The channel is assumed to be disturbed by an additive white Gaussian noise, and a band-limited, time-continuous signal, with an average power not to exceed a given value S, is applied at the input end of the channel.

It is noted that if a random process is said to be a stationary Gaussian process, then the corresponding joint probability density distribution, assumed by the time functions at any finite time interval, is independent of the time origin selected, and it has a Gaussian distribution. If a stationary Gaussian process is said to be white, then the power spectral density must be uniform (constant) over the entire range of the frequency variable. An example, which we encounter later, is thermal noise, which is commonly regarded as having a *stationary* Gaussian distribution and frequently assumed to be white.

To evaluate the capacity of an additive stationary Gaussian channel, we first illustrate a basic property of white Gaussian noise. Let $c(t)$ be a white Gaussian noise; then by the Karhunen–Loéve expansion theorem [14,15], $c(t)$ can be written over a time interval $-T/2 \leq t < T/2$:

$$c(t) = \sum_{i=-\infty}^{\infty} c_i \phi_i(t) \qquad (1.115)$$

where the $\phi_i(t)$'s are *orthonormal functions*, such that

$$\int_{-T/2}^{T/2} \phi_i(t) \phi_j(t) \, dt = \begin{cases} 1 & i = j \\ 0 & i \neq j \end{cases} \qquad (1.116)$$

and c_i are real coefficients commonly known as *orthogonal expansion coefficients*. Furthermore, the c_i's are statistically independent, and the individual probability densities have a stationary Gaussian distribution, with zero mean and the variances equal to $N_0/2T$, where N_0 is the corresponding power spectral density.

Now we consider an input time function $a(t)$, applied to the communication channel, where the frequency spectrum is limited by some bandwidth Δv of the channel. Since the channel noise is assumed to be additive white Gaussian noise, the output response of the channel is

$$b(t) = a(t) + c(t) \qquad (1.117)$$

Such a channel is known as a band-limited channel with additive white Gaussian noise.

Again, by the Karhunen–Loéve expansion theorem, the input and output time functions can be expanded:

$$a(t) = \sum_{i=-\infty}^{\infty} a_i \phi_i(t) \qquad (1.118)$$

and

$$b(t) = \sum_{i=-\infty}^{\infty} b_i \phi_i(t) \qquad (1.119)$$

Thus, we see that

$$b_i = a_i + c_i \qquad (1.120)$$

Since the input function $a(t)$ is band-limited by Δv, only $2T \Delta v$ coefficients a_i, $i = 1, 2, \ldots, 2T \Delta v$, within the passband are considered (we discuss this in Chapter 2). In other words, the input signal ensemble can be represented by a $2T \Delta v$-order product ensemble over a, that is, $A^{2T\Delta v}$. Similarly, the above statement is also true for the output ensemble over b, that is, $B^{2T\Delta v}$. Thus, the average amount of information between the input and output ensembles is

$$I(A^{2T\Delta v}; B^{2T\Delta v}) = H(B^{2T\Delta v}) - H(B^{2T\Delta v}/A^{2T\Delta v}) \qquad (1.121)$$

It is also clear that a, b, and c each form a $2T \Delta v$-dimensional *vector space*. For convenience, we denote by $\mathbf{a}$, $\mathbf{b}$, and $\mathbf{c}$ the respective vectors in the vector space. Thus, we see that

$$\mathbf{b} = \mathbf{a} + \mathbf{c} \qquad (1.122)$$

If we let $p(\mathbf{a})$ and $p(\mathbf{c})$ be the probability density distributions of $\mathbf{a}$ and $\mathbf{c}$, respectively, then the transitional probability density of $p(\mathbf{b/a})$ is

$$p(\mathbf{b/a}) = p(\mathbf{b} - \mathbf{a}) = p(\mathbf{c}) \qquad (1.123)$$

where $\mathbf{a}$ and $\mathbf{c}$ are statistically independent. For simplicity, we let $X \triangleq A^{2T\Delta v}$ be the vector space (the product space) of $\mathbf{a}$. The probability density distribution of $\mathbf{b}$ can be determined:

$$p(\mathbf{b}) = \int_x p(\mathbf{a}) p(\mathbf{c}) \, dX \qquad (1.124)$$

where the integral is over the entire vector space X.

Similarly, $Y \triangleq B^{2T\Delta v}$ and $Z \triangleq C^{2T\Delta v}$ represent the vector space of $\mathbf{b}$ and $\mathbf{c}$, respectively. The average mutual information of Eq. (1.121) can therefore be written as

$$I(X; Y) = H(Y) - H(Z) \qquad (1.125)$$

where

$$H(Y) = -\int_Y p(\mathbf{b}) \log_2 p(\mathbf{b}) \, dY \qquad (1.126)$$

and

$$H(Z) = H(Y/X) = -\int_Z p(\mathbf{c}) \log_2 p(\mathbf{c}) \, dZ \qquad (1.127)$$

In view of Eq. (1.96), we see that the problem of a time-continuous channel has been reduced to a form very similar to that of a time-discrete channel with additive Gaussian noise. Thus, the channel capacity can be defined, as before, as

$$C \triangleq \max_{T, p(a)} \frac{I(X; Y)}{T} \quad \text{bits/time} \qquad (1.128)$$

We can now maximize the average mutual information of $I(X; Y)$ under the constraint that the mean square fluctuation of the input signal ensemble cannot exceed a specified value S:

$$\int_X |\mathbf{a}|^2 p(\mathbf{a}) \, dX \leq S \qquad (1.129)$$

Since each of the vectors $\mathbf{a}$, $\mathbf{b}$, and $\mathbf{c}$ can be represented by $2T \Delta v$ continuous variables, and each c_i is statistically independent, with Gaussian distribution, zero mean, and the variance equal to $N_0/2T$, we quickly see that

$$I(X; Y) = I(A^{2T\Delta v}; B^{2T\Delta v}) = \sum_{i=1}^{2T\Delta v} I(A_i; B_i) \qquad (1.130)$$

Thus, from Eq. (1.105),

$$H(Z) = 2T\Delta v H(C_i) \qquad (1.131)$$

where

$$H(C_i) = \frac{1}{2} \log_2 \left(2\pi e \sigma_{c_i}^2 \right) \qquad (1.132)$$

the entropy of any one of the c_i variables. If we let $N = \sigma_{c_i}^2 = N_0 \Delta v$, then we have

$$H(Z) = T\Delta v \log_2 \left(\frac{\pi e N_0}{T} \right) \qquad (1.133)$$

In view of Eq. (1.104), we see that

$$H(B_i) \leq \frac{1}{2}\log_2\left(2\pi e\sigma_{b_i}^2\right) \qquad (1.134)$$

with the equality holding if and only if b_i is Gaussianly distributed, with zero mean and the variance equal to $\sigma_{b_i}^2$. Since $\mathbf{b} = \mathbf{a} + \mathbf{c}$, we see that for $p(\mathbf{b})$ to have a Gaussian distribution, $p(\mathbf{a})$ also must have a Guassian distribution, with zero mean. Thus, the average mutual information of Eq. (1.130) can be written as

$$\begin{aligned}I(X;Y) &= \sum_{i=1}^{2T\Delta v} H(B_i) - H(Z)\\&= \frac{1}{2}\log_2\left[\prod_{i=1}^{2T\ v}\left(2\pi e\sigma_{b_i}^2\right)\right] - T\Delta v\log_2\left(\frac{\pi e N_0}{T}\right)\end{aligned}$$

$$(1.135)$$

where Π denotes the product ensemble. Since a_i and c_i are statistically independent,

$$\sigma_{b_i}^2 = \sigma_{a_i}^2 + \sigma_{c_i}^2 = \sigma_{a_i}^2 + \frac{N_0}{2T} \qquad (1.136)$$

In view of Eq. (1.129), we see that

$$\sum_{i=1}^{2T\Delta v}\sigma_{b_i}^2 = \sum_{i=1}^{2T\Delta v}\sigma_{a_i}^2 + N_0\,\Delta v \leq S + N \qquad (1.137)$$

where $N = N_0\,\Delta v$. The equality holds for Eq. (1.137) when the input probability density distribution $p(\mathbf{a})$ has a Gaussian distribution, with zero mean and the variance equal to S. Furthermore, from Eq. (1.137), we can write

$$\prod_{i=1}^{2T\ v}\sigma_{b_i}^2 \leq \left(\frac{S+N}{2T\Delta v}\right)^{2T\Delta v} \qquad (1.138)$$

where the equality holds if and only if the $\sigma_{b_i}^2$'s are all equal and $p(\mathbf{a})$ has a Gaussian distribution, with zero mean and the variance equal to S.

Therefore, in the maximization of Eq. (1.135), the corresponding channel capacity is

$$C = \max_{T,p(a)}\frac{I(X;Y)}{T} = \Delta v\log_2\left(1+\frac{S}{N}\right) \quad \text{bits/sec} \quad (1.139)$$

where S and N are the average signal and average noise power, respectively. This is one of the most popular results derived, by Shannon [3] and independently by Wiener [5], for the memoryless additive Gaussian channel. Because of its conceptual and mathematical simplicity, this equation has been frequently used in practice, but it has been occasionally misused. It is noted that this channel capacity is derived under the additive white Gaussian noise regime, and the average input signal power cannot exceed a specified value of S. As noted, we can obtain the capacity of the channel if and only if the input signal is also Gaussianly distributed, with zero mean and the variance equal to S.

Since the average noise power over the specified bandwidth of the channel is $N_0\,\Delta v$, we see that for a fixed value of S/N, as the channel bandwidth increases to infinitely large, the capacity of the channel approaches a definite value:

$$C(\infty) = \lim_{\Delta v\to\infty} C(\Delta v) = \frac{S}{N_0}\log_2 e \qquad (1.140)$$

This equation possesses an important physical significance: the measurement or observation of any physical quantity is practically always limited by *thermal noise agitation*. This thermal agitation can usually be considered white (at least within the bandwidth of interest) additive Gaussian noise. The noise power spectral density N_0 can be related to the thermal temperature T:

$$N_0 = kT \qquad (1.141)$$

where k is Boltzmann's constant, and T is in Kelvin. Moreover, from Eq. (1.140), it follows that the signal energy transmitted through a physical communication channel must be at least kT per nat of information the signal is able to provide. In other words, it takes at least kT energy for a nat of information to be properly transmitted.

In concluding this section, we plot the capacity of the additive Gaussian channel as a function of bandwidth, for a fixed value of S/N_0, as shown in Fig. 1.6. We see that for small values of channel bandwidth Δv, the capacity increases very rapidly with Δv, but that it asymptotically approaches $C(\infty)$ of Eq. (1.140) when Δv becomes infinitely large.

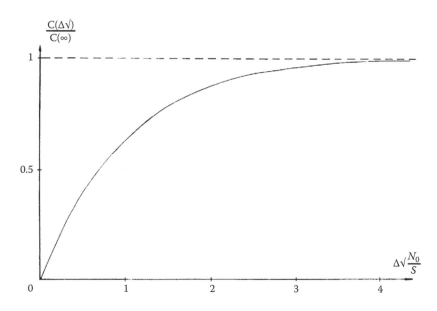

FIG. 1.6   Capacity of an additive white Gaussian channel as a function of bandwidth Δv.

1.6   SUMMARY AND REMARKS

We devoted the earlier part of this chapter to the definition of information measures. With these basic definitions, it is possible to study the important properties of transmission of information. More precisely, the measure of information was defined as the logarithm of the ratio of the *a posteriori* probability to the priori probability of an event. We showed that such an information measure possesses a symmetric property such that the measure of mutual information can be defined.

Moreover, self-information can also be defined as the logarithm of the *a priori* probability of an event. Thus, the measure of self-information of an event is the amount of information required to specify that event *uniquely*. In other words, it is the amount of information the event can provide.

Furthermore, with these primary information measures, the ensemble average of information measures can be determined, namely, the average self-information and the average mutual information. We have shown that the average mutual information can never be negative, in contrast to mutual information about a particular event, which can be either positive or negative. However, the greatest significance of the average mutual information is that it leads us to the basic definition and evaluation of channel capacity.

It is also noted that the information measures we have defined were originally applied on discrete space. However, the definition can be easily extended to continuous space.

With the entropy equations, we have evaluated simple communication channels. In the evaluation of channel capacity, one can maximize the corresponding average mutual information of the channel by searching for an optimum input probability distribution. In the maximization of the mutual information, it is required that the input events be statistically independent. As a result, the output events are also statistically independent.

After evaluation of simple memoryless discrete channels, we turned our attention to memoryless continuous channels with an additive noise regime. However, the capacities of continuous channels were evaluated only in some special, although important, cases, namely, for additive Gaussian noise. The case for additive non-Gaussian noise was also discussed, and we showed that the capacity has a lower and an upper bound.

We also evaluated the capacity of a memoryless time-continuous channel under the additive noise regime and arrived at an important and interesting formula for the channel capacity. This formula was first derived independently by Wiener and by Shannon. Because of its conceptual and mathematical simplicity, this formula has been frequently used by communication engineers. We should remember that this formula was derived under the assumption of an additive white Gaussian noise regime and to approach the capacity only through a Gaussian signal.

It is emphasized that the evaluation of a communication channel with memory is very laborious and extremely difficult. To our knowledge, no satisfactory result has been achieved for a general memory channel. However, researchers in this area are still very active, and certainly important results for memory channels will eventually be obtained.

In concluding this brief summary, we state a very important fundamental theorem without proof: if a communication channel (noisy or nonnoisy) possesses a channel capacity of C bits per second and is assumed to accept an input signal from an information source at a rate of H bits per second, where $H \leq C$, then by properly encoding the signal, it is possible to transmit it through the channel with as little error as we please. However, if the rate of information is greater than the channel capacity, $H > C$, then it is impossible to code the source signal so that it can be transmitted through the channel with as small an error as we please.

Finally, it should be clear that the purpose of this chapter is not to cover the entire domain of information transmission but rather to present a basic introduction. The topics selected for this text were those of general interest. Therefore, readers interested in a detailed discussion on information transmission can refer to the original papers by Shannon [1–3] and the excellent texts by Fano [4] and by Gallager [16].

REFERENCES

1. C. E. Shannon, A Mathematical Theory of Communication, *Bell Syst. Tech. J.*, vol. 27, 379–423, 623–656 (1948).
2. C. E. Shannon, Communication in the Presence of Noise, *Proc. IRE*, vol. 37, 10 (1949).
3. C. E. Shannon and W. Weaver, *The Mathematical Theory of Communication*, University of Illinois Press, Urbana, 1949.
4. R. M. Fano, *Transmission of Information*, MIT Press, Cambridge, MA, 1961.
5. N. Wiener, *Cybernetics*, MIT Press, Cambridge, MA, 1948.
6. N. Wiener, *Extrapolation, Interpolation, and Smoothing of Stationary Time Series*, MIT Press, Cambridge, MA, 1949.
7. R. V. L. Hartley, Transmission of Information, *Bell Syst. Tech. J.*, vol. 7, 535 (1928).
8. L. Brillouin, The Negentropy Principle of Information, *J. Appl. Phys.*, vol. 24, 1152 (1953).
9. L. Brillouin, *Science and Information Theory*, Academic, New York, 1956.
10. L. Brillouin, *Scientific Uncertainty and Information*, Academic, New York, 1964.
11. D. Gabor, Light and Information, in E. Wolf, ed., *Progress in Optics*, vol. I, North-Holland, Amsterdam, 1961.
12. D. Gabor, Informationstheorie in der Optik, *Optik*, vol. 39, 86 (1973).
13. W. J. McGill, Multivariate Information Transmission, *IRE Trans. Inf. Theory*, vol. 4, 93 (1954).
14. W. Davenport and W. Root, *Random Signals and Noise*, McGraw-Hill, New York, 1958.
15. M. Loéve, *Probability Theory*, 3rd ed., Van Nostrand, Princeton, NJ, 1963.
16. R. G. Gallager, *Information Theory and Reliable Communication*, John Wiley, New York, 1968.

2 Diffraction and Signal Analysis

Since the first word in the title of this chapter is *diffraction*, the first question is, naturally, what is diffraction? The answer to this question may be simplified as follows. When light passes the edge of an obstacle, its course deviates from that of a straight line. This deviation effect is known as *diffraction*.

Historically, it was the observation of diffraction that ultimately led to the acceptance of the wave theory of light. Thus, it is possible to explain diffraction on the basis of wave theory. For example, the angle of deviation from straight-line propagation, it can be shown, is proportional to the wavelength of the light source. However, the wavelengths of visible light are extremely small, and thus the diffraction angles are also very small. We see that in geometric optics, light propagation is assumed to occur in a straight line. Therefore, we note that in diffraction theory, straight-line propagation occurs only at the limit where the wavelength is zero.

It is emphasized that the theory of diffraction can be treated on the basis of Huygens' principle, in the form of Kirchhoff's integral. Although rigorous analysis in diffraction should be obtained from Maxwell's equations, the scalar theory of Huygens provides very good predicted results, as in the effect of the actual diffraction. Thus, scalar theory provides a good qualitative analysis of the diffraction phenomenon. However, the greatest significance of the scalar theory approach is that the analysis is much less difficult than analysis by means of Maxwell's equations.

In this chapter, we provide a brief discussion of diffraction and signal analysis. Our discussion is primarily an introductory one. Therefore, interested readers are referred to the books by Born and Wolf [1] and Yu [2] for diffraction theory, and to the book by Davenport and Root [3] for signal analysis.

2.1 INTRODUCTION TO DIFFRACTION

It is customary to divide diffraction into two cases, depending on the distance of the light source and the viewing point from the diffracting screen, and these cases have been named after two early investigators of diffraction. If the source and viewing point are so far from the diffracting screen that lines drawn from the source or viewing point to all points of the apertures do not differ in length by more than a small fraction of a wavelength, the phenomenon is called *Fraunhofer diffraction*. If these conditions do not hold, it is called *Fresnel diffraction*. The boundary between these two cases is somewhat arbitrary and depends on the accuracy desired in the results. In most cases, it is sufficient to use Fraunhofer methods if the difference in distances does not exceed 1/20 of a wavelength. Of course, Fraunhofer diffraction can be achieved without the source being at a great physical distance if a collimating lens is used to make the rays of light from the source nearly parallel.

Figure 2.1 illustrates the above considerations. A source of monochromatic light is located at S, and a viewing point at P, and between them is an opaque screen having a finite number of apertures. Let a circle C be drawn on the plane of the screen, and let it be as small as possible while still enclosing all the apertures. Let C be the base of cones, with S and P as vertices. Also draw spherical surfaces Σ_1 and Σ_2, having S and P as centers, and r_1 and r_2 as radii, the shortest distances from S and P to the base C. If the longest distances from C to Σ_1 and to Σ_2 are not more than 1/20 of the wavelength of the light used, then the diffraction is Fraunhofer, and the light falling on the observing screen at P forms a *Fraunhofer diffraction pattern*.

However, if, by reason of the large size of C or the shortness of the distances to S and P, the distances between C and Σ_1 or Σ_2 are greater than 1/20 of the wavelength, then we have Fresnel diffraction and a *Fresnel diffraction pattern* at P.

The radius of the circle C in Fig. 2.2 is denoted by p, the shortest distance to S from the screen is l, and the greatest separation of the sphere and the screen is Δl. From the definition of Fraunhofer diffraction, Δl must be a small fraction of a wavelength. However, ρ may be many wavelengths long (as it can also be in the Fresnel case). In the right triangle in Fig. 2.2, we have

$$(l + \Delta l)^2 = l^2 + \rho^2 \tag{2.1}$$

and because of the small size of $(\Delta l)^2$ in comparison with the other quantities, we make the approximation

$$l \simeq \frac{\rho^2}{2\Delta l} \tag{2.2}$$

As an example, suppose that ρ is 1 cm, and that the source has the longest wavelength in the visible red, about 8×10^{-5} cm. Let Δl be 1/20 of this, or 0.4×10^{-5} cm. The distance l is then approximately 1.25 km. If violet light of half the wavelength is used, l will be 2.5 km. The requirements for Fraunhofer diffraction can thus be rather stringent.

We now turn our attention to one of the most significant principles in the theory of diffraction, namely, Huygens' principle. By means of Huygens' principle, it is possible to obtain by graphical methods the shape of a wave front at any instant if the wave front at an earlier instant is known. The principle may be stated as follows. Every point of a wave front may be considered the source of a small secondary wavelet that spreads in all directions from the point at the wave propagation velocity. A new wave front is found by constructing a surface tangent to all the second wavelets. If the velocity of propagation is not constant for all parts of the wave front, then each wavelet must be given an appropriate velocity.

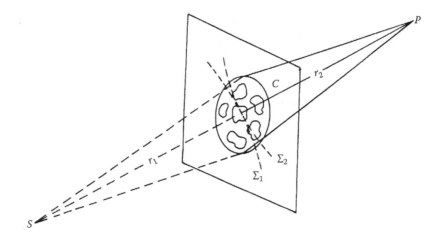

FIG. 2.1 Geometry for defining Fraunhofer and Fresnel diffraction.

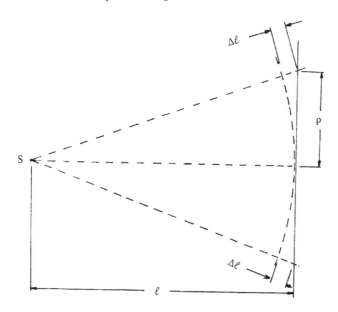

FIG. 2.2 Geometry for determining the Fraunhofer diffraction.

An illustration of the use of Huygens' principle is given in Fig. 2.3. The known wave front is shown by the arc Σ, and the directions of propagation are indicated by small arrows. To determine the wave front after an interval Δt, with a wave velocity v, simply construct a series of spheres of radius $r = v\Delta t$

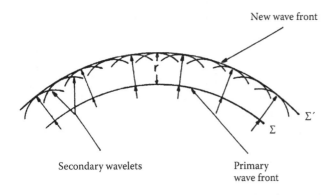

FIG. 2.3 Huygens' principle.

from each point of the original front Σ. These spheres represent the secondary wavelets. The envelope enclosing the surfaces of the spheres represents the new wave front. This is the surface marked Σ' in the figure. In this example, the magnitude of the wave velocity is considered the same at all points.

It is noted that Huygens' principle predicts the existence of a backward wave that, however, is never observed. An explanation for this discrepancy can be found by examining the interference of the secondary wavelets throughout the space surrounding Σ. The main use of Huygens' principle is in predicting diffraction patterns, which we see in Section 2.2. When first stated, the principle was a useful method for finding the shape of a new wave front; little physical significance was attached to the secondary wavelets at that time. Later, as the wave nature of light came to be more fully understood, Huygens' principle took on deeper significance.

2.2 FRESNEL–KIRCHHOFF THEORY

In Section 2.1, we discussed the basic concept of diffraction. We defined far-field Fraunhofer diffraction and near-field Fresnel diffraction. The basic concept of Huygens' principle was also illustrated.

In this section, we derive one of the most important theories in diffraction, namely, the *Fresnel–Kirchhoff theory*. We show that the Fresnel–Kirchhoff theory essentially coincides with Huygens' principle.

A rigorous approach to the Fresnel–Kirchhoff theory should start with the application of *Green's theorem to the scalar wave theory*. However, such a derivation is in general more mathematically involved, so we derive the Fresnel–Kirchhoff theory from a *system theory* viewpoint. In beginning the derivation, we recall Huygens' principle from Section 2.1. Thus, according to Huygens' principle, the amplitude observed at a point P' of a given coordinate system $\sigma(\alpha, \beta, \gamma)$, due to a light source located in a given coordinate system $\rho(x, y, z)$, as shown in Fig. 2.4, can be calculated by assuming that each point of the light source is an infinitesimal spherical radiator. Thus, the complex light amplitude $u(r)$ contributed by a point

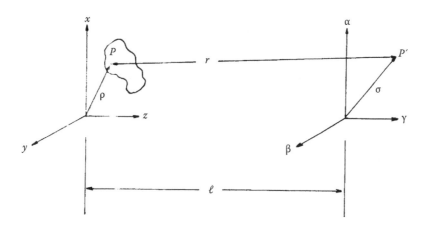

FIG. 2.4   Fresnel–Kirchhoff theory.

P in the ρ coordinate system can be considered that from an unpolarized monochromatic point source, such that

$$u(r) = \frac{1}{r}\exp[i(kr - \omega t)] \qquad (2.3)$$

where k and ω are the wave number and angular frequency, respectively, of the point source, and r is the distance between the point source and the point of observation:

$$r = [(l + \gamma - z)^2 + (\alpha - x)^2 + (\beta - y)^2]^{1/2} \qquad (2.4)$$

If the separation l of the two coordinate systems is assumed to be large compared with the magnitude of ρ and σ, then r can be approximated by l in the denominator of Eq. (2.3), and by

$$r = \left[l' + \frac{(\alpha - x)^2}{2l'} + \frac{(\beta - y)^2}{2l'} \right] \qquad (2.5)$$

in the exponent of Eq. (2.3). Since $l' = l + \gamma - z$, Eq. (2.3) can be written as

$$u(\sigma - \rho) \simeq \frac{1}{l}\exp\left\{ ik\left[l' + \frac{(\alpha - x)^2}{2l'} + \frac{(\beta - y)^2}{2l'} \right] \right\} \qquad (2.6)$$

where the time-dependent exponent has been omitted for convenience.

Furthermore, if the point of radiation and the point of observation are interchanged in the two coordinate systems, then the complex light amplitude observed at $\rho(x, y, z)$ is

$$u(\rho - \sigma) = \frac{1}{l}\exp\left\{ ik\left[l' + \frac{(\alpha - x)^2}{2l'} + \frac{(\beta + y)^2}{2l'} \right] \right\} \qquad (2.7)$$

where $l'' = l + z - \gamma$.

It is clear that Eqs. (2.6) and (2.7) represent the free-space radiation from a monochromatic point source. They are also called free-space *impulse responses*.

Therefore, the complex amplitude produced at the σ coordinate system by a monochromatic radiating surface located in the ρ coordinate system can be written as

$$U(\sigma) = \int\int_{\Sigma} T(\rho)u(\sigma - p)\,d\Sigma \qquad (2.8)$$

where $T(\rho)$ is the complex light field of the monochromatic radiating surface, Σ denotes the surface integral, and $d\Sigma$ is the incremental surface element. It is noted that Eq. (2.8) represents the Fresnel–Kirchhoff theory.

Furthermore, by replacing $u(\sigma - \rho)$ by $u(r)$, from Eq. (2.8) we see that

$$T(\rho)u(\sigma - p)d\Sigma = T(\rho)u(r)d\Sigma = \frac{C\,d\Sigma}{r}e^{i(kr - \omega t)} \qquad (2.9)$$

where $C = T(\rho)$. Equation (2.9) is essentially the secondary-point radiation proposed by Huygens. Thus, Eq. (2.8) also represents Huygens' principle in the form of Kirchhoff's integral.

2.3   LINEAR SYSTEMS AND FOURIER ANALYSIS

The concept of a linear system is very important in the analysis of optical communication and information processing systems for at least two major reasons:

1. A great number of applications in optical communication and processing systems are assumed to be linear, at least within specified ranges.
2. An exact solution in the analysis of linear information transmission can be obtained by standard techniques.

Except for a very few special cases, there is no general procedure for analyzing nonlinear problems. Of course, there are practical ways of solving nonlinear problems that may involve graphical or experimental approaches. Approximations are often necessary in solving nonlinear problems, and each situation may require special handling techniques. In practice, fortunately, a great number of optical communication and processing problems are linear, and these are generally solvable. However, we emphasize that in practice, no physical system is strictly linear unless certain restrictions are imposed.

It is a common practice to study the behavior of a physical system by the input excitation and output response. Both excitation and response may be some physically measurable quantity, depending on the nature of the system. Suppose an input excitation of $f_1(t)$ produces an output response of $g_1(t)$, and that a second excitation $f_2(t)$ produces a second response $g_2(t)$ of a physical system, as shown in Fig. 2.5. Symbolically, we write

$$f_1(t) \rightarrow g_1(t) \tag{2.10}$$

and

$$f_2(t) \rightarrow g_2(t) \tag{2.11}$$

Then for a linear physical system, we have

$$f_1(t) + f_2(t) \rightarrow g_1(t) + g_2(t) \tag{2.12}$$

Equation (2.12), in conjunction with Eqs. (2.10) and (2.11), represents the *additivity* property of the linear system. Thus, a necessary condition for a system to be linear is that the *principle of superposition* hold. The principle of superposition implies that the presence of one excitation does not affect the response due to other excitations.

Now, if the input excitation of the same physical system is $Cf_1(t)$, where C is an arbitrary constant, then the output response is $Cg_1(t)$:

$$Cf_1(t) \rightarrow Cg_1(t) \tag{2.13}$$

Equation (2.13) represents the *homogeneity* characteristic of the linear system. Thus, a property of a linear system is preserving the magnitude scale factor. Therefore, it may be concluded that a physical system is *linear* if and only if Eqs. (2.12) and (2.13) are satisfied. In other words, if a system possesses the additivity and homogeneity properties, then it is a *linear system*.

There is, however, another important physical aspect that characterizes a linear system with constant parameters. If the input excitation $f(t)$ applied to such a system is an alternating function of t with frequency v, and if the output response $g(t)$ appears to be alternating with the same frequency v, then the system can be said to have *time invariance*. In other words, a time-invariant system does not generate new frequencies. Time invariance implies that if

$$f(t) \rightarrow g(t)$$

then

$$f(t - t_0) \rightarrow g(t - t_0) \tag{2.14}$$

where t_0 is an arbitrary time delay. Therefore, if a linear system possesses the time invariance property, then it is a *linear time-invariant system*.

We now turn to Fourier analysis. Fourier transforms are particularly important in the analysis of optical information processing and communication. We consider first a class of real functions $f(t)$ that satisfy the following sufficient conditions:

1. $f(t)$ must be sectionally continuous in every finite region over the t domain, except for a finite number of discontinuities.
2. $f(t)$ must be absolutely integrable over the entire t domain:

$$\int_{-\infty}^{\infty} |f(t)| \, dt < \infty \tag{2.15}$$

These functions can be represented by the equation

$$f(t) = \int_{-\infty}^{\infty} F(v) \exp(i2\pi vt) \, dv \tag{2.16}$$

where

$$F(v) = \int_{-\infty}^{\infty} f(t) \exp(-i2\pi vt) \, dt \tag{2.17}$$

and v is the frequency variable.

Equations (2.16) and (2.17) are known as a *Fourier transform pair*. Equation (2.17) is often called the *Fourier transform*, and Eq. (2.16) is known as the *inverse Fourier transform*.

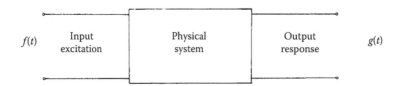

FIG. 2.5 Input–output physical system.

Symbolically, Eqs. (2.16) and (2.17) can be written, respectively, as

$$f(t) = F^{-1}[F(v)] \quad\quad (2.18)$$

and

$$F(v) = F[f(t)] \quad\quad (2.19)$$

where F^{-1} and F denote the inverse and direct Fourier transformations, respectively. It is noted that $F(v)$ is generally a complex function:

$$F(v) = |F(v)| \exp[i\phi(v)] \quad\quad (2.20)$$

where $|F(v)|$ and $\phi(v)$ are referred to as the *amplitude spectrum* and the *phase spectrum*, respectively, and $F(v)$ is also known as the *Fourier spectrum* or *frequency spectrum*. However, if $f(t)$ is a periodic function such that (1) $f(t) = f(t + T)$, where T is the period, and (2) $f(t)$ is sectionally continuous over the period T, then $f(t)$ can be expanded into a *Fourier series*:

$$f(t) = \sum_{n=-\infty}^{\infty} C_n \exp(i2\pi n v_0 t) \quad\quad (2.21)$$

where

$$v_0 = \frac{1}{T}$$

and

$$C_n = \frac{1}{T} \int_0^T f(t) \exp(-i2\pi n v_0 t)\, dt \quad\quad (2.22)$$

C_n is known as the *complex Fourier coefficient*, and $|C_n|$ is known as the *amplitude spectrum*. It should be noted that if $f(t)$ is a real function, then

$$C_{-n} = C_n^* \quad\quad (2.23)$$

where the superscripts asterisk denotes the *complex conjugate*.

In concluding this section, we refer the reader to the excellent texts by Papoulis [4] and by Bracewell [5] for a detailed treatment of Fourier analysis.

2.4   FINITE BANDWIDTH ANALYSIS

Strictly speaking, optical communication channels, optical information processors, and physical systems are restricted to a finite bandwidth. For example, a physical system can be a *low-pass*, *band-pass*, or *discrete band-pass* system. But a strictly high-pass system can never exist in practice, since

every physical system has an upper frequency limit. It may be pointed out that the term *frequency* can be applied to temporal and spatial coordinates [2], as we briefly note at the end of this section. A low-pass system is defined as a system that possesses a nonzero transfer characteristic from zero frequency to a definite limit of maximum frequency v_m. Thus, the bandwidth Δv of a low-pass system is equal to the maximum frequency limit:

$$\Delta v = v_m \quad\quad (2.24)$$

However, if the system possesses a nonzero transfer characteristic from a lower-frequency limit v_1 to a higher-frequency limit v_2, then it is a band-pass system. Thus, the corresponding bandwidth is

$$\Delta v = v_2 - v_1 \quad\quad (2.25)$$

Similarly, one can generalize the band-pass system to a finite discrete band-pass system. Thus, the bandwidths must be the corresponding passbands.

Since, as is well known, the analysis of a band-pass system can be easily reduced to the case of an equivalent low-pass system [6], we restrict our discussion to only the low-pass problem.

Before beginning a detailed discussion of the analysis, one may ask a very fundamental question: Given a low-pass communication system of bandwidth $\Delta v = v_m$, what sort of output response can be expected? In other words, from the frequency domain standpoint, what would happen if the Fourier spectrum of the input signal were extended beyond the passband?

To answer this basic question, we present a very simple but important example. For simplicity, we let the low-pass system, an ideal filter, as shown in Fig. 2.6, have the corresponding transfer function:

$$H(v) = \begin{cases} 1 & |v| \le v_m \\ 0 & |v| > v_m \end{cases} \quad\quad (2.26)$$

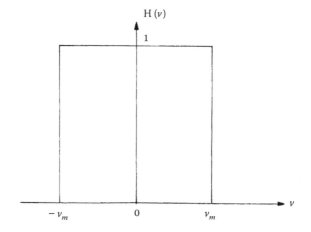

FIG. 2.6   Ideal low-pass filter.

If the input signal to this low-pass system has a finite duration of Δt, then we see that to have good output reproduction of the input signal, it is required that the system bandwidth Δv be greater than or at least equal to $1/\Delta t$:

$$\Delta v \geq \frac{1}{\Delta t} \qquad (2.27)$$

where $1/\Delta t$ is known as the input signal bandwidth. Thus, we have the basic relationship

$$\Delta t \, \Delta v \geq 1 \qquad (2.28)$$

where $\Delta v = {}^2v_m$. We see that Eq. (2.28), the duration bandwidth product, has a definite lower bound. The basic significance is that if the Fourier spectrum of the input signal is more or less concentrated in the passband of the system, that is, if $|v| \leq v_m$, then the output response will quite faithfully reproduce the input signal. However, if the Fourier spectrum of the input signal spreads beyond the passband of the system, then the output response will be severely distorted; that is, it will fail to reproduce the input signal. It is noted that Eq. (2.28) possesses an intimate relationship to Heisenberg's uncertainty principle in quantum mechanics, as we elaborate shortly.

We now give a few more examples in which we show that the inequality of Eq. (2.28) is indeed relevant in information transmission. Let us consider a short symmetric pulse with a duration of Δt, which has its maximum value at $t = 0$:

$$f(t) = \begin{cases} f(t) & |t| \leq \dfrac{\Delta t}{2} \\[2ex] 0 & |t| > \dfrac{\Delta t}{2} \end{cases} \qquad (2.29)$$

$$f(-t) = f(t) \qquad (2.30)$$

and

$$f(0) \geq f(t) \qquad (2.31)$$

Since the symmetric pulse $f(t)$ is Fourier transformable, we have the corresponding Fourier spectrum of $f(t)$:

$$F(v) = F[f(t)] \qquad (2.32)$$

where F denotes the direct Fourier transformation.

We define a *nominal duration* of Δt equivalent to the duration of a rectangular pulse of amplitude $f(0)$:

$$\Delta t f(0) \triangleq \int_{-\infty}^{\infty} f(t)\, dt \qquad (2.33)$$

Similarly, one can also define a *nominal bandwidth* of Δv:

$$\Delta v F(0) \triangleq \int_{-\infty}^{\infty} F(v)\, dv \qquad (2.34)$$

which is equivalent to the bandwidth of a flat, rectangular Fourier spectrum.

From the definition of the Fourier transform pair (Eqs. (2.16) and (2.17)), we see that Eqs. (2.33) and (2.34) can be written as

$$\Delta t = \frac{F(0)}{f(0)} \qquad (2.35)$$

and

$$\Delta v = \frac{f(0)}{F(0)} \qquad (2.36)$$

Thus,

$$\Delta t \Delta v = 1 \qquad (2.37)$$

which gives the lower bound condition of Eq. (2.28).

It should be noted that if the symmetric pulse $f(t)$ contains *negative* values, as shown in Fig. 2.7, then the definitions of nominal duration and nominal bandwidth must be modified:

$$\Delta t \,|\, f(0)\,| \triangleq \int_{-\infty}^{\infty} |\, f(t)\,|\, dt \geq \left| \int_{-\infty}^{\infty} f(t)\, dt \right| = |\, F(0)\,| \qquad (2.38)$$

and

$$\Delta v \,|\, F(0)\,| \triangleq \int_{-\infty}^{\infty} |\, F(v)\,|\, dv \geq \left| \int_{-\infty}^{\infty} F(v)\, dv \right| = |\, f(0)\,| \qquad (2.39)$$

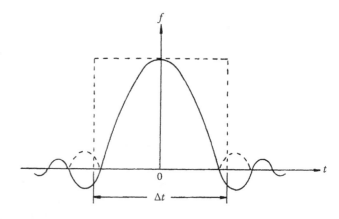

FIG. 2.7   Symmetric pulse with negative values.

These definitions give rise to the uncertainty relation

$$\Delta t\, \Delta v \geq 1 \qquad (2.40)$$

which is essentially the condition of Eq. (2.28). From Fig. 2.7, we see that the nominal duration was determined by equating the area under the rectangular pulse function to the area under the curve of $|f(t)|$. It is evident that the nominal duration Δt is *wider* under this new definition of Eq. (2.38), provided $f(t)$ contains negative values. Similarly, the nominal bandwidth Δv of Eq. (2.39) can be interpreted in the same manner.

We now turn our attention to the inequality of Eq. (2.28) and show that it is essentially identical to *Heisenberg's uncertainty principle* in quantum mechanics [7]. In order to do so, let us denote by x a position variable for a particle, and by p the corresponding momentum variable. The uncertainty principle states that the position variable x and its momentum variable p cannot be observed or measured simultaneously with arbitrary accuracy:

$$\Delta x\, \Delta p \geq h \qquad (2.41)$$

where Δx and Δp are the position and momentum errors, respectively, and h is Planck's constant. The Heisenberg uncertainty relation of Eq. (2.41) can also be written in the form of energy and time variables:

$$\Delta E\, \Delta t \geq h \qquad (2.42)$$

where ΔE and Δt are the corresponding energy and time errors. But since $E = hv$, and hence $\Delta E = h\, \Delta v$, we see that

$$\Delta E\, \Delta t = h\, \Delta v\, \Delta t \geq h \qquad (2.43)$$

Therefore, one concludes that $\Delta v\, \Delta t \geq 1$ is in fact the *Heisenberg uncertainty relation*. The reader may realize how unrealistic it would be to select smaller values of Δv and Δt, which would violate the uncertainty relation.

In concluding this section, we emphasize that time and frequency variables, as mentioned on several occasions, can be easily applied to *spatial coordinates*, namely, *spatial* and *spatial frequency* domains. For example, in most optical information processing [2], we use spatial domains, instead of the time variable.

2.5 DEGREES OF FREEDOM OF A SIGNAL

We now consider the problem of degrees of freedom of a band-limited signal. Let us denote by $f(t)$ a band-limited signal whose spectrum extends from zero frequency up to a definite maximum limit of v_m. We let $f(t)$ extend over a time interval of T, where $v_m T \gg 1$. Now the basic question is, how many sampling points or degrees of freedom are required in order to describe the function $f(t)$, over T, uniquely? To answer this fundamental question, we present a simple example. However,

before going into a detailed discussion, it is noted that $f(t)$ is not *completely* defined when only the value of $f(t)$, over T, is given. For simplicity, however, we let the value of $f(t)$ repeat for every interval of T:

$$f(t) = f(t + T) \qquad (2.44)$$

which can be considered periodic for every T. Thus, the function $f(t)$, over the period T, can be expanded in a Fourier series:

$$f(t) = \sum_{n=-M}^{M} C_n \exp(i2\pi n v_0 t) \qquad (2.45)$$

where $v_0 = 1/T$, and $M = v_m T$.

From this Fourier expansion of Eq. (2.45), we see that $f(t)$ contains a finite number of terms:

$$N = 2M + 1 = 2v_m T + 1 \qquad (2.46)$$

which includes the zero-frequency Fourier coefficient C_0. Thus, if the average value of $f(t)$, over T, is zero, then we see that $C_0 = 0$, and the total number of terms of the corresponding Fourier series reduces to

$$N = 2v_m T \quad \text{for } C_0 = 0 \qquad (2.47)$$

If the duration T is sufficiently large, we see that Eq. (2.46) reduces to

$$N \simeq 2v_m T \qquad (2.48)$$

which is the number of degrees of freedom required to specify $f(t)$, over T. In other words, it requires a total of N equidistant sampling points of $f(t)$, over T, to describe the function

$$t_s = \frac{T}{N} = \frac{1}{2v_m} \qquad (2.49)$$

where t_s is the *Nyquist sampling interval*. Thus, the corresponding *sampling frequency* is

$$f_s = \frac{1}{t_s} = 2v_m \qquad (2.50)$$

which is called the *Nyquist sampling rate*. We see that

$$fs \geq 2v_m \qquad (2.51)$$

that is, the sampling frequency is at least equal to twice the highest frequency limit of $f(t)$. Now we consider reconstruction of the original function $f(t)$ with the corresponding N equidistant sample points, that is, the N degrees of freedom.

First, let us take the Fourier transform of $f(t)$:

$$F(v) = \int_{-\infty}^{\infty} f(t) \exp(-i2\pi nvt)\, dt \qquad (2.52)$$

By virtue of the band-limited nature of $f(t)$, we see that

$$F(v) = 0 \quad \text{for } v > v_m \qquad (2.53)$$

Let $F(v)$ be made arbitrarily periodic in the frequency coordinates, as shown in Fig. 2.8, with a period of $2v_m$. Thus, in the frequency domain, $F(v)$ can be expanded in a Fourier series:

$$F(v) = \sum_{n=-\infty}^{\infty} K_n \exp\left(\frac{i\pi nv}{v_m}\right) \quad \text{for } |v| \le v_m \qquad (2.54)$$

and

$$F(v) = 0 \quad \text{for } |v| > v_m \qquad (2.55)$$

with the corresponding Fourier coefficient K_n defined by

$$K_n = \int_{-v_m}^{v_m} F(v) \exp\left(-i\frac{\pi nv}{v_n}\right) dv \qquad (2.56)$$

But since $F(v)$ is the Fourier transform of $f(t)$, then $f(t)$ can be written as

$$f(t) = \int_{-v_m}^{v_m} F(v) \exp(i2\pi vt)\, dv \qquad (2.57)$$

In particular, at sampling points $t = -n/2v_m$, we have

$$f\left(-\frac{n}{2v_m}\right) = \int_{-v_m}^{v_m} F(v) \exp\left(-i\frac{\pi nv}{v_m}\right) dv = K_n \qquad (2.58)$$

which is equal to the Fourier coefficient K_n of Eq. (2.56). Thus, we see that $f(t)$ is given at various Nyquist intervals ($t = n/2v_m$); then the corresponding Fourier coefficient K_n can be obtained. But, from Eq. (2.54) we see that $F(v)$ can in

turn be determined, and from Eq. (2.57) that knowledge of $F(v)$ implies a knowledge of $f(t)$. Therefore, if we substitute Eq. (2.54) into Eq. (2.57), we have

$$f(t) = \int_{-v_m}^{v_m} \sum_{n=-\infty}^{\infty} K_n \exp\left(\frac{i\pi nv}{v_m}\right) \exp(i2\pi vt)\, dv \qquad (2.59)$$

By interchanging the integration and summation of Eq. (2.59), we obtain

$$f(t) = \sum_{n=-\infty}^{\infty} K_n \frac{\sin 2\pi v_m(t + n/2v_m)}{2\pi v_m(t + n/2v_m)} \qquad (2.60)$$

By substituting Eq. (2.58) into Eq. (2.60), we have

$$f(t) = \sum_{n=-\infty}^{\infty} f\left(\frac{n}{2v_m}\right) \frac{\sin 2\pi v_m(t - n/2v_m)}{2\pi v_m(t - n/2v_m)} \qquad (2.61)$$

where all the positive and negative signs of n have been adjusted.

From Eq. (2.61), we see that each sample of $f(t)$ is multiplied by a weighting factor $[(\sin x)/x]$, called a *sampling function*, which located every sample point of occurrence. This is in fact the output response of an ideal low-pass filter having a cutoff frequency at v_m (Fig. 2.6) when the samples $f(n/2v_m)$ are applied at the input end of the filter.

It is also noted that Shannon essentially used this concept, degrees of freedom, in the application of his sampling theory.

In concluding this section, we point out that the sampling theorem can be extended to higher-dimensional band-limited functions. For example, with a two-dimensional band-limited function $f(x, y)$, it can be shown that

$$f(x, y) = \sum_{n=-\infty}^{\infty} \sum_{m=-\infty}^{\infty} f(nx_0, my_0) \frac{\sin p_0(x - nx_0) \sin q_0(y - my_0)}{p_0(x - nx_0) q_0(y - my_0)}$$

$$(2.62)$$

where p_0 and q_0 are the highest angular frequencies with respect to the x and y coordinates of $f(x, y)$, and $x_0 = \pi/p_0$ and $y_0 = \pi/q_0$ are the respective Nyquist sampling intervals.

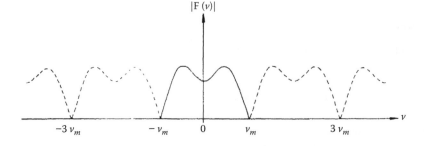

FIG. 2.8 Periodic representation of the Fourier spectrum.

2.6 GABOR'S INFORMATION CELL

In 1946, Gabor [8] published a paper entitled "Theory of Communication" in the *Journal of the Institute of Electrical Engineers*. This was about 2 years before Shannon's [9] classical article, "A Mathematical Theory of Communication," appeared in the *Bell System Technical Journal*. Several of Gabor's concepts of information were quite consistent with Shannon's theory of information. In this section, we illustrate briefly one of his concepts, the information cell, which is related to the uncertainty principle in quantum mechanics. However, for a detailed discussion of Gabor's [8] work on communication theory, we refer to his original papers.

Let us first take the frequency and the time variable to form a finite two-dimensional *Euclidean space*, as shown in Fig. 2.9, with v_m the maximum frequency limit and T the finite time sample of the signal function $f(t)$. This frequency–time space can be subdivided into elementary information elements or cells, which Gabor called *logons*, such as

$$\Delta v \Delta t = 1 \qquad (2.63)$$

We quickly recognize that Eq. (2.63) is essentially the lower bound of the uncertainty relation of Eq. (2.28). However, it is noted that the signal in each of the information cells has two possible *elementary signals*, a symmetric one and an *antisymmetric* one, with the same bandwidth Δv and the same duration Δt. Furthermore, the amplitudes of both these elementary signals should be given so that the signal function $f(t)$ can be uniquely described. From Fig. 2.9, we see that for a given highest-frequency content v_m and a finite time sample, the total number of information cells is

$$N_1 = v_m T \qquad (2.64)$$

However, since each of the information cells contains two elementary functions, the total number of elementary functions is

$$N = 2N_1 = 2v_m T \qquad (2.65)$$

which is the result of Eq. (2.48).

The *shapes* of the information cells are not particularly critical, but their unit area $\Delta v \Delta t = 1$ is. Moreover, from sampling technique we see that the information cells are on the horizontal axis of the time coordinate, and as in Fourier analysis, the information cells are on the vertical axis of the frequency coordinate. For the elementary signals, Gabor [8,10] suggested the use of *Gaussian cosine* and *Gaussian sine* signals, also called wavelets, as shown in Fig. 2.10.

It is seen from Eq. (2.63) that the elementary information cell suggested by Gabor is in fact the lower bound of the Heisenberg uncertainty principle in quantum mechanics:

$$\Delta E \, \Delta t = h \qquad (2.66)$$

Since $E = h_v$, we see that Eq. (2.66) is the same as Eq. (2.63).

It is emphasized that the band-limited signal must be a very special type of function; that is, the function must be well behaved. It contains no discontinuity or sharp angles and has only rounded-off features, as shown in Fig. 2.11. Thus, this signal must be *analytic* over T.

In concluding this section, we note that the concept of Gabor's information cell can be applied to functions of a spatial domain. For example, given a spatial band-limited signal $f(x)$, the corresponding elementary information cell, in the angular spatial frequency p and spatial coordinate x, is

$$\frac{1}{2\pi} \Delta p \, \Delta x = 1 \qquad (2.67)$$

The relation of Eq. (2.67) is used in calculating the spatial channel capacity in Chapter 3.

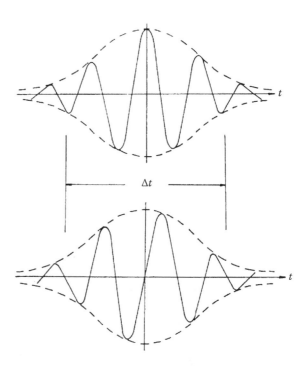

FIG. 2.10 Cosine and sine elementary signals, with Gaussian envelope.

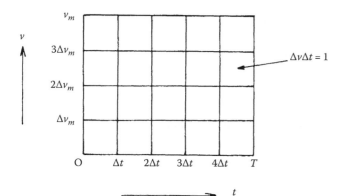

FIG. 2.9 Gabor's information cell.

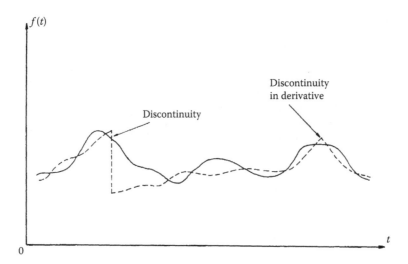

FIG. 2.11 Typical example of a band-limited signal.

2.7 SIGNAL DETECTION

A problem of considerable importance in optical communication and information processing is the detection of a signal embedded in random noise. Therefore, in this section, we discuss a special type of optimum linear filter known as a *matched filter*, which is remarkably useful in optical signal detection and in optical pattern or character recognition. We derive an expression for the filter transfer function on a somewhat general basis, namely, for a stationary additive noise.

It is well known in communication theory that the signal-to-noise ratio at the output end of a correlator can be improved to a certain extent. Let us consider the input excitation to a linear filtering system to be an additive mixture of a signal $s(t)$ and a stationary random noise $n(t)$:

$$f(t) = s(t) + n(t) \qquad (2.68)$$

Let the output response of the linear filter due to the signal $s(t)$ alone be $s_0(t)$, and that due to the random noise $n(t)$ alone be $n_0(t)$. The figure of merit on which the filter design is based is the output signal-to-noise ratio at $t = 0$:

$$\frac{S}{N} \triangleq \frac{|s_0(0)|^2}{\sigma^2} \qquad (2.69)$$

where σ^2 is the mean square value of the output noise.

In terms of the filter transfer function $H(v)$ and the Fourier transform $S(v)$ of the input signal $s(t)$, these quantities can be written as

$$s_0(0) = \int_{-\infty}^{\infty} H(v)S(v)\,dv \qquad (2.70)$$

and

$$\sigma^2 = \int_{-\infty}^{\infty} |H(v)|^2\,N(v)\,dv \qquad (2.71)$$

where $|H(v)|^2 N(v)$ is the power spectral density of the noise at the output end of the filter, and $N(v)$ is the power spectral density of the noise at the input end. Thus, the output signal-to-noise ratio can be expressed explicitly in terms of the filter function $H(v)$:

$$\frac{S}{N} = \frac{\left| \displaystyle\int_{-\infty}^{\infty} H(v)S(v)\,dv \right|^2}{\displaystyle\int_{-\infty}^{\infty} |H(v)|^2\,N(v)\,dv} \qquad (2.72)$$

The objective of the filter design is to specify a filter function such that the output signal-to-noise ratio is maximum. To obtain such a filter transfer function, we can apply the *Schwarz inequality*, which states that

$$\frac{\left| \displaystyle\int_{-\infty}^{\infty} u(t)v^*(t)\,dt \right|^2}{\displaystyle\int_{-\infty}^{\infty} |u(t)|^2\,dt} \leq \int_{-\infty}^{\infty} |v^*(t)|^2\,dt \qquad (2.73)$$

where $u(t)$ and $v(t)$ are arbitrary functions, and the asterisk denotes the complex conjugate. The equality in Eq. (2.73) holds if and only if $u(t)$ is proportional to $v(t)$.

To make the Schwarz inequality applicable to Eq. (2.72), it may be expedient to express the output noise spectral density as the product of the two conjugate factors:

$$N(v) = N_1(v)N_1^*(v) \qquad (2.74)$$

Then Eq. (2.72) can be written as

$$\frac{S}{N} = \frac{\left| \displaystyle\int_{-\infty}^{\infty} [H(v)N_1(v)][S^*(v)/N_1^*(v)]^*\,dv \right|^2}{\displaystyle\int_{-\infty}^{\infty} |H(v)N_1(v)|^2\,dv} \qquad (2.75)$$

If we identify the bracketed quantities of Eq. (2.75) as $u(t)$ and $v(t)$, in view of the Schwarz inequality, we then have

$$\frac{S}{N} \le \int_{-\infty}^{\infty} \frac{|S(v)|^2}{N(v)} \, dv \tag{2.76}$$

The equality in Eq. (2.76) holds if and only if the filter function is

$$H(v) = K \frac{S*(v)}{N(v)} \tag{2.77}$$

where K is a proportional constant. The corresponding value of the output signal-to-noise ratio is therefore

$$\frac{S}{N} = \int_{-\infty}^{\infty} \frac{|S(v)|^2}{N(v)} \, dv \tag{2.78}$$

It is interesting to note that if the stationary additive noise is white (i.e., the noise spectral density is uniform over all the frequency domain), then the optimum filter function is

$$H(v) = KS*(v) \tag{2.79}$$

which is proportional to the conjugate of the signal spectrum. This optimum filter is then said to be matched to the input signal $s(t)$. The output spectrum of the matched filter is therefore proportional to the power spectral density of the input signal:

$$G(v) = K|S(v)|^2 \tag{2.80}$$

Consequently, we see that the phase variation at the output end of the matched filter vanishes. In other words, the matched filter is capable of eliminating all the phase variations of $S(v)$ over the frequency domain.

2.8   STATISTICAL SIGNAL DETECTION

In Section 2.7, we illustrated that one of the objectives in signal filtering is to increase the signal-to-noise ratio. In fact, the increase in signal-to-noise ratio is purposely to minimize the probability of error in signal detection. It is noted, however, that in certain signal detections, an increase in the signal-to-noise ratio does not necessarily guarantee minimizing the probability of error. But minimization of the probability of error can *always* be achieved by using the *decision process*. That is, the detection of signals can be achieved by making decisions about the presence or absence of the signals. This decision making gives rise to the basic concept of *threshold decision level*. Thus, the decision must be made with respect to a threshold level chosen under the minimum probability of

error criterion. We see that if a decision is made that a signal was present but that it was caused by noise, then this event is called a *false alarm*. However, if a decision is made that no signal was present when a signal was actually present, then this event is called a *miss*. Therefore, for an optimum decision, the probability of error must be minimized; if the signal-to-noise ratio is high, *a priori*, then the threshold level can be raised higher, and the probability of error can be reduced. Thus, the minimization of error probability can be achieved at a cost of high signal energy.

Let us consider the detection of *binary signals*. It is possible for us to establish a *Bayes' decision rule*. Through Bayes' *theorem*, the conditional probabilities have the relation

$$P(a/b) = \frac{P(a)P(b/a)}{P(b)} \tag{2.81}$$

One can write

$$\frac{P(a=0/b)}{P(a=1/b)} = \frac{P(a=0)P(b/a=1)}{P(a=1)P(b/a=1)} \tag{2.82}$$

where $P(a)$ is the a priori probability of a; that is, $a = 1$ corresponds to the signal presented and $a = 0$ corresponds to no signal.

We see that a logical decision rule is that if $P(a = 0/b) > P(a = 1/b)$, then we decide that there is no signal ($a = 0$) for a given b. However, if $P(a = 0/b) < P(a = 1/b)$, then we decide that there is a signal ($a = 1$) for a given b. Thus, from Eq. (2.82), Bayes' decision rule can be written as follows:

Accept $a = 0$ if

$$\frac{P(b/a=0)}{P(b/a=1)} > \frac{P(a=1)}{P(a=0)} \tag{2.83}$$

Accept $a = 1$ if

$$\frac{P(b/a=0)}{P(b/a=1)} < \frac{P(a=1)}{P(a=0)} \tag{2.84}$$

It is clear that two possible errors can occur: if we accept that the received event b contains no signal ($a = 0$), but the signal does in fact present ($a = 1$), and vice versa, if we accept that event b contains a signal ($a = 1$), but the signal in fact is not present ($a = 0$). In other words, the error of accepting $a = 0$ when $a = 1$ has actually occurred is a miss, and the error of accepting $a = 1$ when $a = 1$ has actually not occurred is a false alarm.

Let us assign the cost values C_{00}, C_{01}, C_{10}, and C_{11}, respectively, to the following cases: (1) $a = 0$ is actually true, and the decision is to accept it; (2) $a = 0$ is actually true, and the decision is to reject it; (3) $a = 1$ is actually true, and the decision is to reject it; and (4) $a = 1$ is actually true, and the decision is to accept it. Thus, the overall *average cost* is

$$\bar{C} = \sum_{i=0}^{1} \sum_{j=0}^{1} C_{ij} P(a_i) P(B_j / a_i) \qquad (2.85)$$

where $a_0 = 0$, $a_1 = 1$, $P(a_i)$ is the *a priori* probability of a_i, and $P(B_j/a_i)$ is the conditional probability that b falls in B_j if a_i is actually true, as shown in Fig. 2.12.

To minimize the average cost, it is desirable to select a region B_0, where $B_1 = B - B_0$, in such a manner that the average cost is minimum, in other words, to place certain restrictions on the cost values C_{ij} so that $\bar{C}$ will be minimum for a desirable B_0. For example, a miss or a false alarm may be costlier than correct decisions:

$$C_{10} > C_{11} \qquad (2.86)$$

and

$$C_{01} > C_{00} \qquad (2.87)$$

But since

$$P(B_j / a_i) = \int_{B_j} p(b/a_i) \, db \quad i, j = 0, 1 \qquad (2.88)$$

and

$$P(B_1 / a_i) = 1 - P(B_0 / a_i) \quad i = 0, 1 \qquad (2.89)$$

by substituting Eqs. (2.86) and (2.87) into Eq. (2.85), we have

$$\begin{aligned} \bar{C} = {} & C_{01} P(a = 0) + C_{11} P(a = 1) \\ & + \int_{B_0} P(a = 1)(C_{10} - C_{11}) p(b/a = 1) \, db \\ & - \int_{B_0} P(a = 0)(C_{01} - C_{00}) p(b/a = 0) \, db \end{aligned} \qquad (2.90)$$

where the first and second terms are positive constants.

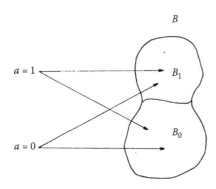

FIG. 2.12   Hypothesis of the received events for two possible transmitted events.

In the minimization of C, we select an optimum region of B_0. But because of Eqs. (2.86) and (2.87), we see that it is sufficient to select region B_0 such that the second integral of Eq. (2.90) is larger than the first integral. Thus, equivalently,

$$\frac{P(b/a = 0)}{P(b/a = 1)} > \frac{P(a = 1)(C_{10} - C_{11})}{P(a = 0)(C_{01} - C_{00})} \qquad (2.91)$$

Now let us write

$$\alpha \triangleq \frac{P(b/a = 0)}{P(b/a = 1)} \qquad (2.92)$$

which is the *likelihood ratio*, and

$$\beta \triangleq \frac{P(a = 1)(C_{10} - C_{11})}{P(a = 0)(C_{01} - C_{00})} \qquad (2.93)$$

which is simply a constant incorporated with the *a priori* probabilities and the error costs. The decision rule is to select the hypothesis for which the signal is actually absent, if $\alpha > \beta$.

If the inequality of Eq. (2.91) is reversed ($\alpha < \beta$), then one chooses B_1 instead. In other words, Bayes' decision rule (Eq. (2.91)) ensures a minimum average cost for the decision making.

Furthermore, if the costs of the errors are equal, $C_{10} = C_{01}$, then the decision rule reduces to Eqs. (2.83) and (2.84). If the decision making has sufficient information on the error costs, then one uses Bayes' decision rule of Eq. (2.91) to begin with. However, if information on the error costs is not provided, then one uses the decision rule as given by Eqs. (2.83) and (2.84).

It is also noted that the Bayesian decision process depends on the *a priori* probabilities $P(a = 0)$ and $P(a = 1)$. However, if the *a priori* probabilities are not provided but one wishes to proceed with decision making alone, then the likelihood ratio test can be applied. That is, if

$$\frac{P(b/a = 0)}{P(b/a = 1)} > 1 \qquad (2.94)$$

then one accepts $a = 0$ for the received event of b. But, if

$$\frac{P(b/a = 0)}{P(b/a = 1)} < 1 \qquad (2.95)$$

then one accepts $a = 1$.

From Eq. (2.91), we see that Eq. (2.94) implies that

$$P(a = 1)(C_{10} - C_{11}) = P(a = 0)(C_{01} - C_{00})$$

Thus, if $C_{10} - C_{11} = C_{01} - C_{00}$, then the *a priori* probabilities of $P(a)$ are equal. It is noted that applications of the likelihood ratio test are limited, since the decision making takes places without knowledge of the *a priori* probabilities. Thus, the results frequently are quite different from those from the minimum error criterion.

Although there are times when there is no satisfactory way to assign appropriate *a priori* probabilities and error costs, there is a way of establishing an optimum decision criterion, namely, the *Neyman–Pearson criterion* [11]. One allows a fixed false-alarm probability and then makes the decision in such a way as to minimize the miss probability.

Since it is not our present objective to discuss this subject in detail, we refer interested readers to the texts by Middleton [12], Selin [13], and Van Trees [14].

2.9   SIGNAL RECOVERING

An interesting problem in optical processing is signal recovering (e.g., image restoration). The type of optimum filter is the solution based on minimization of the mean square error [15], as given by

$$\min \left\langle \overline{\varepsilon^2(t)} \right\rangle = \min \left\langle \lim_{T \to \infty} \frac{1}{2T} \int_{-\infty}^{\infty} [f_o(t) - f_d(t)]^2 \, dt \right\rangle \quad (2.96)$$

where $f_o(t)$ and $f_d(t)$ are the actual output and the desired output response of the filter. A linear filter that is synthesized by using the minimum mean square error criterion is known as the *optimum linear filter*.

Let us look at the mean square error performance of a linear filter function as given by

$$\overline{\varepsilon^2(t)} = \lim_{T \to \infty} \frac{1}{2T} \int_{-T}^{T} \left[\int_{-\infty}^{\infty} h(\tau) f_i(t - \tau) \, d\tau - f_d(t) \right]^2 dt \quad (2.97)$$

where $h(t)$ is impulse response of the filter and $f_i(t)$ is the input signal function. By expanding the preceding equation, it can be shown that the mean square error can be written as

$$\overline{\varepsilon^2(t)} = \int_{-\infty}^{\infty} h(\tau) \, d\tau \int_{-\infty}^{\infty} h(\sigma) \, d\sigma R_{ii}(\tau - \sigma)$$
$$-2 \int_{-\infty}^{\infty} h(\tau) \, d\tau R_{id}(\tau) + R_{dd}(0) \quad (2.98)$$

where $h(t)$ is the impulse response of the filter, $R_{ii}(x)$ is the autocorrelation function of the input signal, $R_{id}(\tau)$ is the cross-correlation of the input with the desired output function, and $R_{dd}(0)$ is the mean square value of the desired output function, as given by

$$R_{ii}(\tau) = \int_{-\infty}^{\infty} f_i(t) f_i(t + \tau) \, dt \quad (2.99)$$

$$R_{id}(\tau) = \int_{-\infty}^{\infty} f_i(t) f_d(t + \tau) \, dt \quad (2.100)$$

$$R_{dd}(\tau) = \int_{-\infty}^{\infty} f_d(t) f_d(t + \tau) \, dt \quad (2.101)$$

$$R_{dd}(0) = \int_{-\infty}^{\infty} \left[f_d^2(t) \right] dt \quad (2.102)$$

It is trivial to see that the mean square error is a function of $h(t)$, which is the center problem to determine $h(t')$ by minimizing the error. Since the minimization of functionals is a subject to *calculus of variation*, which leads to the *Wiener–Hopf* integral equation as given by

$$\int_{-\infty}^{\infty} h(\sigma) R_{ii}(\tau - \sigma) \, d\sigma - R_{id}(\tau) = 0 \quad (2.103)$$

we note that by following the realizable constraint of a physical filter, strictly speaking the preceding result should be written as

$$\int_{-\infty}^{\infty} h(\sigma) R_{ii}(\tau - \sigma) \, d\sigma - R_{id}(\tau) \neq 0 \quad \text{for } \tau < 0 \quad (2.104)$$

$$\int_{-\infty}^{\infty} h(\sigma) R_{ii}(\tau - \sigma) \, d\sigma - R_{id}(\tau) = 0 \quad \text{for } \tau \geq 0 \quad (2.105)$$

An interesting point of the Wiener–Hopf integral equation is the fact that it corresponds to the *Euler–Lagrange* differential equation in the calculus of variation. To further emphasize the Wiener–Hopf equation such as

$$R_{id}(\tau) = \int_{-\infty}^{\infty} h_{opt}(\tau) R_{ii}(\tau - \sigma) \, d\sigma \quad (2.106)$$

in which the optimum filter $h_{opt}(\tau)$ is remained to be determined.

If the desired output could be obtained by assuming without the physical realizable constraint, the input–output cross-correlation of the optimum filter is, in fact, the input and the desired output cross-correlation itself, which holds not only for $\tau \geq 0$ but also for $\tau < 0$. Now the problem becomes a trivial one; by taking the Fourier transform of Eq. (2.106), we have

$$R_{id}(v) = H_{opt}(v) R_{ii}(v) \quad (2.107)$$

where $H_{opt}(v)$ is the transform function of the optimum filter, which can be written as

$$H_{opt}(v) = \frac{R_{id}(\omega)}{R_{ii}(\omega)} \quad (2.108)$$

That is, in fact, the solution of our problem.

Mention must be made that we do not exactly know the desired output response, but we have been demanding the best filter by minimizing the mean square error, which gives us the *best* approximation. In other words, with this method we try to get the best result we can, similar to bargaining with a used car salesman. Because we demanded that more than a filter can be synthesized, the input–output cross-correlation *cannot* be equal to the cross-correlation of the input and the desired output. According to the Wiener–Hopf equation, the filter design should be based on the minimum error criterion and the impulse response should be compelled to the inequality of $\tau \geq 0$, but allow it to vary somewhat for $\tau < 0$, such as

$$q(\tau) = R_{\text{id}}(\tau) - \int_{-\infty}^{\infty} h_{\text{opt}}(\sigma) R_{\text{ii}}(\tau - \sigma) \, d\sigma \qquad (2.109)$$

where $q(\tau) \neq 0$ for $\tau < 0$ but vanishes for $\tau > 0$.

For the readers who are interested in the Wiener–Hopf solution, see the text by Lee [15].

2.10 SIGNAL AMBIGUITY

The choice of a transmitted waveform for radar seems more trivial than it is for communication. However, it is actually not so fundamental, as it does not contain any information. Nevertheless, it is the reflected (or echo) wave from the target that provides the temporal and the Doppler shift (i.e., frequency), which determines the range and radial velocity information of the target. Whenever time and frequency are mentioned, we would anticipate to relate the quantities to the uncertainty relationship. In other words, no waveform can occupy a very short time and have a very narrow bandwidth. Thus, there is a limitation imposed on the combined resolution in time and frequency [16].

To simplify the analysis, we set scaler wave field $u(t)$ to be constrained by the following equation:

$$\int |u(t)|^2 \, dt = \int |U(v)|^2 \, dv = 1 \qquad (2.110)$$

In view of the Wiener–Khinchin and Parseval theorems, we see that in the time domain we have

$$R(\tau) = \int u(t)u^*(t + \tau) \, dt = \int |U(v)|^2 \, e^{i2\pi v \tau} \, dv \quad (2.111)$$

and

$$T = \int |R(\tau)|^2 \, d\tau = \int |u(v)|^4 \, dv$$

where the superscript asterisk denotes the complex conjugate. Similarly, in frequency domain we have

$$K(v) = \int U^*(v')U(v' + v) \, dv' = \int |u(t)|^2 \, e^{-i2\pi v t} \, dt \quad (2.112)$$

and

$$F = \int |K(v)|^2 \, dv = \int |u(t)|^4 \, dt \qquad (2.113)$$

where F is defined as the frequency resolution cell, which provides the well-known uncertainty relationship as given by (see Eq. (2.63))

$$(\text{Frequency resolution cell}) \times (\text{Time span}) = 1 \quad (2.114)$$

Notice that Eq. (2.111) applies frequently to the resolution of detecting stationary targets at different ranges, while Eq. (2.112) applies to the resolution of detecting moving targets at different radial velocities. However, when the targets are both at different ranges, and moving with different radial velocities, neither quantity being known in advance, the separation of time (e.g., range) and frequency (i.e., radial velocity) resolution does not always provide the actual resolving power of the target. Thus, we need a more general description of time and frequency shift to interpret the resolving power of the signal, which is known as the *ambiguity function*, as given by

$$\chi(\tau, v) = \int u(t)u^*(t + \tau) e^{-i2\pi v \tau} \, dt = \int U^*(v')U(v' + v) e^{-i2\pi v' \tau} \, dv'$$

$$(2.115)$$

The significance of $\chi(\tau, v)$ is that the range and (radial) velocity of a target cannot be resolved at $(\tau_0 + \tau, v_0 + v)$, where τ_0 and v_0 are the mean range and velocity of the target. By virtue of normalization, we let

$$\int \int |\chi(\tau, v)|^2 \, d\tau \, dv = 1 \qquad (2.116)$$

which is called the *area of ambiguity* in the time–frequency domain.

Example

One of the simplest examples of ambiguity function is obtained from a single Gaussian pulse as given by

$$\mu(t) = \sqrt[4]{2} e^{-\pi t^2}$$

By substituting into Eq. (2.115), we have

$$\chi(\tau, v) = \exp\left(-\frac{1}{2}\pi\tau^2\right)\exp\left(-\frac{1}{2}\pi v^2\right)\exp(i\pi v \tau)$$

From the preceding equation, we see that the magnitude of ambiguity distribution, as given by

$$|\chi(\tau, v)|^2 = e^{-\pi(\tau^2 + v^2)}$$

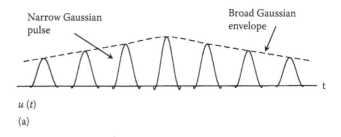

Narrow Gaussian pulse   Broad Gaussian envelope

$u(t)$

(a)

FIG. 2.13   Ambiguity of a Gaussian pulse.

describes a circular pattern in the (τ, v) domain, as can be seen in Fig. 2.13. If the pulse is shortened, it can be seen that the circular ambiguity pattern would become elliptical, as illustrated in Fig. 2.14.

A less trivial example is that $u(t)$ represents a train of Gaussian pulses, modulated by a broad Gaussian envelope, as shown in Fig. 2.15a. The corresponding ambiguity $|\chi(\tau, v)|^2$ is sketched in Fig. 2.15b, in which the ambiguity in frequency resolution is split up into narrow bands, whereas in time it is repeated itself. Notice that the shaded areas of the redistributed ambiguity are in fact equal to the area of the single-pulse ambiguity, i.e.,

$$\int\int |\chi(\tau, v)|^2 \, d\tau \, dv = 1 \qquad (2.117)$$

Mention must be made that in actual application, it is not our interest to resolve all the ranges of the target for which the redundant parts of the ambiguity diagram remote from the origin have no practical usage.

For example, a target 10 mi away should not be confused as being at 1010 mi. Therefore, in practice, we may draw a rectangular region around the origin of the (τ, v) coordinate plane to discriminate the ambiguity outside the region. Thus, we see that the pulse train example will be a more acceptable waveform for range and (radial) velocity discrimination than the ambiguity distribution obtained from

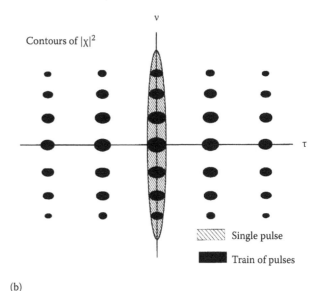

Contours of $|\chi|^2$

▨ Single pulse

■ Train of pulses

(b)

FIG. 2.15   (a) Gaussian pulse train. (b) Ambiguity diagram of (a).

single pulse. However, when interpreting the ambiguity diagram, one should remember that τ and v do not represent the actual range and (radial) velocity, but rather the difference between the ranges and velocities of any two targets that need to be resolved.

2.11   WIGNER SIGNAL REPRESENTATION

There is another form of time–frequency signal representation, similar to the ambiguity function, defined by [17,18]

$$W(\tau, v) = \int u(t)u^*(\tau - t)e^{-i2\pi vt} \, dt = \int U^*(v')U(v - v')e^{i2\pi v'\tau} \, dv'$$

$$(2.118)$$

which is known as the *Wigner distribution function* (WDF). Instead of using correlation operator [i.e., $u(t)u^*(t + \tau)$], Wigner used the convolution operator (i.e., $u(t)u^*(\tau - t)$) for his transformation. The physical properties of the WDF can be shown as

$$\int W(\tau, v) \, dv = |u(v)|^2 \qquad (2.119)$$

$$\int W(\tau, v) \, d\tau = |U(v)|^2 \qquad (2.120)$$

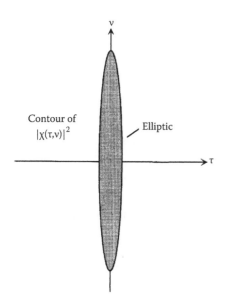

Contour of $|\chi(\tau, v)|^2$ Elliptic

FIG. 2.14   Ambiguity of a very short Gaussian pulse.

$$\int W(\tau, v)\, d\tau\, dv = 1 \qquad (2.121)$$

in which the WDF has been normalized to unity for simplicity.

One of the interesting features of WDF is the inversion. Apart from the association with a constant phase factor, the transformation is unique, as can be shown:

$$\int W(\tau, v) e^{-i4\pi v\tau}\, d\tau = U(2v)U^*(0) \qquad (2.122)$$

$$\int W(\tau, v) e^{i4\pi v\tau}\, dv = u(2\tau)u^*(0) \qquad (2.123)$$

If the constant phase factors $u^*(0)$ or $U^*(0)$ happen to be very small or even zero, then we can reformulate the transformations by using the maximum value of $|u(\tau)|^2$ or $|U(\tau)|^2$, which occurs at τ_{max} or f_{max}, respectively, as given by

$$\int W(\tau, v) \exp[-4\pi(v - v_{max})\tau]\, d\tau = U(2f - v_{max})U^*(v_{max}) \qquad (2.124)$$

$$\int W(\tau, v) \exp[i4\pi v(\tau - \tau_{max})]\, dv = u(2\tau - \tau_{max})u^*(\tau_{max}) \qquad (2.125)$$

In other words, WDF does provide the frequency content of a time-varying signal, instead of providing the time–frequency resolution as the ambiguity function does.

Examples

In the following, we provide a few simple examples of the WDF. First, let us consider a single-tone signal as given by

$$u(t) = e^{i2\pi v_0 t}$$

By substition into Eq. (2.118), we have

$$W(\tau, v) = \delta(v - v_0)$$

which is shown in Fig. 2.16a.

If the signal is a single-pulse function, such as

$$u(t) = \delta(t - t_0)$$

then the corresponding WDF would be

$$W(\tau, v) = \delta(\tau - t_0)$$

which is plotted in Fig. 2.16b.

(a)

(b)

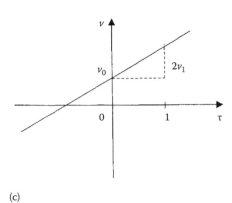

(c)

FIG. 2.16 Samples of Wigner distributions (a) for a single-tone signal, (b) for a pulse signal, and (c) for a chirp signal.

On the other hand, if it is a chirp signal as given by

$$u(t) = \exp(i2\pi v_1 t^2 + v_0 t)$$

then the corresponding WDF can be shown as

$$W(\tau, v) = \delta(v - 2v_1\tau - v_0)$$

as illustrated in Fig. 2.16c.

2.12 FOURIER TRANSFORM PROPERTIES OF LENSES

It is extremely useful to obtain a two-dimensional Fourier transformation from a positive lens [19,20]. Fourier transform operations usually bring to mind complicated electronic spectrum analyzers or digital computers. However, this complicated transformation can be performed extremely simply in a coherent optical system, and because the optical transform is two-dimensional, it has a greater information capacity than transforms carried out by means of electronic circuitry.

In the derivation of the Fourier transform properties of a lens, we call attention to the Fresnel–Kirchhoff theory discussed in Section 2.2. The complex light field on a planar surface (α, β) due to a point source at the (x, y) plane, as shown in Fig. 2.4, can be described by

$$h_l(\sigma - \rho) = C \exp\left\{i\frac{\pi}{\lambda l}[(\alpha - x)^2 + (\beta - y)^2]\right\} \quad (2.126)$$

where $C = \exp(ikl)/l$ is a complex constant. Equation (2.126) is known as a *spatial impulse response* (Section 2.2).

Thus, the complex light field due to a coherent source $f(x, y)$ can be written as

$$g(\alpha, \beta) = \int\limits_{-\infty}^{\infty}\int\limits_{\infty} f(x, y) h_l(\sigma - \rho)\, dx\, dy \quad (2.127)$$

which is, equivalently,

$$g(\alpha, \beta) = f(x, y) * h_l(\sigma - \rho)\, dx\, dy \quad (2.128)$$

where the asterisk denotes the convolution operation.

However, for Fourier transformation, a lens is needed for the operation. Now let us consider Fig. 2.17, a simple optical system. If the complex light field at P_1 is $f(\xi, \eta)$, then the complex light distribution at P_2 can be written

$$g(\alpha, \beta) = C\{[f(\xi, \eta) * h_l(\xi, \eta)]T(x, y)\} * h_f(x, y) \quad (2.129)$$

where C is an arbitrary complex constant, $h_l(\xi, \eta)$ and $h_f(x, y)$ are the corresponding spatial impulse responses, and $T(x, y)$ is the *phase transformation* of the lens. Equation (2.129) can be written in the integral form:

$$g(\alpha, \beta) = C\int\int\limits_{S_1}\left[\int\int\limits_{S_2} \exp\left(i\frac{k}{2}\Delta\right) dx\, dy\right] f(\xi, \eta)\, d\xi\, d\eta \quad (2.130)$$

where S_1 and S_2 denote the surface integrals of the light field P_1 and the lens T, respectively, and

$$\Delta = \left\{\frac{1}{l}[(x - \xi)^2 + (y - \eta)^2] + \frac{1}{f}[(\alpha - x)^2 + (\beta - y)^2 - (x^2 - y^2)]\right\} \quad (2.131)$$

It is noted that the surface integral of the lens may be assumed to be of infinite extent, since the lens is very large compared with the spatial apertures at P_1 and P_2 paraxiality.

Equation (2.131) can be written as

$$\Delta = \frac{1}{f}(\mu\xi^2 + \mu x^2 + \alpha^2 - 2\mu\xi x - 2x\alpha - \mu\eta^2 \\ + \mu y^2 + \beta^2 - 2\mu\eta y - 2y\beta) \quad (2.132)$$

where $\mu = f/l$.

By completing the square, Eq. (2.132) can be written as

$$\Delta = \frac{1}{f}[\mu^{1/2}x - \mu^{1/2}\xi - \mu^{-1/2}\alpha)^2 - \alpha^2\frac{1-\mu}{\mu}2\xi\alpha \\ + (\mu^{1/2}y - \mu^{1/2}\eta - \mu^{-1/2}\beta)^2 - \beta^2\frac{1-\mu}{\mu} - 2\eta\beta] \quad (2.133)$$

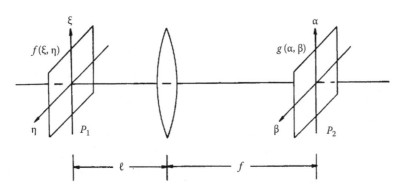

FIG. 2.17 Geometry for determination of the optical Fourier transformation.

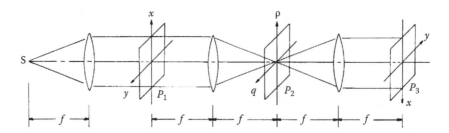

FIG. 2.18 Successive Fourier transformations of lenses. A monochromatic point source is located at S.

By substituting in Eq. (2.130), we have

$$g(\alpha,\beta) = C \exp\left[-i\frac{k}{2f}\left(\frac{1-\mu}{\mu}\right)(\alpha^2 + \beta^2)\right]$$
$$\cdot \iint_{S_1} f(\xi,\eta) \exp\left[-i\frac{k}{f}(\alpha\xi + \beta\eta)\right] d\xi\, d\eta \qquad (2.134)$$
$$\cdot \iint_{S_2} \exp\left\{i\frac{k}{2f}[(\mu^{1/2}x - \mu^{1/2}\xi - \mu^{-1/2}\alpha)^2\right.$$
$$\left. + (\mu^{1/2}y - \mu^{1/2}\eta - \mu^{-1/2}\beta)^2]\right\} dx\, dy$$

Since the integrations over S_2 are assumed to be taken from $-\infty$ to ∞, we obtain a complex constant that can be incorporated with C. Thus, we have

$$g(\alpha,\beta) = C \exp\left[-i\frac{k}{2f}\left(\frac{1-\mu}{\mu}\right)(\alpha^2 + \beta^2)\right]$$
$$\iint_{S_1} f(\xi,\eta) \exp\left[-i\frac{k}{f}(\alpha\xi + \beta\eta)\right] d\xi\, d\eta \qquad (2.135)$$

From this result, we see that except for a spatial quadratic phase variation, $g(\alpha, \beta)$ is the Fourier transform of $f(\xi, \eta)$. In fact, the quadratic phase factor vanishes for $l = f$. Evidently, if the signal plane P_1 is placed at the front focal plane of the lens, the quadratic phase factor disappears, which leaves an exact Fourier transform relation. Thus, Eq. (2.135) can be written as

$$G(p,q) = C_1 \iint_{S_1} f(\xi,\eta) \exp[-i(p\xi + q\eta)]\, d\xi, d\eta \quad \text{for } \mu = 1$$

$$(2.136)$$

where $p = k\alpha/f$ and $q = k\beta/f$ are the spatial frequency coordinates.

It is emphasized that the exact Fourier transform relation takes place under the conditions $l = f$. Under the conditions $l \neq f$, a quadratic phase factor is included. Furthermore, it can easily be shown that a quadratic phase factor also results if the signal plane P_1 is placed behind the lens.

In conventional Fourier transform theory, transformation from the spatial domain to the spatial frequency domain requires the kernel exp $[-i(px + pq)]$, and transformation from the spatial frequency domain to the spatial domain requires the

conjugate kernel exp $[i(px + qy)]$. Obviously, a positive lens always introduces the kernel exp $[-i(px + qy)]$. Since an optical system takes only successive transforms, to perform the inverse transform one can simply reverse the coordinates axes at the output plane, as shown in Fig. 2.18.

REFERENCES

1. M. Born and E. Wolf, *Principle of Optics*, 2nd rev. ed., Pergamon, New York, 1964.
2. F. T. S. Yu, *Introduction to Diffraction, Information Processing, and Holography*, MIT Press, Cambridge, MA, 1973.
3. W. B. Davenport Jr. and W. L. Root, *Random Signals and Noise*, McGraw-Hill, New York, 1958.
4. A. Papoulis, *The Fourier Integral and Its Applications*, McGraw-Hill, New York, 1962.
5. R. N. Bracewell, *The Fourier Transform and Its Applications*, McGraw-Hill, New York, 1965.
6. E. A. Guillemin, *Synthesis of Passive Networks*, John Wiley, New York, 1957.
7. J. I. Powell and B. Crasemann, *Quantum Mechanics*, Addison-Wesley, Reading, MA, 1961.
8. D. Gabor, Theory of Communication, *J. Inst. Electr. Eng.*, vol. 93, 429 (1946).
9. C. E. Shannon, A Mathematical Theory of Communication, *Bell Syst. Tech. J.*, vol. 27, 379, 623 (1948).
10. D. Gabor, Communication Theory and Physics, *Phil. Mag.*, vol. 41, no. 7, 1161 (1950).
11. J. Neyman and E. S. Pearson, On the Use and Interpretation of Certain Test Criteria for Purposes of Statistical Inference, *Biometrika*, vol. 20A, 175, 263 (1928).
12. D. Middleton, *Statistical Communication Theory*, McGraw-Hill, New York, 1960.
13. I. Selin, *Detection Theory*, Princeton University Press, Princeton, NJ, 1965.
14. H. L. Van Trees, *Detection, Estimation, and Modulation Theory*, John Wiley, New York, 1968.
15. Y. W. Lee, *Statistical Theory of Communication*, Wiley, New York, 1961.
16. P. M. Woodward, *Probability and Information Theory, with Application to Radar*, Pergamon Press, London, 1953.
17. E. Wigner, On the Quantum Correction for Thermodynamic Equilibrium, *Phys Rev.*, vol. 40, 49–759 (1932).
18. H. O. Bartelt, K. H. Brenner, and A. W. Lohmann, The Wigner Distribution Function and Its Optical Production, *Opt. Commun.*, vol. 32, 32 (1980).
19. L. J. Cutrono et al., Optical Data Processing and Filtering Systems, *IRE Trans. Inform. Theory*, IT-6, 386 (1960).
20. F. T. S. Yu and S. Jutamulia, *Optical Signal Processing, Computing and Neural Networks*, Wiley-Interscience, New York, 1992.

3 Optical Spatial Channel and Encoding Principles

In Chapter 1, we defined the basic concept of information theory: if H is the average information per unit resolution cell provided by an optical source, and if the source is emitting at a rate of m resolution cells per second, then the rate of information emitted by the optical source is

$$I = mH \quad \text{bits/sec} \tag{3.1}$$

It is noted that in most of the optical information sources, there is some degree of *temporal* and *spatial* redundancy. To reduce this redundancy, it is convenient to replace the original signal (or image) by a different set of new signal elements, that is, optical or image codes. The general procedure for replacing original signals is to perform a certain encoding operation. If the decoding procedure at the receiving end is assumed to be unique, then the encoding and decoding processes will result in no loss of information. However, if the decoding is not a unique process, then some information will be lost. Thus, one of the required restrictions on the basic coding theory is a uniquely decodable code.

It can be seen that numerous uniquely decodable codes exist. However, we consider the set of uniquely decodable codes with which the highest efficiency in information transmission can be achieved. There is no general method of obtaining the most efficient code, since coding depends on the intention to transfer the original signals. In this connection, we center our attention on the optical channel in which the encoded signals (or images) are transmitted or stored.

3.1 OPTICAL SPATIAL COMMUNICATION CHANNEL

In general, an optical channel may be considered a space- and time-variable channel. The channel may be described as a function of spatial coordinates that vary with respect to time, that is, $f(x, y; t)$, where x and y are the orthogonal spatial coordinates. In the following, we discuss the optical channel separately with respect to spatial and temporal domains. General optical channel capacity with respect to this concept is also treated.

If it is assumed that we want the encoded spatial signal (the image) to be transmitted through an optical spatial channel or to be recorded (stored) on a photosensitive plate (photographic plate), then we will search for the smallest spatial bandwidth Δp of the optical spatial channel. The spatial signal transmitted through the channel with a resolution element Δx is

$$\frac{1}{2\pi} \Delta p \, \Delta x = 1 \tag{3.2}$$

where Δp is the spatial bandwidth in radians per unit distance, and Δx is the resolution element in distance.

It is noted that the spatial constraints of Eq. (3.2) may not be sufficient to specify a general optical channel, since a physical channel is usually restricted by a certain time imposition. It is the time constraint that must also be considered. If it is assumed that the optical spatial channel possesses a system bandwidth of $\Delta \omega$ in radians per second, then the shortest time interval Δt required for the channel to respond efficiently will be

$$\Delta t \geq \frac{2\pi}{\Delta \omega} \tag{3.3}$$

Equation (3.3) implies that the smallest bandwidth required for the optical channel for a resolution time interval Δt to be effectively transmitted is

$$\frac{1}{2\pi} \Delta \omega \, \Delta t = 1 \tag{3.4}$$

Now let us consider a set of resolution cells in the spatial domain of the channel. For simplicity, we assume there is an $m \times n$ array of resolution cells, as shown in Fig. 3.1. And we let each cell be centered at each (x_i, y_j), where $i = 1, 2, \ldots, m$, and $j = 1, 2, \ldots, n$. We also select a set of signal levels (i.e., different levels of irradiance) for the optical spatial communication. Once we have adopted these sets of cells and signal levels, an optical spatial communication channel can be defined.

Based on restrictions of Eq. (3.2), we can select the set of resolution cells, and with the constraints of Eq. (3.4), we can consider statistically a set of signal levels quantized into k discrete irradiances, $E_1, E_2, \ldots, E_k$. If we assume that the incremental time duration of the spatial signal is Δt, then we can use it to make the random selection of each of the signal levels within a cell, Thus, the optical spatial channel can be defined, within the time interval Δt, by the $m \times n$ array of resolution cells and the different signal levels E_i.

Now if we assume that the E_i's, within the resolution cells, distributed in the spatial domain are statistically independent, then an arbitrary distribution of the signal levels across the $m \times n$ array of resolution cells can represent a spatially coded message. Thus, for a number of $m \times n$ arrays, a spatially coded message is simply the combination of the individual signal levels E_i over the $m \times n$ array of cells.

Now let N be the total number of these combinations. If these N combinations are assumed, *a priori*, to be equiprobable,

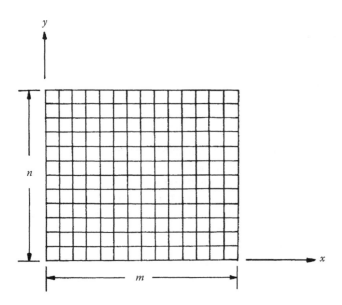

FIG. 3.1 $m \times n$ array of resolution cells.

then the information provided by each of the spatially coded messages (i.e., the combination of E_i over $m \times n$ arrays) is

$$I = \log_2 N \quad \text{bits/message} \tag{3.5}$$

But since the total number of combinations can be written as

$$N = k^{mn} \tag{3.6}$$

the information provided by each of the spatially coded messages can be written as

$$I = mn \log_2 k \quad \text{bits/message} \tag{3.7}$$

The corresponding rate of information transmission through the optical spatial channel is therefore

$$R = \frac{I}{\Delta t} = \frac{mn}{\Delta t} \log_2 k \quad \text{bits/sec} \tag{3.8}$$

It is noted that if the arrays of $m \times n$ resolution cells are not the same size, then the total combinations of the cells N, strictly speaking, is a function over all the spatial plane S, that is, $N(S)$. If we let $\sigma_1, \sigma_2, \ldots, \sigma_q$ be the spatial area elements of corresponding resolution cells, then S can be approximated:

$$S \simeq \sum_{i=1}^{q} \sigma_i \tag{3.9}$$

Now the information provided by each of the spatially coded messages can be approximated:

$$I(S) = \log_2 N(S) \quad \text{bits/message} \tag{3.10}$$

which is a function of S.

The rate of information per unit spatial element transmitted through the spatial channel is therefore

$$R(S) = \frac{I(S)}{S \Delta t} \quad \text{bits/unit area/sec} \tag{3.11}$$

It is interesting to note that the rate changes as S varies. But it can be shown, for all practical cases, that the rate approaches a certain limit as S becomes sufficiently large. It is this limit, originally used by Shannon [1–3], that we adopt to define the capacity of an optical spatial channel:

$$C \triangleq \lim_{S \to \infty} \frac{\log_2 N(S)}{S \Delta t} \quad \text{bits/unit area/sec} \tag{3.12}$$

This is the capacity that corresponds to the maximum average rate of information the spatial channel is capable of conveying under the constraints of Δt of Eq. (3.4). The spatial channel capacity in bits per unit spatial area can also be written as

$$C \triangleq \lim_{S \to \infty} \frac{\log_2 N(S)}{S} \quad \text{bits/unit area} \tag{3.13}$$

This is the limit we adopted for the spatial channel capacity as defined by Shannon [1–3].

It is noted that if the optical channel is reduced to a single resolution cell (e.g., a single optical fiber), then the capacity of the channel can be calculated in a similar manner. For example, let us define a set of quantized signal levels E_i (irradiance levels) of duration t_i, where $i = 1, 2, \ldots, k$, and $t_i \geq \Delta t$ of Eq. (3.4). It is noted that we do not impose any restriction on the time duration; in other words, the duration may be identical. If we assume that the levels E_i are statistically independent over the time variable, then an arbitrary succession of signal levels can represent a coded message in the time domain. Thus, for a long duration of T (i.e., where T is the sum of some t_i), a timely coded message transmitted through a channel element is simply the combination of all possible E_i's. Now let $N(T)$ be a total possible distinct combination of E_i's over T. If all the $N(T)$ possible combinations are assumed, *a priori*, to be equiprobable, then the information provided by each of the timely coded messages (i.e., the combination of E_i's) over T is

$$I(T) = \log_2 N(T) \quad \text{bits/message} \tag{3.14}$$

The corresponding rate at which information is transmitted through a single channel element is

$$R(T) = \frac{I(T)}{T} \quad \text{bits/sec} \tag{3.15}$$

It is clear that the rate varies with T. However, in all practical cases, the rate has a certain limit as T becomes increasingly large. This is the limit, as defined by Shannon [1–3], we used to define the channel capacity of a single resolution cell:

$$C = \lim_{T \to \infty} \frac{\log_2 N(T)}{T} \quad \text{bits/sec} \qquad (3.16)$$

This is the capacity that corresponds to the maximum average rate of information it is possible to convey through a single-element optical channel.

It is also noted that if we have $m \times n$ independent arrays of optical channels whose capacities are defined by Eq. (3.16), then the overall channel capacity is

$$C = \sum_{i=1}^{mn} C_i \quad \text{bits/sec} \qquad (3.17)$$

It is also clear that if all these $m \times n$ channels have an identical time duration of Δt over E_i ($t_1 = t_2 = \cdots = t_R = \Delta t$), then Eq. (3.16) can be reduced to Eq. (3.8), which is identical to the optical spatial channel. Since the time durations are all identical to the lowest limit Δt of the channel, the rate of information is maximum under the constraints of Eq. (3.4). This maximum rate of information transmission is the rate we have obtained for an optical channel with $m \times n$ arrays. It is noted that the result of Eq. (3.8) is also the capacity of the optical spatial channel under the constraints of Δt of Eq. (3.4).

We discuss the definition of channel capacity more thoroughly in Section 3.2, and conditions for its existence are illustrated.

3.2 OPTICAL MESSAGE IN SPATIAL CODING

Let us assume that we have made a selection of discrete signal levels (irradiances) and we want to apply these levels in synthesizing a spatially coded message. This is a typical example in optical spatial encoding. For instance, the distinct irradiance levels may correspond to a basic alphabet (e.g., the English alphabet), and their corresponding possible combinations are coded words. The irradiance levels could also represent basic digits whose sequences give rise to numbers.

It is noted that each of the signal levels used in information transmission represents a certain amount of energy (the cost) required for the transmission. The minimum amount of energy required for efficient information transmission through an optical channel depends on the physical nature of the channel. We investigate only the cost corresponding to the spatial elements of the optical signal. The basic problem in optical spatial communication is how to transmit as many messages as possible over a certain spatial domain S, and keep the required cost at a minimum.

Note that each signal level E_i occupies a set of spatial elements over S, and that many irradiance levels may have spatial cells of identical size. Let us also assume that no constraints or restrictions are imposed on the possible distribution of the signal levels across the spatial domain (i.e., the levels are

statistically independent). We now separate the signal levels according to the size of the spatial elements. Let h_i denote the number of signal levels that have spatial elements σ_i of the same size, where we assume $\sigma_1 < \sigma_2 < \ldots < \sigma_q$, with σ_q representing the largest possible spatial element. Let M denote the total number of h_i:

$$M = \sum_{i=1}^{q} h_i \qquad (3.18)$$

We now seek the total number $N(S)$ of distinct spatial combinations of signal levels over the entire spatial domain S, and obtain

$$N(S) = \sum_{i=1}^{q} h_i N(S - \sigma_i) \qquad (3.19)$$

The above equation implies that the spatially coded messages corresponding to the spatial elements $S - \sigma_1$ can be completed by any one of h_1 signal levels to construct a message S, and in a similar manner for h_2, h_3, ..., h_q.

For simplicity, we assume that the σ_i's are integral values. It can be seen that Eq. (3.19) is a linear difference equation with constant coefficients of h_i. Thus, it has σ_q independent solutions of the following type [4] (see Appendix A):

$$N_n(S) = Z_n^S \quad n = 1, 2, \ldots, \sigma_q \qquad (3.20)$$

and the general solution is the linear combination of Eq. (3.20):

$$N(S) = \sum_{n=1}^{q} K_n Z_n^S \qquad (3.21)$$

where the K_n's are arbitrary constant coefficients.

We can now calculate the capacity of an optical spatial channel, which was defined in Section 3.1,

$$C = \lim_{S \to \infty} \frac{\log_2 N(S)}{S} \quad \text{bits/unit area} \qquad (3.22)$$

Let us consider an optical spatial channel in the case of spatial encoding. For a sufficiently large spatial domain S, the solution of Eq. (3.21) reduces to a dominant term:

$$N(S) = \lim_{S \to \infty} \sum_{n=1}^{q} K_n Z_n^S = K_1 Z_1^S \qquad (3.23)$$

Alternatively, Eq. (3.23) can be written as

$$N(S) = K_1 2^{\beta S} \quad \text{as } S \to \infty \qquad (3.24)$$

where $\beta = \log_2 Z_1$. By substituting Eq. (3.24) into Eq. (3.22), the capacity of the optical spatial channel is found:

$$C = \lim_{S \to \infty} \left(\frac{\log_2 K_1}{S} + \beta \right) = \beta \quad \text{bits/unit area} \quad (3.25)$$

We noted that if the optical spatial channel has a capacity C, and it is used to convey a spatial signal of I bits, then the best possible spatial encoding system that will enable us to utilize the channel efficiently has a spatial rate of $r = C/I$ signal levels per unit area.

In order to prove this general result, we discuss more precisely the physical aspects of a fundamental theorem regarding a noiseless spatial channel. We specify the basic properties of the encoding system so that the spatial channel will be more efficiently utilized. In the result, we show that the best spatial encoding techniques are the ones yielding the most probable distribution of signal levels. Moreover, certain basic encoding techniques are required for obtaining these optimum coding processes.

Let k be the total number of distinguishable signal levels for the spatial optical communication channel, and $n_1, n_2, \ldots, n_k$ the numbers of the respective signal levels $E_1, E_2, \ldots, E_k$ used in the spatially encoded messages. Then the total number of the coded messages is the total number of possible different combinations of the n_i's:

$$N = \frac{\left(\sum_{i=1}^{k} n_i \right)!}{n_1! \, n_2! \cdots n_k!} \quad (3.26)$$

where $\sum_{i=1}^{k} n_i$ is the total number of signal levels used in the encoding over S, which is equivalently equal to the number of resolution cells.

Now if we assume that the signal levels across the spatial domain are statistically independent, then their probability distribution is

$$p_j = \frac{n_j}{\sum_{i=1}^{k} n_i} \quad j = 1, 2, \ldots, K \quad (3.27)$$

with

$$\sum_{j=1}^{k} P_j = 1 \quad (3.28)$$

By means of Stirling's approximation [5], Eq. (3.25) can be approximated by

$$\log_2 N \approx -\left(\sum_{i=1}^{k} n_i \right) \sum_{j=1}^{k} P_j \log_2 P_j \quad \text{bits/message} \quad (3.29)$$

Since Eq. (3.29) is used to define the average amount of information provided by any one of the N possible spatially encoded messages, the average amount of information provided per signal is

$$I = \frac{\log_2 N}{\sum_{i=1}^{k} n_i} = -\sum_{j=1}^{k} P_j \log_2 P_j \quad \text{bits/signal level} \quad (3.30)$$

We can now present a simple example. We first consider that all the resolution cells over the spatial domain are equal in size, but that there are no restrictions on the signal levels. After this simple example, we treat a more general problem, namely, unequal resolution cells.

Let us now consider the case of resolution cells of equal size, $\sigma_1 = \sigma_2 = \cdots = \sigma$. In a large spatial domain S, the spatially encoded messages contain a total number of S/σ resolution cells. In each of the resolution cells, it is possible to encode any of the 1 to k signal levels. Thus, the total number of possible spatially encoded messages (i.e., the total possible combinations) is

$$N(S) = k^{S/\sigma} \quad (3.31)$$

Then the corresponding capacity of the optical spatial channel of resolution cells of equal size is

$$C = \lim_{S \to \infty} \frac{\log_2 N(S)}{S} = \frac{\log_2 k}{\sigma} \quad \text{bits/unit area} \quad (3.32)$$

It is noted that in defining the spatial channel capacity, *a priori* probability distributions of the signal levels across the spatial domain are assumed equiprobable. It is clear that under this equiprobability assumption, the average amount of information provided by the spatial information source is maximum:

$$I_{\max} = -\sum_{i=1}^{k} P_i \log_2 P_i = \log_2 k \quad \text{bits/signal level} \quad (3.33)$$

Since each signal level is assumed to encode in a resolution cell of the same size, the rate at which information is provided by the source is

$$R = \frac{I_{\max}}{\sigma} = \frac{\log_2 k}{\sigma} \quad \text{bits/unit area} \quad (3.34)$$

It can be seen that Eq. (3.34) essentially results from the optical spatial channel capacity of Eq. (3.32). It is noted that this result basically comes from the apparent optimization of the spatial information source, in which the signal levels are taken to be equal in probability. Furthermore, the rate of transmission is

$$r = \frac{C}{I_{\max}} = \frac{1}{\sigma} \quad \text{signal levels/unit area} \quad (3.35)$$

The basic significance of Eq. (3.35) is that we simply cannot transmit more than r signal levels per unit area without introducing significant error.

It is also noted that for a finite spatial domain S, the spatial channel capacity can be written as

$$C(S) = \frac{S}{\sigma} \log_2 k \quad \text{bits} \qquad (3.36)$$

Equation (3.36) indicates the maximum allowable rate at which spatial information can be transmitted through the channel without causing significant error.

With this simple illustration, it can be seen that the best spatial encoding scheme involves application of the equiprobability of the signal levels at the input end of the channel. This is a significant result, based on the basic information theory of Shannon [1–3], for efficient utilization of the channel.

In concluding this example, we note that the equiprobability distribution of the signal levels across the spatial domain yields the maximum spatial information. This maximum spatial information, in fact, is the asymptotic result of obtaining the spatial channel capacity of Eq. (3.32). It is noted that in the course of the calculation, we did not impose an *a priori* probability distribution of the signal levels. But it should be cautioned that the asymptotic method of determining the capacity of the channel is an *average notion*, which is applied to a large spatial domain. This asymptotic approximation is the basic feature in calculation of the spatial channel capacity. We show in the following, a more general case for resolution cells of different sizes, that the spatial averaging notion is still applicable in the calculation of channel capacity.

3.3 SPATIAL CHANNEL WITH RESOLUTION CELLS OF DIFFERENT SIZES

For the more general case of resolution cells of different sizes, the illustration is more involved. Therefore, we treat this example in a separate section.

We discuss this case in detail and show that the result obtained in Section 3.2 still applies. The most efficient spatial encoding is the one that yields the most probable *a priori* probability distribution for the different signal levels across the spatial domain.

Let us now assume an encoded message that occupies the spatial domain S of the channel. Each signal level (irradiance) E_i, where $i = 1, 2, ..., k$, has resolution cells of a certain size, and many signal levels have cells of the same size. Let n_j, where $j = 1, 2, ..., k$, be the number of signal levels having a resolution cell of the same size σ_j. The total spatial domain is then

$$S = \sum_{j=1}^{k} n_j \sigma_j \qquad (3.37)$$

where $\sigma_1 < \sigma_2 < \cdots < \sigma_k$. The total number of signal levels across the spatial domain S is

$$M = \sum_{j=1}^{k} n_j \qquad (3.38)$$

The probability distribution of the signal levels over the spatial domain S is therefore

$$P_j = \frac{n_j}{M} \qquad (3.39)$$

with

$$\sum_{j=1}^{k} p_j = 1 \qquad (3.40)$$

By substituting $n_j = MP_j$ in Eq. (3.37), we have

$$S = M \sum_{j=1}^{k} \sigma_j P_j \qquad (3.41)$$

Since the total number of different spatially coded messages (i.e., combinations of n_j over M) is

$$N = \frac{M!}{n_1! n_2! \cdots n_k!} \qquad (3.42)$$

the logarithm of N, by Stirling's approximation [5], can be approximated by

$$\log_2 N = -M \sum_{j=1}^{k} P_j \log_2 P_j \qquad (3.43)$$

If we assume that the spatially encoded messages over the spatial domain are statistically independent, the average amount of information provided by one of the N encoded messages is

$$I = \log_2 N \quad \text{bits/message} \qquad (3.44)$$

It is noted that all the spatially encoded N messages occupy the same spatial domain S, since they are assumed to have fixed numbers of n_j of size σ_j over S, where $j = 1, 2, ..., k$, and they are assumed, *a priori*, to be equiprobable across the spatial domain of S.

Now in the search for the most probable distribution of the input signal ensemble for the maximum amount of information per encoded message, we consider the following variational technique [6]:

$$d(\log_2 N) - \alpha d \left(\sum_{j=1}^{k} P_j \right) - \beta d(S) = 0 \qquad (3.45)$$

where the parameters α and β are known as Lagrange multipliers. We note that M, p_1, p_2, ..., p_k are the variables for consideration. Thus, by differentiating Eq. (3.43) with respect to these variables, we have

$$d(\log_2 N) = -dM \sum_{j=1}^{k} P_j \log_2 P_j - M \sum_{j=1}^{k} (\log_2 e + \log_2 P_j) dP_j$$

(3.46)

By substituting Eq. (3.46) into Eq. (3.45), we have

$$-dM \left[\sum_{j=1}^{k} P_j (\log_2 P_j + \beta \sigma_j) \right]$$
$$+ \sum_{j=1}^{k} dP_j [-\alpha - \beta M \sigma_j - M(\log_2 e + \log_2 P_j)] = 0$$

(3.47)

With regard to the independent variables dM and dP_j, it can be seen that

$$\sum_{j=1}^{k} P_j (\log_2 P_j + \beta \sigma_j) = 0$$

(3.48)

and

$$\alpha + \beta M \sigma_j + M(\log_2 e + \log_2 P_j) = 0$$

(3.49)

Equation (3.49) can be written as

$$-\log_2 P_j = \log_2 e + \beta \sigma_j + \frac{\alpha}{M}$$

(3.50)

By substituting Eq. (3.50) into Eq. (3.48), we have

$$\left(\log_2 e + \frac{\alpha}{M} \right) \sum_{j=1}^{k} P_j = 0$$

(3.51)

But since Eq. (3.40), we conclude that

$$\alpha = -M \log_2 e$$

(3.52)

Now Eq. (3.50) can be reduced to

$$-\log_2 P_j = \beta \sigma_j$$

(3.53)

P_j can be written as

$$P_j = 2^{-\beta \sigma_j}$$

(3.54)

and β can be determined by

$$\sum_{j=1}^{k} P_j = \sum_{j=1}^{k} 2^{-\beta \sigma_j} = 1$$

(3.55)

Let

$$Z = 2^{\beta}$$

(3.56)

Then Eq. (3.55) can be written as

$$\sum_{j=1}^{k} Z^{-\sigma_j} = 1$$

(3.57)

which is essentially the same as Eq. (3.21), with all $K_n = 1$.

If the spatial encoding is over a very large spatial domain, then the input signal ensemble will approach asymptotically the most probable distribution at which a maximum amount of spatial information will be provided by the encoded messages. Substitution of Eq. (3.54) into Eq. (3.43) yields

$$\log_2 N = M \sum_{j=1}^{k} \beta \sigma_j 2^{-\beta \sigma_j} - M\beta \sum_{j=1}^{k} \sigma_j P_j$$

(3.58)

By definition, the average size of the resolution cells is

$$\bar{\sigma} = \sum_{j=1}^{k} \sigma_j P_j$$

(3.59)

and

$$S = \bar{\sigma} M$$

(3.60)

Then Eq. (3.58) can be written as

$$\log_2 N = \beta S$$

(3.61)

which is equivalent to

$$N = 2^{\beta S} = Z_1^S$$

(3.62)

where $\beta = \log_2 Z_1$.

It is noted that with the most probable distribution of the input signal ensemble, N is essentially the same as $N(S)$ of Eq. (3.24). Therefore, the spatial channel capacity is

$$C = \lim_{S \to \infty} \frac{\log_2 N}{S} = \beta \quad \text{bits/spatial area}$$

(3.63)

From this result, it can be seen that for spatial encoding over large domains, the coded message is probably used more frequently than any other encoding. Therefore, all the other possible distributions can simply be disregarded. The best coding is the one that achieves the most efficient distribution of signal levels:

$$\frac{\log_2 P_j}{\sigma_j} = -\beta \qquad (3.64)$$

where β is a constant.

From Eq. (3.55), we learn that the largest real root of β corresponds to the largest value of Z. In view of Eq. (3.61), the largest root of β gives rise to the largest value of N. Under these conditions, it yields the most probable distribution of the signal levels.

Now if we use the most efficient coding and the most probable distribution of the signal levels, then the amount of information per signal level provided by this spatial encoding over S will be

$$I = \sum_{j=1}^{k} P_j \log_2 P_j = \beta\bar{\sigma} \quad \text{bits/signal level} \qquad (3.65)$$

where $\bar{\sigma}$ is the average size of the resolution cells, as defined by Eq. (3.59). The corresponding rate of spatial information in a signal level per unit area is

$$r = \frac{C}{I} = \frac{1}{\bar{\sigma}} \qquad (3.66)$$

which is essentially the same as the rate given by Eq. (3.35).

In concluding this section, we point out that the coincidence of the average and most probable distribution of the signal levels is a well-known result based on the law of large numbers in statistics [7]. This is also a basic and important concept of entropy theory in statistical thermodynamics [8].

3.4 MATCHING A CODE WITH A SPATIAL CHANNEL

In Section 3.3, we arrived at the conclusion that for the most efficient spatial encoding, the most probable distribution of the input signal ensemble over the spatial domain is that of Eq. (3.54) or, equivalently, Eq. (3.64). This spatial encoding is just as important as the problem of matching the refractive index of an optical component. However, the result we obtained was rather general. In principle, it can be applied to all coding problems. In order to demonstrate the applicability of the result to a general spatial channel, we discuss a few simple examples.

Let us recall the efficient encoding conditions:

$$P_j = 2^{-\beta\sigma_j} \quad j = 1, 2, \ldots, k \qquad (3.67)$$

where P_j can be considered, *a priori*, the probability of the spatial signal, σ_j is the corresponding spatial cell, and β is an arbitrary constant.

If the most probable input spatial signal has a probability of P_j, then $-\log_2 P_j$ will be relatively small. In the most efficient spatial encoding, the corresponding σ_j should also be made smaller. However, for the most improbable one, σ should be made larger.

In view of the probability distribution of the input spatial signals, we can obtain in principle a corresponding variation in spatial encoded messages, for example, in a code of k levels. This would apply to a spatial channel capable of transmitting k quantized signal levels, namely, E_i, $i = 1, 2, \ldots, k$, irradiance levels.

Since our spatial encoding may involve resolution cells of different sizes, it is customary to provide some means for identifying any one of the input spatial signals. For simplicity, we assume a rectangular spatial domain with square array of resolution cells of the same size. It is also assumed that k levels of encoding take place on a row-by-row basis by sequentially scanning across the spatial domain. In this case, a specified signature to characterize the end of each code word is added. Therefore, in a long encoded message, the average size of the resolution cell per input signal can be obtained, such as

$$\bar{\sigma} = \sum_{j=1}^{k} \sigma_j P_j \qquad (3.68)$$

In order to illustrate this basic technique, we present two simple examples.

We first consider the encoding of a set of letters in the English alphabet sequentially in a rectangular binary spatial domain ($k = 2$). In this encoding process, we use a spatial channel capable of transmitting only one type of spatial irradiance in each of the resolution cells; with no irradiance, the cell is regarded as a 0, and with irradiance, it is regarded as a 1.

Since the spatial coding corresponds to different lengths for each input signal, we must provide some method for recognizing signal terminations. For example, a double, triple, or quadruple of 1 may be used for the termination of each spatial encoded letter. It is also noted that a single 1 is regarded as a signal proper but not as a signature for termination. We now consider the *a priori* probability distribution of a set of English letters derived from a book, as shown in Table 3.1 [9]. The corresponding binary codes are also tabulated. It can be seen from this table that the spatial coding varies from three to eight cells per letter. In a long message sequentially encoded on a spatial rectangular domain, the average length per letter is

$$\bar{\sigma} = \sum_{j=1}^{27} \sigma_j P_j \simeq 4.65 \,\text{cells/letter} \qquad (3.69)$$

TABLE 3.1

Probability of Occurrence of English Letters

Letter	Probability	Code	Letter	Probability	Code
Space	0.1859	111			
A	0.0642	01111	N	0.0574	01011
B	0.0127	0001111	O	0.0632	0011
C	0.0218	000111	P	0.0152	0100111
D	0.0317	001111	Q	0.0008	01010011
E	0.1031	011	R	0.0484	00011
F	0.0208	000011	S	0.0514	010111
G	0.0152	0010011	T	0.0796	0111
H	0.0467	010011	U	0.0228	0101111
I	0.0575	00111	V	0.0083	0001011
J	0.0008	01010111	W	0.0175	0010111
K	0.0049	0000111	X	0.0013	000011
L	0.0321	001011	Y	010164	0100011
M	0.0198	0101011	Z	0.0005	01001111

Source: F. M. Reza, *An Introduction to Information Theory*, McGraw-Hill, New York, 1961.

TABLE 3.2

Possible Ternary Code

Letters	Code	Letters	Code
Space	20		
A	210	N	220
B	12210	O	120
C	2210	P	11210
D	2110	Q	21220
E	10	R	1120
F	2220	S	1210
G	12120	T	110
H	1220	U	11110
I	1110	V	12220
J	21210	W	12110
K	21110	X	21120
L	2120	Y	11220
M	11120	Z	22110

Since we are using binary coding, $\bar{\sigma}$ is also the average information provided by the encoded letter:

$$I' = \bar{\sigma} \log_2 2 = 4.65 \text{ bits/letter} \tag{3.70}$$

It is also noted that the average amount of information provided by the letters is

$$I = \sum_{j=1}^{27} P_j \log_2 P_j = 4.03 \text{ bits/letter} \tag{3.71}$$

It can be seen from Eq. (3.70) that the entropy of our binary coding is still larger than the entropy of the source of Eq. (3.71). If we took the binary coding process to be equal in word length, then it would require five cells for each letter. In this case, we can see that the coding is even worse than the one chosen at the beginning. Nevertheless, there exists an optimum technique so that the encoding may approach the theoretical limit of Eq. (3.71). Obviously, in our example the coding is by no means optimum.

In the second example, we adopt the same probability distribution of the English alphabet as in the previous example, but we use a ternary code ($k = 3$). In this encoding process, we employ a spatial channel transmitting three levels of irradiance, namely, E_0, E_1, and E_2. Let E_0 correspond to 0, E_1 to 1, and E_2 to 2. The ternary codes are listed in Table 3.2. We used E_0 (i.e., zero) as a termination signature for the encoded

letters. It is noted that without this signature, we are left with a binary coding for the letters. The average length of the encoding is

$$\bar{\sigma} = \sum_{j=1}^{27} \sigma_j P_j \simeq 3.3 \text{ cells/letter} \tag{3.72}$$

The average information provided by this encoding procedure is

$$I' = \bar{\sigma} \log_2 3 = 5.25 \text{ bits/letter} \tag{3.73}$$

which is higher than in the previous example. Therefore, it may be seen that for multi-signal-level encodings, it does not necessarily reflect the merit of coding. It is the optimum encoding process that reflects the merit.

It is true for most sequential encoding techniques that, for different lengths, a special signature is required for the termination of each encoded signal. This is the price we pay for lengthening the encoding scheme. For example, in our case, zero (i.e., E_0) appears less frequently in the encoded signals. It is for this reason that the average information provided by the encoded letters is greater than that of the previous example. Nevertheless, a more elaborate encoding procedure may improve the result.

In concluding this section, we also note that the *coding efficiency* can be defined:

$$\eta = \frac{1}{I'} \tag{3.74}$$

where I and I' are the average information provided by the signals and encoded signals, respectively. It is also possible to define the redundancy of the encoding:

$$\text{Redundancy} = 1 - \eta \qquad (3.75)$$

Therefore, in the previous examples the coding efficiencies for the binary and ternary cases are

$$\eta = \frac{4.03}{4.65} = 0.867 \quad \text{or } 86.7\% \qquad (3.76)$$

and

$$\eta = \frac{4.03}{5.25} = 0.768 \quad \text{or } 76.8\% \qquad (3.77)$$

The corresponding redundancies are

$$\text{Redundancy} = (1 - 0.867)\% = 13.3\% \qquad (3.78)$$

and

$$\text{Redundancy} = (1 - 0.768)\% = 23.2\% \qquad (3.79)$$

Since optimum encoding processes are beyond the scope of this book, we refer the reader to the excellent text by Peterson [10].

REFERENCES

1. C. E. Shannon, A Mathematical Theory of Communication, *Bell Syst. Tech. J.*, vol. 27, 379–423, 623–656 (1948).
2. C. E. Shannon, Communication in the Presence of Noise, *Proc. IRE*, vol. 37, 10 (1949).
3. C. E. Shannon and W. Weaver, *The Mathematical Theory of Communication*, University of Illinois Press, Urbana, 1949.
4. C. H. Richardson, *An Introduction to the Calculus of Finite Differences*, Van Nostrand, Princeton, NJ, 1954.
5. W. Kaplan, *Advanced Calculus*, Addison-Wesley, Reading, MA, 1952.
6. N. I. Akhiezer, *The Calculus of Variations*, Blaisdell, New York, 1962.
7. E. Parzen, *Modern Probability Theory and Its Applications*, John Wiley, New York, 1960.
8. F. W. Sears, *Thermodynamics, the Kinetic Theory of Gases, and Statistical Mechanics*, Addison-Wesley, Reading, MA, 1953.
9. F. M. Reza, *An Introduction to Information Theory*, McGraw-Hill, New York, 1961.
10. W. W. Peterson, *Error-Correcting Codes*, MIT Press, Cambridge, MA, 1961.

4 Entropy and Information

In the preceding chapters, we discussed the basic definition of information, spatial channel capacity, and the principles of coding. The fundamental constraints of these definitions were also discussed. In this chapter, we deal with the relationship between information theory and physical science.

The measure of information, as Shannon [1–3] defined it, has an intimate relationship with entropy theory in statistical thermodynamics. Therefore, information theory and thermodynamics must have some common points of interest. In order to discuss the basic aspects of entropy as related to information, we first review the two fundamental laws or principles of thermodynamics [4,5].

4.1 FUNDAMENTAL LAWS OF THERMODYNAMICS

We state, without proof, the two fundamental laws of thermodynamics. The *first law* of thermodynamics may be regarded as the law of *energy conservation*. In very general terms, the first law asserts that the net flow of energy across the boundary of a system is equal to the change in energy of the system. For this reason, in thermodynamics it is sufficient to consider only two types of energy flow across the boundary, namely, the work done on or by the system, and the heat flow supplied by radiation or conduction to or by the system.

The *second law* of thermodynamics can be regarded as Carnot's principle [4,5], which asserts that in every process that takes place in an *isolated* system, the entropy, as defined by Clausius [4,5], of the system either increases or remains unchanged. In other words, the entropy of an isolated system never decreases.

According to Kelvin [4,5], the second law is essentially the law of *energy degradation*. That is, although the total energy in a closed isolated system remains constant, energy degradation for all *irreversible* processes results. We say the energy is degraded because it tends to a common level, and hence cannot perform any work by flowing from a higher to a lower (temperature) level.

Furthermore, if an isolated system is in such a state that its entropy is at a maximum, then a change from that maximum entropy state to a lower entropy state is *not* possible, since the system is closed and isolated.

Now if a system receives an amount of heat ΔQ (by conduction, radiation, or any physical means), then the increase in entropy in the system can be written as [4,5]

$$\Delta S = \frac{\Delta Q}{T} \tag{4.1}$$

where $T = C + 273$ is the absolute temperature in Kelvin, and C is the temperature in degrees Celsius.

The following examples will clarify the reader's definition of entropy. Let us consider two bodies that are in contact, but

as a whole are isolated, as shown in Fig. 4.1. These two bodies may be assumed to conduct heat exchange and work done between one another. For convenience, we denote these two bodies 1 and 2. Then by the first law (energy conservation) of thermodynamics, we conclude that

$$\Delta Q_1 - \Delta W_1 + \Delta Q_2 - \Delta W_2 = 0 \tag{4.2}$$

where ΔQ and ΔW are the corresponding heat input and work done by the bodies, and the subscripts 1 and 2 denote the respective bodies. It is noted from Eq. (4.2) that there is no change in the overall energy. However, the changes in entropy within the two bodies are

$$\Delta S_1 = \frac{\Delta Q_1}{T_1} \tag{4.3}$$

and

$$\Delta S_2 = \frac{\Delta Q_2}{T_2} \tag{4.4}$$

Since the two bodies as a whole are isolated, by the second law of thermodynamics we have

$$\Delta S = \Delta S_1 + \Delta S_2 > 0 \tag{4.5}$$

Thus, the entropy of the whole system increases.

By a slight change in the above example, we let the temperature of body 1 be higher than that of body 2, $T_1 > T_2$. Then the heat flow is conducted from body 1 to body 2. It is assumed that no work is performed in either of these bodies, $\Delta W_1 = \Delta W_2 = 0$, so the heat transfer from body 1 to body 2 is therefore

$$\Delta Q_2 - \Delta Q_1 = 0 \tag{4.6}$$

where ΔQ_2 is the amount of heat *input* to body 2, a positive quantity, and $-\Delta Q_1$ is the amount of heat *output* from body 1, a negative quantity. Thus, the corresponding entropy changes of these bodies are

$$\Delta S_1 = -\frac{\Delta Q_1}{T_1} = -\frac{\Delta Q_2}{T_1} \tag{4.7}$$

and

$$\Delta S_2 = -\frac{\Delta Q_2}{T_2} = -\frac{\Delta Q_1}{T_2} \tag{4.8}$$

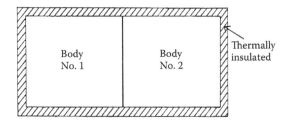

FIG. 4.1 Irreversible thermodynamic process.

Although the entropy of body 1 decreases, we can see that the overall entropy, that is, the combined entropy of these two bodies, increases.

Since by Eq. (4.6), $\Delta Q_2 = \Delta Q_1$,

$$\Delta S = \Delta S_1 + \Delta S_2 = \Delta Q_1 \left(\frac{1}{T_2} + \frac{1}{T_1} \right) > 0 \qquad (4.9)$$

where $T_1 > T_2$. This is an example of an irreversible process of an isolated system in the thermodynamic sense, for heat cannot flow from a lower to a higher temperature.

With respect to the same example, if we let the temperature of both bodies remain constant, $T_1 = T_2 = T$, and we assume that there is heat input but no work done in body 1, $\Delta W_1 = 0$, and $\Delta Q_1 > 0$, and there is work done but no heat exchange in body 2, $\Delta W_2 > 0$, and $\Delta Q_2 = 0$, then by the first law of thermodynamics we see that

$$\Delta Q_1 - \Delta W_2 = 0 \qquad (4.10)$$

In this case, the entropy changes within these two bodies are

$$\Delta S_1 = \frac{\Delta Q_1}{T} = \frac{\Delta W_2}{T} \qquad (4.11)$$

and

$$\Delta S_2 = 0 \qquad (4.12)$$

Once again, as a whole, the entropy of the isolated system increases:

$$\Delta S = \Delta S_1 + \Delta S_2 = \frac{\Delta Q_1}{T} = \frac{\Delta W_2}{T} > 0 \qquad (4.13)$$

So one sees that the exchange of heat and work done is an irreversible process that results in an entropy increase in an isolated system.

It is noted that the preceding illustrations are typical in thermodynamics. We have shown that it is impossible for the entropy to decrease in an isolated system. The second example shows that the entropy within one of the bodies essentially decreases (i.e., the problem of heat flow but no work

done), $\Delta S_1 < 0$, but the overall entropy increases, $\Delta S = \Delta S_1 + \Delta S_2 > 0$.

The increase in entropy has been regarded as the degradation of energy by Kelvin. It is noted that the word *entropy* was originally introduced by Clausius, and he may have intended that it be used for a negative interpretation of the entropy concept. Therefore, it is occasionally found to be much easier to discuss the problem from the point of view of negative entropy, or simply *negentropy* ($N \triangleq -S$). Historically, the importance of the negentropy concept was first recognized by Tait [6], a close associate of Kelvin. Negentropy represents the quality or grade of energy of an isolated system and must always decrease or remain constant. In other words, negentropy may be regarded as the opposite interpretation of the second law of thermodynamics.

From the above examples, it is noted that for an isolated system (i.e., the combination of the two bodies in our example) to perform work (e.g., electrical or mechanical) or even thermal dissipation, it must contain a source of negentropy. This source of negentropy is used to do electrical or mechanical work or is simply lost by thermal dissipation. An isolated system containing a charged battery and a direct current (dc) motor with a load is a good example. We can consider the charged battery a source of negentropy (i.e., the entropy in the battery must increase).

The negentropy principle of information was first emphasized by Brillouin [7] and is fully discussed in his excellent book *Science and Information Theory* [8]. Historically, the negative entropy concept may be due to Schrödinger [9] and his interesting book *What Is Life?* He pointed out that for a living organism to sustain life, it needs to draw negative entropy from its environment continually. However, in our version, if a living organism is searching for food, then it requires a certain amount of negentropy from some source for it to reach for the food. The amount of negentropy required makes up for the work it has performed in searching for the food. Since energy as a whole is conserved, the energy provided by the food is not what enables the organism to reach for it—the negentropy required is important.

4.2 PHYSICAL ENTROPY AND INFORMATION

The basic definition of information was given in earlier chapters from a strictly statistical point of view. We considered an example of N possible symbols (or outcomes) of an information source (or experiment). If these N possible symbols are assumed, *a priori*, to be equiprobable, then the average amount of information provided by the source will be

$$I_0 = -\sum_{j=1}^{N} P_j \log_2 P_j = \log_2 N \quad \text{bits/symbol} \qquad (4.14)$$

where $P_j = 1/N$.

A certain amount of information I is acquired, with which it may be possible to reduce the N symbols to a smaller set of symbols M. If it is also assumed, *a priori*, that the M symbols are equiprobable, then the average amount of information provided by the M symbols will be

$$I_1 = \sum_{j=1}^{M} P_j \log_2 P_j = \log_2 M \quad \text{bits/symbol} \qquad (4.15)$$

where $P_j = 1/M$, and $N > M$. The amount of information required for this information reduction is

$$I = I_0 - I_1 = \log_2 \frac{N}{M} \quad \text{bits/symbol} \qquad (4.16)$$

We are seeking the relationship between information and entropy. For this reason, we consider mainly only information connected with certain physical problems in which entropy theory can be easily treated. In order to derive the relationship between information and entropy, we consider a physical system in which equiprobability in complexity of the structures has been established *a priori*. It is noted in the previous example (where the N possible outcomes are reduced to M) that the system *cannot* be isolated, since the second law of thermodynamics prevails. The corresponding entropies for the N and M equiprobable states are

$$S_0 = k \ln N \qquad (4.17)$$

and

$$S_1 = k \ln M \qquad (4.18)$$

where $N > M$, and k is Boltzmann's constant. Thus,

$$S_0 > S_1 \qquad (4.19)$$

It can be seen that the entropy decreases when information I is obtained. However, this amount of information I must be furnished by some external source whose entropy increases. The decrease in entropy in our system is obviously derived from the information I we have acquired from some external sources:

$$\Delta S = S_1 - S_0 = -kI \ln 2 \qquad (4.20)$$

Or we write, equivalently,

$$S_1 = S_0 - kI \ln 2 \qquad (4.21)$$

where $S_0 > S_1$.

However, the information I is required to be proportional to the decrease in entropy ΔS in our system. This

is the basic connection between entropy and information theory. Equation (4.20) or, equivalently, Eq. (4.21) can be regarded as an expression of the *entropy theory of information*. We discuss this relationship in more detail as we apply Eq. (4.21) to some physical problems. We show that the information I and the change in entropy ΔS can simply be traded.

4.3 TRADING ENTROPY WITH INFORMATION

We have established an intimate relationship between information and entropy (Eq. (4.21)). The amount of information acquired in reducing the entropy of a physical system must be derived from some external source.

By the second law of thermodynamics, if we isolate the entire system, which includes sources of I, then for any further evolution within the whole system the entropy will increase or remain constant:

$$\Delta S_1 = \Delta(S_0 - kI \ln 2) \geq 0 \qquad (4.22)$$

From Eq. (4.22), we see that any further increase in entropy ΔS_1 can be due to ΔS_0 or ΔI, or both. Although in principle it is possible to distinguish the changes in ΔS_0 and ΔI separately, in some cases, the separation of the changes due to ΔS_0 and ΔI may be difficult to discern.

It is interesting to note that if the initial entropy S_0 of the system corresponds to some complexity of a structure but not the maximum, and if S_0 remains constant ($\Delta S_0 = 0$), then after a certain free evolution without the influence of external sources, from Eq. (4.22) we will have

$$\Delta I \leq 0 \qquad (4.23)$$

since $\Delta S_0 = 0$.

From Eq. (4.23), we see that the changes in information ΔI are negative, or decreasing. The interpretation is that when we have no *a priori* knowledge of the system complexity, the entropy S_0 is assumed to be maximum (i.e., the equiprobable case). Hence, the information provided by the system structure is maximum. Therefore, $\Delta I \leq 0$ is due to the fact that in order to increase the entropy of the system, $\Delta S_1 > 0$, a certain decrease in information is needed. In other words, information can be provided or transmitted (a source of negentropy) only by increasing the entropy of the system. However, if the initial entropy S_0 is at a maximum, then $\Delta I = 0$ and the system cannot be used as a source of negentropy.

In Section 4.4, we present several examples illustrating that entropy and information can be interchanged. Information can be obtained *only* by increasing the entropy of certain physical devices. This statement is expressed by the simple symbolic equation

$$\Delta I \rightleftharpoons \Delta S \qquad (4.24)$$

We summarize the relationship between entropy and information theory as follows.

1. The entropy corresponds to the average amount of information provided, as defined by Shannon [1–3].
2. The increase in entropy in a physical system may be used as a source of negentropy to provide information, or vice versa.
3. Work done or energy can be shown to be related to information:

$$\Delta W = \Delta Q = T\,\Delta S = IkT \ln 2 \qquad (4.25)$$

where T is the thermal noise temperature. Thus, from the above equation, we see that with higher thermal noise T, the amount of energy required for the transmission of information I is higher.

From Eq. (4.22), the equality holds only for a reversible process; otherwise, the entropy increases.

4.4   TYPICAL EXAMPLES

In order to provide examples of the relationship between entropy and information, we first discuss a general information problem.

For simplicity, let us begin with a communication example, say an oral conversation. An individual possesses a certain piece of information that is to be shared with a friend. We illustrate with a step-by-step process how this piece of information may be transmitted.

First, an appropriate coding (e.g., the right language) must be selected, and then this piece of information must be transformed, or encoded, into an appropriate speech signal suitable for the acoustical communication channel. However, in the signal transformation process certain errors may be introduced. These errors may represent certain information loss due to the encoding process. It is noted that the loss of information occurs before the information is transmitted through the channel.

Second, a certain loss of information may also occur as a result of noise disturbances in the channel, for example, acoustical noise, which depends on the nature of the communication channel.

Third, the receiver (the friend) may have difficulty in decoding the information. For example, because of a defect in the receiving system (the ears), the coding may be misinterpreted or even missed by the receiver. This represents additional information loss.

Fourth, after a period of time elapses, the receiver (the friend) may forget some of the information that has been received.

So it can be seen that there is a certain amount of information loss at each stage. Of course, the most efficient communication we are aiming at is one without any loss during information transmission.

This example is a very common one and applies to our daily communication. Another trivial example involves a broadcaster reporting daily news events on the radio. Not all the information conveyed by the broadcaster is equally valuable to the listener, so some of it may be ignored or overlooked. It is noted that our definition of information was not originally intended to include this value aspect, but it is possible to extend the theory in this way. However, the discussion of such a problem is beyond the scope of this book. Interested readers may refer to the existing literature.

We now discuss a few typical examples from the statistical thermodynamics point of view, to which the relationship between entropy and information can be easily applied.

Let us consider, for example, an ideal monatomic gas, assumed to exist in a thermally isolated container. Under steady-state conditions, the familiar entropy equation can be written as [10,11]

$$S = \frac{5Nk}{2} + Nk \ln\left[\frac{Vg}{N}\left(\frac{4\pi mE}{3Nh^2}\right)^{3/2}\right] \qquad (4.26)$$

where N is the total number of atoms in the container, k is Boltzmann's constant, V is the volume of the container, m is the atomic mass, E is the total energy, h is Planck's constant, $g = 2j + 1$ is the number of indistinguishable ground states of the atom, and j is the momentum moment.

The value of g is worth mentioning. If the ground state is not degenerate, then $j = 0$, and $g = 1$. In this case, the atom has no momentum moment in the ground state. However, if $j \neq 0$, then there are $g = 2j + 1$ ground states, and we assume *a priori* that these ground states are equiprobable.

We now discuss the first example. We assume that we had *a priori* knowledge of the state of the gas. For instance, we may know that at a precise moment the gas occupies a volume V_0, and that at another moment it occupied a larger volume V_1:

$$V_1 = V_0 + \Delta V \qquad (4.27)$$

where ΔV is the enlarging portion.

By the entropy equation (4.26), the corresponding entropies can be written, respectively,

$$S_0 = Nk(\ln V_0 + K) \qquad (4.28)$$

and

$$S_1 = Nk(\ln V_1 + K) \qquad (4.29)$$

where

$$K = \frac{5}{2} + \ln\left[\frac{g}{N}\left(\frac{4\pi mE}{3Nh^2}\right)^{3/2}\right] \quad \text{and} \quad S_0 < S_1$$

The initial entropy S_0 is somewhat smaller than the final entropy S_1 after the gas expansion. The corresponding increase in entropy of this process is, from Eq. (4.20),

$$\Delta S = S_1 - S_0 = Nk(\ln V_1 - \ln V_0) > 0 \qquad (4.30)$$

This increase in entropy may be considered the loss in information of the process:

$$I = \frac{\Delta S}{k \ln 2} \qquad (4.31)$$

Evaluation of the above results is made when the steady state of the gas is reached, that is, after a uniform density distribution is established throughout the final volume V_1.

Again, we see that an increase in entropy of a process corresponds to certain information loss. In this example, it may be said that as the volume expanded, the gas progressively lost or forgot the information.

However, we can see that if the volume progressively shrinks, the gas obtains additional information, since the entropy progressively decreases. But we have seen from Eq. (4.20) that this is possible only if a certain amount of information is acquired from some external source.

We use the same monatomic gas for the second example, and consider the problem of diffusion. For simplicity, we assume that two different types of gases are involved, but that they have an identical mass m and the same g factor. We let N_1 atoms of type 1 occupy volume V_1, and N_2 atoms of type 2 occupy volume V_2. Thus, the total atoms and total volume are

$$N = N_1 + N_2 \qquad (4.32)$$

and

$$V = V_1 + V_2 \qquad (4.33)$$

Let us define the corresponding probabilities:

$$P_1 = \frac{N_1}{N} = \frac{V_1}{V} \qquad (4.34)$$

and

$$P_2 = \frac{N_2}{N} = \frac{V_2}{V} \qquad (4.35)$$

where $P_1 + P_2 = 1$.

From Eqs. (4.34) and (4.35), we have

$$\frac{V_1}{N_1} = \frac{V_2}{N_2} = \frac{V}{N} \qquad (4.36)$$

Let us also assume that the law of *energy equipartition* is applied. Then we conclude that

$$\frac{E_1}{N_1} = \frac{E_2}{N_2} = \frac{E}{N} \qquad (4.37)$$

where

$$E = E_1 + E_2$$

It is noted from Eqs. (4.36) and (4.37) that the two gases are under the same pressure and temperature. For simplicity, we let volume V be constant throughout the gas diffusion process. Thus, the entropies before and after the diffusion process are

$$S_0 = Nk\left(\ln \frac{V}{N} + G \right) \qquad (4.38)$$

and

$$S_1 = N_1 k\left(\ln \frac{V}{N_1} + G \right) + N_2 k\left(\ln \frac{V}{N_2} + G \right) \qquad (4.39)$$

where

$$G = \frac{5}{2} + \ln\left[g\left(\frac{4\pi mE}{3Nh^2} \right)^{3/2} \right]$$

From Eqs. (4.36) and (4.38), Eq. (4.39) can be written as

$$S_1 = S_0 + k\left(N_1 \ln \frac{N}{N_1} + N_2 \ln \frac{N}{N_2} \right) \qquad (4.40)$$

By substituting Eqs. (4.34) and (4.35) into Eq. (4.40), we have

$$S_1 = S_0 - Nk(P_1 \ln P_1 + P_2 \ln P_2) \qquad (4.41)$$

Therefore, the corresponding increase in entropy after the gas diffusion process is

$$\Delta S = S_1 - S_0 = -Nk(P_1 \ln P_1 + P_2 \ln P_3) \qquad (4.42)$$

where $P_1 < 1$, and $P_2 < 1$. Since

$$I = \frac{\Delta S}{k \ln 2} = -N(P_1 \log_2 P_1 + P_2 \log_2 P_2) \qquad (4.43)$$

it corresponds to information *loss* in this diffusion process.

It is interesting to compare the information measure provided by Shannon [1–3]:

$$I = -\sum_{j=1}^{N} P_j \log_2 P_j \tag{4.44}$$

For a *binary* information source, Shannon's formula is

$$I = -(P_1 \log_2 P_1 + P_2 \log_2 P_2) \tag{4.45}$$

Moreover, for *N independent binary* sources, the formula becomes

$$I = -N(P_1 \log_2 P_1 + P_2 \log_2 P_2) \tag{4.46}$$

Equation (4.46) is essentially the same as Eq. (4.43).

The interpretation of our example is that after the diffusion process, it requires at least the amount of information I (Eq. (4.43)) to return the N_1 and N_2 atoms (if we could) to the original state. This amount of information is related to the amount of entropy increase (Eq. (4.42)). In fact, this amount of information is related to the energy required to reverse the process (Eq. (4.25)), if we could.

For our third example, we use the second example—the diffusion problem—and include the effect due to the g factor in the entropy equation (4.26). Let us consider for a moment the quantity $Nk \ln g$, where $g = 2j + 1$. If we assume that the atoms of both gases have electronic spin $\frac{1}{2}$, for type 1 the electronic spin will be $\frac{1}{2}$ and for type 2 it will be $-\frac{1}{2}$. In the diffusion process, the atomic collisions make possible transition between the two kinds of atomic spins; the final state is assumed to be one with equiprobable spin orientations of the atoms. Thus, this corresponds to $g = 2$, where $j = \frac{1}{2}$.

With the inclusion of the second example, after the diffusion process, the overall entropy equation becomes

$$S_2 = S_0 - Nk(P_1 \ln P_1 + P_2 \ln P_2 + \ln 2) \tag{4.47}$$

The overall increase in entropy is

$$\Delta S = S_2 - S_0 = -Nk(P_1 \ln P_1 + P_2 \ln P_2 + \ln 2) \tag{4.48}$$

This increase in entropy in the process is again recognized as information loss in the system:

$$I = \frac{\Delta S}{k \ln 2} = -N(P_1 \log_2 P_1 + P_2 \log_2 P_2 + 1) \tag{4.49}$$

In comparison with Eq. (4.43), there is a net change in information loss due to $g = 2$, which is equal to N bits, where N is the total number of atoms.

4.5 REMARKS

A low-entropy system is an unstable system. Eventually, it follows the normal evolutionary process leading to a more stable high-entropy state. From our examples, we see that information can be obtained from a physical system if and only if the entropy of the system progressively increases. In other words, a low-entropy system can be used to provide information, and is also known as a source of negentropy.

The information stored in a system can be regarded as a decrease in entropy of the system. The stored information is progressively destroyed or erased by the increase in entropy due to the normal evolution of the system. In practice, it may take anywhere from a small fraction of a second to years, or even centuries, to erase the information, depending on the nature of the system (i.e., the recording material).

In statistical thermodynamics, entropy is defined as a measure of the disorder of a system. However, in information theory entropy is a measure of the *lack* of information about the actual structure of the system.

Historically, the idea of an entropy theory of information was first presented by Szilard [12,13] in his famous paper "On the Decrease of Entropy in a Thermodynamic System by the Intervention of Intelligent Beings," but at that time, the concept was not well understood or accepted. It was not until the late 1940s that the relationship between entropy and information was rediscovered by Shannon [1–3].

Finally, for further reading on the entropy theory of information, we refer the reader to the excellent article and book by Brillouin [7,8] and the papers by Rothstein [14,15].

REFERENCES

1. C. E. Shannon, A Mathematical Theory of Communication, *Bell Syst. Tech. J.*, vol. 27, 379–423, 623–656 (1948).
2. C. E. Shannon, Communication in the Presence of Noise, *Proc. IRE*, vol. 37, 10 (1949).
3. C. E. Shannon and W. Weaver, *The Mathematical Theory of Communication*, University of Illinois Press, Urbana, 1962.
4. F. W. Sears, *Thermodynamics, the Kinetic Theory of Gases, and Statistical Mechanics*, Addison-Wesley, Reading, MA, 1953.
5. M. W. Zemansky, *Heat and Thermodynamics*, 3rd ed., McGraw-Hill, New York, 1951.
6. P. G. Tait, *Sketch of Thermodynamics*, Edmonston and Douglas, Edinburgh, 1868, p. 100.
7. L. Brillouin, The Negentropy Principle of Information, *J. Appl. Phys.*, vol. 24, 1152 (1953).
8. L. Brillouin, *Science and Information Theory*, 2nd ed., Academic, New York, 1962.
9. E. Schrödinger, *What Is Life?* Cambridge University Press, New York, 1945.

10. J. E. Mayer and M. G. Mayer, *Statistical Mechanics*, John Wiley, New York, 1940, p. 115.

11. A. Katz, *Principles of Statistical Mechanics*, W. F. Freeman, San Francisco, 1967, p. 97.

12. L. Szilard, Über die Entropieverminderung in Einem Thermodynamishen System bei Eingriffen Intelligenter Wesen, *Z. Phys.*, vol. 53, 840 (1929).

13. L. Szilard, On the Decrease of Entropy in a Thermodynamic System by the Intervention of Intelligent Beings (translated by A. Rapoport and M. Knoller), *Behav. Sci.*, vol. 9, 301 (1964).

14. J. Rothstein, Information, Measurement, and Quantum Mechanics, *Science*, vol. 114, 171 (1951).

15. J. Rothstein, Information and Thermodynamics, *Phys. Rev.*, vol. 85, 135 (1952).

5 Demon Exorcist and Cost of Entropy

We discussed in detail the entropy theory of information in Chapter 4. The relationship between physical entropy and information theory was derived strictly from the statistical thermodynamic approach. We showed that information and entropy can be traded. However, information is gained inevitably at the expense of increasing entropy from a different source. This relationship of entropy and information was first recognized by Szilard [1,2] in 1929 and later successfully applied by Brillouin [3,4] in 1951. In practice, the amount of increasing entropy from a different source is generally excessive. In principle, the amount of information gain at best can only approach the equivalent amount of entropy traded off:

$$I \le \frac{\Delta S}{k \ln 2} \qquad (5.1)$$

In this chapter, we discuss one of the most interesting applications of the entropy theory of information, namely, the demon's operation and information theory.

Of all the demons in the physical sciences, Maxwell's demon may be regarded as the most famous of all. The problem of Maxwell's demon and the theory of heat may be the best example for application of the entropy theory of information. What follows illustrates the impossibility of a perpetual motion machine. Subsequent sections demonstrate the demon's operation and information theory.

5.1 PERPETUAL MOTION MACHINE

According to the first law of thermodynamics, the law of energy conservation, a perpetual motion machine is impossible. Other perpetual motion machines, those of the second kind [5], used to convert the disorder of heat (e.g., by conduction or radiation) into work under the same temperature or pressure, are also known to be impossible, since they violate the second law of thermodynamics. For example, in a refrigerator the transfer of heat from a cold body to a hot body requires an external source of energy.

As illustrated in Chapter 4, a system can perform work if two bodies or subsystems with different temperatures or pressures are in contact. As an example, we denote these two bodies as 1 and 2, and

$$T_2 > T_1 \qquad (5.2)$$

where T_1 and T_2 are the temperatures of body 1 and body 2, respectively. When these two bodies are in contact, a certain amount of heat is expected to flow between them. Let the input heat into a body be a positive quantity; then we have

$$\Delta Q_2 = -K \Delta Q_1 \qquad (5.3)$$

where K is a positive proportional constant; ΔQ_1 is the amount of heat input to body 1, a positive quantity; and ΔQ_2 is the amount of heat output from body 2, a negative quantity. Let W be the work done by the system, and then by the energy conservation law we have

$$\Delta Q_1 + \Delta Q_2 + \Delta W = 0 \qquad (5.4)$$

Let us assume the thermodynamic process of the system is *reversible*; hence, there is no increase in total entropy S:

$$\Delta S = \Delta S_1 + \Delta S_2 = 0 \qquad (5.5)$$

By the definition of entropy, Eq. (5.5) can be written as

$$\frac{\Delta Q_1}{T_1} + \frac{\Delta Q_2}{T_2} = 0 \qquad (5.6)$$

or equivalently,

$$\Delta Q_2 = -\frac{T_2}{T_1} \Delta Q_1 \qquad (5.7)$$

Thus,

$$\Delta Q_1 + \Delta Q_2 = (T_1 - T_2)\Delta S_1 < 0 \qquad (5.8)$$

which is a negative quantity. In Eq. (5.4), this amount equals the work done by the system. Therefore, the system appears to perform work. The corresponding thermal efficiency [5] is

$$\eta = \frac{\Delta W}{\Delta Q_2} = \frac{T_2 - T_1}{T_2} < 1 \qquad (5.9)$$

From Eq. (5.9), it can be seen that heat can be converted to work, but not completely. However, work can always be totally converted to heat. Thus, from this heat–work transformation problem it can be seen that it is not possible to realize a perpetual motion machine using a refrigerating system as an example. We illustrate in the following sections a perpetual motion machine of the second kind, from an information theory point of view. The information aspect of perpetual motion was first recognized by Szilard [1,2] well before modern information theory was introduced by Shannon [6–8] in 1948. However, it was not until Shannon's work appeared that the earlier work of Szilard was recognized and understood. We discuss these problems from Maxwell's demon approach, but with the use of information theory. In short, a perpetual motion machine is again shown to be unrealizable in practice.

5.2 MAXWELL'S DEMON

The perpetual motion machine of the second kind was created by James Clerk Maxwell as early as 1871. The sorting demon, known as Maxwell's demon, first appeared in Maxwell's *Theory of Heat* [9, p. 328]. Since then, it has intrigued investigators of the physical world and also provided an excellent example of the application of the entropy theory of information. Maxwell's sorting demon is some kind of intelligent being capable of tracing every molecule's motion. In other words, he is able to perform certain tasks beyond the current physical constraints.

In order to illustrate the demon's operation, let us suppose we have a thermally insulated chamber filled with gas molecules. Let the chamber be divided into two parts, chamber 1 and chamber 2, as shown in Fig. 5.1. Also, let the partition wall be equipped with a small trapdoor that the demon is able to open and close as he wishes. The demon is able, we assume, to see every individual molecule, so he can allow a molecule to pass from one chamber to the other. Now the demon decides to let the fast molecules in chamber 1 pass into chamber 2, and the slower molecules in chamber 2 to pass into chamber 1. In this manner, the demon is eventually able to collect all the fast molecules in chamber 2 and all the slower molecules in chamber 1. Thus, without any expenditure of work, he is able to raise the temperature in one chamber (chamber 2) and lower the temperature in the other. Hence, work can be performed by the difference in temperature between the two chambers. Therefore, we see that the demon is indeed able to create a perpetual motion machine of the second kind. This famous paradox has been explained by many physicists since the invention of Maxwell's demon. In fact, he has been known as a *temperature demon* in the theory of heat.

There is a similar paradox involving *Brownian motion*, discussed by von Smoluchowski [10], in which a simple trapdoor device is used. The trapping procedure is based on the random opening and closing of a spring valve by the random bombardment of particles. In this manner, the spring valve creates high- and low-pressure chambers, so work can be performed, again, by pressure differences. This paradox was

recognized by von Smoluchowski. Because of the random unpredictability and short duration of particle motion, he concluded that Brownian motion constitutes only an apparent contradiction to the second law of thermodynamics, so that permanent operation of such a system is not possible. That is, the system may move irregularly, but not in a manner that allows partition of the particles. Thus, a perpetual motion machine is again shown to be impossible.

A case similar to Brownian motion, but in which thermal agitation occurs in a rectifier, was also described by Brillouin [4]. He concluded that no rectified current can be produced by a physical rectifier. Once again, the second law of thermodynamics is unchallenged.

Now we come to the application of information theory to the problem of Maxwell's demon. The basic question is, "Are we sure the demon is actually able to see the individual molecules?" This basic question was raised by Demers [11,12] and Brillouin [3,4], and the connection between information theory and the demon's operation was first fully discussed in their publications. We have adopted their basic explanation in the following discussion.

We again assume the problem of a thermally isolated chamber. Since the demon is within the enclosure, he is therefore in an equilibrium temperature state. The radiation within the chamber is only blackbody radiation; thus, it is impossible for the demon to see anything inside the chamber. Raising the temperature of the chamber would not help, since the demon would be able to see only the thermal radiation and its random fluctuations, but not the molecules. Thus, under these isolated conditions, the demon is unable to operate the door. Separation of temperatures in the partitioned chamber is impossible.

However, one might ask, although the demon cannot see the molecules, can he detect them by some other means? Yes, he can. For example, he might detect the molecules by the electric field caused by the electric dipoles, or a magnetic field caused by the electronic spin, or even by their van der Waals forces. But all these fields are detectable only at a very short distance, since their field strength varies with the inverse square of the distance. Thus, the demon can detect the presence of a molecule only when it is very close to the trapdoor. At this very short distance, it is already too late for him to open the door without performing any work. Thus, the demon must rely on detection of the molecule long before it approaches the trapdoor. Nevertheless, this long-distance detection requires a certain compensation for the external source of radiation. In our case, an optical radiator (e.g., a flashlight) may be the simplest example. A detailed discussion of this example follows in Section 5.3.

5.3 INFORMATION AND DEMON EXORCIST

Let us now provide the demon with an electric flashlight to see the individual molecules within the isolated chamber. As pointed out in Chapter 4, the flashlight acts as an external source providing information for the demon. This is essentially what was called the source of negative entropy, or simply

Thermally insulated

FIG. 5.1 Operation of Maxwell's demon.

negentropy, by Brillouin [4,13]. Then, by using the amount of information provided by the flashlight, the demon is able to operate the trapdoor. In this manner, he is eventually able to decrease the entropy of the chamber to a lower state. Thus, work can be performed by this lower entropy state. However, the entropy increase in the flashlight can be shown to be larger than or at least equal to the entropy decrease in the chamber. The total entropy (i.e., the combined entropy of the chamber and the flashlight) increases or remains constant, and as a result, the second law of thermodynamics is still satisfied.

We now discuss in detail the problem of a temperature demon. We use the same thermally isolated chamber that is filled with gas molecules and in which the temperature is uniform throughout. Now we provide the demon with a flashlight with freshly charged batteries. In order to perceive the molecules, the demon turns on the flashlight. The batteries heat up the filament of the light bulb to a temperature T_α higher than that of the gas:

$$T_\alpha > T_0 \qquad (5.10)$$

where T_0 is the temperature of the gas. In order to obtain visible light within the chamber, the flashlight is required to emit a quantum of light:

$$h\nu_\alpha > kT_0 \qquad (5.11)$$

where h is Planck's constant, ν_α is the frequency of the visible light from the flashlight, k is Boltzmann's constant, and T_0 is the temperature of the gas in degrees Kelvin.

In a period of time, total energy ΔE is radiated from the flashlight. Then the corresponding amount of entropy increase is

$$\Delta S_\alpha = \frac{\Delta E}{T_\alpha} \qquad (5.12)$$

This is also the reduction in entropy that the demon, *at best*, is able to obtain in the chamber by operating the trapdoor.

If the demon does not intervene, then the energy ΔE will be totally absorbed by the gas at temperature T_0. This corresponds to an increase in entropy in the gas

$$\Delta S_g = \frac{\Delta E}{T_0} \qquad (5.13)$$

Now let us illustrate the demon's intervention. We have noticed that the demon is able to see an approaching molecule if and only if that molecule scatters at least a quantum of light $h\nu_\alpha$. The quantum of light is absorbed by the eye of the demon, or by means of a photodetector. This quantum of absorbed light represents an increase in entropy in the demon:

$$\Delta S_d = \frac{h\nu_\alpha}{T_0} \qquad (5.14)$$

On the basis of the entropy theory of information, Eq. (5.14) also provides the demon with an amount of information:

$$I_d = \frac{\Delta S_d}{k \ln 2} \qquad (5.15)$$

Once the information is obtained, the demon can use it to decrease the entropy in the chamber. To compute the amount of entropy reduction, we let the initial entropy in the chamber be

$$S_0 = k \ln N_0 \qquad (5.16)$$

where N_0 represents the total initial microscopic complexity of the gas. After receiving I_d, the demon is able to reduce the initial complexity N_0 to N_1:

$$N_1 = N_0 - \Delta N \qquad (5.17)$$

where ΔN is the net change in complexity. Thus, the final entropy in the chamber is

$$S_1 = k \ln N_1 \qquad (5.18)$$

The corresponding amount of decreased entropy is

$$\Delta S_1 = S_1 - S_0 = k \ln \left(1 - \frac{\Delta N}{N_0} \right) \qquad (5.19)$$

Since

$$\ln \left(1 - \frac{\Delta N}{N_0} \right) = - \left[\frac{\Delta N}{N_0} + \frac{1}{2} \left(\frac{\Delta N}{N_0} \right)^2 + \frac{1}{3} \left(\frac{\Delta N}{N_0} \right)^3 + \cdots \right] \qquad (5.20)$$

Eq. (5.19) can be approximated by

$$\Delta S_1 \simeq -k \frac{\Delta N}{N_0} \quad \text{for} \quad \frac{\Delta N}{N_0} \ll 1 \qquad (5.21)$$

In conjunction with Eq. (5.14), one sees that the net entropy in the entire chamber is increased:

$$\Delta S = \Delta S_d + \Delta S_1 = k \left[\frac{h\nu_\alpha}{kT_0} - \frac{\Delta N}{N_0} \right] > 0 \qquad (5.22)$$

since $h\nu_\alpha > kT_0$. It can be seen that the demon is not able to violate the second law of thermodynamics. All he can do is convert only a small part of the total entropy increase in the flashlight into information; from this amount of information gain, he is able to reduce the entropy in the chamber. But the entropy reduction is much smaller than the cost in entropy incurred by the flashlight.

Let us proceed with the problem in more detail. We see that after a certain period, the demon is able to establish different temperatures in the partitioned chamber:

$$\Delta T = T_2 - T_1 \qquad (5.23)$$

where T_1 and T_2 are the corresponding temperatures in chambers 1 and 2. It can also be seen that

$$T_0 = T_2 - \frac{\Delta T}{2} = T_1 + \frac{\Delta T}{2} \qquad (5.24)$$

In order for the demon to establish these two temperatures, he must divert the fast molecules from chamber 1 to chamber 2, and the slower molecules from chamber 2 to chamber 1. However, for the demon to be able to do this, he needs at least two quanta of light: one for the fast molecule and another for the slower one. The two quanta of radiation must be provided by an external source, in our example, the flashlight. Therefore, the actual increase in entropy in the flashlight per two molecular exchanges is

$$\Delta S_d = 2\frac{h\nu_\alpha}{T_0} \qquad (5.25)$$

instead of Eq. (5.14). The corresponding amount of information gained by the demon is

$$I_d = \frac{2h\nu_\alpha}{T_0 k \ln 2} \qquad (5.26)$$

If the kinetic energy of the faster and slower molecules is assumed to be

$$E_f = \frac{3}{2}kT_0(1+\varepsilon_1) \qquad (5.27)$$

and

$$E_s = \frac{3}{2}kT_0(1-\varepsilon_2) \qquad (5.28)$$

respectively, where ε_1 and ε_2 are arbitrary small positive constants, then the net transfer of energy due to these two molecules, from chamber 1 to chamber 2, and from chamber 2 to chamber 1, is

$$\Delta E = \frac{3}{2}kT_0(\varepsilon_1 + \varepsilon_2) \qquad (5.29)$$

Thus, the corresponding entropy decrease per operation is

$$\Delta S_c = \Delta E\left(\frac{1}{T_2} - \frac{1}{T_1}\right) = -\frac{3}{2}k(\varepsilon_1 + \varepsilon_2)\frac{\Delta T}{T_0} \qquad (5.30)$$

Thus, the net entropy change in the entire process per operation is

$$\Delta S = \Delta S_d + \Delta S_c = k\left[\frac{2h\nu_\alpha}{kT_0} - \frac{3}{2}(\varepsilon_1 + \varepsilon_2)\frac{\Delta T}{T_0}\right] > 0 \qquad (5.31)$$

where $(\varepsilon_1 + \varepsilon_2) \ll 1$, and $\Delta T \ll T_0$. Once again, it is seen that violation of the second law of thermodynamics is impossible.

It would be interesting to calculate the efficiency of the demon. This can be obtained by using the ratio of the entropy decrease to the amount of information obtained by the demon:

$$\eta = -\frac{\Delta S_c}{I_d k \ln 2} = -\frac{\Delta S_c}{\Delta S_d} \qquad (5.32)$$

Therefore, by substituting Eqs. (5.25) and (5.30) into Eq. (5.32), we have

$$\eta = \frac{3k(\varepsilon_1 + \varepsilon_2)\Delta T}{4h\nu_\alpha} < 1 \qquad (5.33)$$

The demon's efficiency is *always* less than 1; the most he can do is approach unit efficiency. However, for him to do so would require an extreme difference in temperature ($\Delta T \to \infty$).

One may question whether the problem of Maxwell's temperature demon can be rigorously explained, since occasionally we assume that $\Delta N \ll N_0$, $\Delta T \ll T_0$, and $\varepsilon_1 + \varepsilon_2 \ll 1$ 1. However, we discuss Maxwell's pressure demon in Section 5.4 and illustrate a more rigorous demonstration of this problem.

We have achieved a significant result from an optical viewpoint in this section. Every observation must be compensated for by an increase in the entropy of another system. However, this increase in entropy does not compensate for the decrease in entropy in the other system, so the second law of thermodynamics is not violated. Beyond this limit, the observation (or use as a measurement) cannot be achieved in practice, since it basically violates the second law. The important result we have established is the basic connection of observation and information, which is presented in Chapter 6. This result also presents the idea that one cannot get something for nothing; there is always a price to pay. And there is always the question, "Can we afford it?" This is even truer in the information sciences—as Gabor [14] pointed out in his article "Light and Information": "We cannot get something for nothing, not even an observation, far less a law of nature!"

Terminating this section, we emphasize that our treatment of the demon's effectiveness and information is primarily derived from the quantum condition, that is, $h\nu_\alpha$. A similar explanation can also be achieved by means of the classical approach. We refer the interested reader to the following articles: von Smoluchowski [10], Lewis [15], Slater [16], Demers [11,12], and finally, the most interesting work by Brillouin [3,4,13], from whom we benefitted most.

5.4 DEMON EXORCIST, A REVISIT

In this section, we discuss Maxwell's pressure demon. Instead of sorting out faster and slower molecules, this demon operates a trapdoor, allowing the molecules to pass into one chamber but not in the opposite direction. Then, after a time, the demon is able to build up high pressure in one chamber and low pressure in the other. Thus, work can be performed by the difference in pressure between the two connected chambers. One of the interesting facts under these conditions is that the demon does not even need to see the molecules.

In order to discuss the pressure demon's work, we adopt the model proposed by Brillouin [4], as shown in Fig. 5.2. We use a conduit to replace the trapdoor opening and to connect the two chambers. A beam of light from an external source is assumed to pass through a transparent section of the conduit near the trapdoor. The scattered light, if any, is used for detection of the molecules in the conduit, approaching the trapdoor. However, the appearance of molecules in the conduit is not sufficient for the demon to open the trapdoor, unless he knows the approximate average velocity of the molecules. In order for him to determine the molecular velocity, a set of revolving shutters is employed in the conduit, farther away from the trapdoor, as shown in the figure. By using an appropriate shutter speed, it is possible to allow the passage of molecules having a certain average velocity. Molecules possessing either higher or lower velocities are reflected back into the chamber. Now with the photodetector and the revolving shutters, the demon is able to determine not only the presence of the molecules but also their approximate velocity.

We are now able to determine the amount of information the demon requires to operate the trapdoor. We let τ be a long interval of the pressure demon's operation. He is now able to operate the trapdoor at every subinterval Δt_i of τ. To simplify, let every Δt_i be equal to Δt. At each Δt, the demon needs to know whether any molecules are arriving from chamber 1 approaching the trapdoor at an average known speed. If there are, he will open the trapdoor to allow the molecules to pass through, and then immediately close it. If none are arriving, he will remain stationary. This problem is essentially a binary yes or no problem. Therefore, the *a priori* probability of the binary source can be calculated for a long period of τ. Let Δt be sufficiently small so that the probability of opening the trapdoor more than once can be neglected. We denote by N_1 the total number of openings of the trapdoor, over τ, and by

N_2 the total number of intervals during which the trapdoor remains closed, over τ. Then the corresponding *a priori* probabilities of the trapdoor opening or remaining closed at every Δt are

$$P_1 = \frac{N_1}{N} \tag{5.34}$$

and

$$P_2 = \frac{N_2}{N} \tag{5.35}$$

where $N = N_1 + N_2$, and $N = \tau/\Delta t$. Now the amount of information the demon needs is

$$I_d = -(P_1 \log_2 P_1 + P_2 \log P_2) \quad \text{bits/}\Delta t \tag{5.36}$$

To determine the *a priori* probabilities P_1 and P_2, we let $\bar{n}$ be the average number of molecules bombarding the trapdoor from chamber 1. Then $\bar{m} = \bar{n}Z_2/Z_1$ is the average number of molecules from chamber 2 striking the trapdoor. Z_1 and Z_2 are the pressures in chamber 1 and chamber 2, respectively. From Appendix B, the *a priori* probabilities of P_1 and P_2 are

$$P_1 = 1 - e^{-\bar{n}} \tag{5.37}$$

and

$$P_2 = e^{-\bar{n}} \tag{5.38}$$

By substituting Eqs. (5.37) and (5.38) into Eq. (5.36), we have

$$I_d = -\left[(1 - e^{-\bar{n}}) \log_2 (1 - e^{-\bar{n}}) + e^{-\bar{n}} \log_2 e^{-\bar{n}} \right] \tag{5.39}$$

Since Δt is assumed to be very small, it is also justifiable to assume that $\bar{n} \ll 1$. Therefore, Eq. (5.39) can be approximated by

$$I_d \simeq \bar{n} \log_2 \left(\frac{e}{\bar{n}} \right) \tag{5.40}$$

since $1 - e^{-\bar{n}} \simeq \bar{n}$, and $e^{-\bar{n}} \simeq 1 - \bar{n}$. The net (average) number of molecules entering chamber 2, per Δt, is

$$\Delta N = \bar{n} - \bar{m}P_1 \tag{5.41}$$

which is equal to the average number of molecules entering minus the average number of molecules leaving, per Δt. Since $\bar{n} \ll 1$, Eq. (5.41) can be approximated by

$$\Delta N \simeq \bar{n} \left(1 - \bar{n} \frac{Z_2}{Z_1} \right) \tag{5.42}$$

FIG. 5.2 Intervention by Maxwell's pressure demon. L, light beam; P, photodetector; S, rotating shutters.

where Z_1 and Z_2 are the respective pressures of chambers 1 and 2. Thus, the amount of entropy reduction per Δt accomplished by the demon is

$$\Delta S_c = -\Delta Nk \ln \frac{Z_2}{N_1} = -k\overline{n}\left(1 - \overline{n}\frac{Z_2}{Z_1}\right) \ln \frac{Z_2}{Z_1} \qquad (5.43)$$

But from Eq. (5.40), one can see that the lowest entropy increase in the external sources (i.e., the light source and the revolving shutter) is

$$\Delta S_d = I_d k \ln 2 \simeq k\overline{n} \ln\left(\frac{e}{n}\right) \qquad (5.44)$$

Therefore, the net entropy change per Δt is

$$\Delta S = \Delta S_d + \Delta S_c = k\overline{n}\left[\ln\left(\frac{e}{n}\right) - \left(1 - \overline{n}\frac{Z_2}{Z_1}\right)\ln\frac{Z_2}{Z_1}\right] > 0 \qquad (5.45)$$

From Eq. (5.45), we see that the second law of thermodynamics has not been violated.

From Eqs. (5.43) and (5.44), the pressure demon's efficiency is

$$\eta = -\frac{\Delta S_c}{\Delta S_d} = \frac{\left[1 - \overline{n}(Z_2/Z_1)\right]\ln(Z_2/Z_1)}{\ln(e/\overline{n})} < 1 \qquad (5.46)$$

which is always smaller than 1. The best the demon can do is to make the efficiency approach unity, but in order for him to do so, there must be an extremely large pressure difference between the chambers ($Z_2/Z_1 \to \infty$).

In concluding this section, we emphasize that in order for Maxwell's demons (i.e., the temperature and pressure demons) to operate a perpetual motion machine, a certain amount of information must be obtained and must be derived from some external source for which the entropy is increased. As a result of the information gained by the demon, the entropy of the thermodynamic process is reduced and work can be performed. However, the entropy decrease created by the demon is still *smaller*, even under ideal conditions, than the amount of information provided by the external sources. Therefore, even with this most powerful Maxwell's demon, from an information theory point of view, the second law of thermodynamics has been proved to remain unchallenged.

5.5 SZILARD'S DEMON

In the earlier days, before Shannon's information theory, most physicists thought that information had nothing to do with the physical laws. However, the basic relationship between physical entropy and information was discovered by Szilard [1,2] as early as 1929 and is recorded in one of his remarkable works. Unfortunately, this interesting work was neglected for many

years, and not until the development of information theory by Shannon [6–8] in 1948 was it brought to light.

The work on the entropy theory of information by Szilard is, to our knowledge, one of the earliest on this subject. It is certainly the most interesting and rigorous demonstration involving Maxwell's demon and information theory. Because of its importance and historical significance in relation to the entropy theory of information, we devote a separate section of this chapter to Szilard's machine.

Szilard's machine consists of a cylinder of volume V and a frictionless piston acting as a partition wall dividing the cylinder into two chambers, 1 and 2, as shown in Fig. 5.3. As the piston moves either to the left or the right, it raises one of the pans and lowers the other. For convenience, the piston is equipped with a trapdoor that can be opened or closed. Szilard assumed that the cylinder contains only *one* molecule, and that the whole cylinder maintains a constant temperature T. The molecule continuously gains and loses kinetic energy as it randomly bounces off the walls, and it has an average kinetic energy of $3kT/2$. When the trapdoor is opened, no work is performed by simply sliding the piston sideways.

Now let us assume that the piston is moved to a position near the center of the cylinder. In this location, the demon clamps the trapdoor shut, and at this moment he somehow knows where the molecule is located. Let us suppose that the molecule is in chamber 1. Then the frictionless piston moves slowly from left to right, as a result of the random movement of the molecule against the piston walls. In this manner, work is performed by Szilard's machine. After chamber 1 has expanded to the whole volume of the cylinder, the demon opens the trapdoor and slides the piston back to its original position. In this fashion, another cycle of work is done by the machine. Thus, a perpetual motion machine of the second kind can be realized through the demon's intervention. But a puzzle similar to the previous one remains: How can the demon locate the molecule?

As in the previous examples, the demon needs to obtain a certain amount of information in order to locate the molecule. From this information, he is able to reduce the entropy of the system, but the amount of information provided will be shown to be in excess of the entropy reduction.

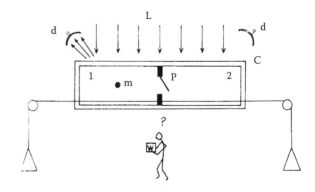

FIG. 5.3 Szilard's machine with intervention of the demon, d, photodetector. L, light beams; C, cylinder; b, pan; w, weight; P, piston; m, molecule.

To illustrate the demon's work, we let V_1 and V_2 be the respective volumes of chambers 1 and 2. The first step is to locate the molecule. For simplicity, we let the cylinder be transparent. We equip the demon with two beams of light, one through chamber 1 and the other through chamber 2, as shown in Fig. 5.3. If light is scattered from one of these chambers, it will show where the molecule is located. However, the absorption of the light quanta hv by one of the photodetectors represents an entropy increase in the process:

$$\Delta S_c = \frac{hv}{T} > k \tag{5.47}$$

where $hv > kT$, h is Planck's constant, v is the frequency of the light beams, T is the Kelvin temperature in the cylinder, and k is Boltzmann's constant. It was noted by Szilard [1,2] that a slightly lower entropy limit can be calculated:

$$\Delta S_c \geq k \ln 2 = 0.7k \tag{5.48}$$

We can now compute the amount of information the demon needs for its intervention. Let P_1 and P_2 be the *a priori* probability of locating the molecule in chamber 1 and in chamber 2, respectively:

$$P_1 = \frac{V_1}{V} \tag{5.49}$$

and

$$P_2 = \frac{V_2}{V} \tag{5.50}$$

where $P_1 + P_2 = 1$ and $V_1 + V_2 = V$.

Now the amount of information required by the demon is

$$I_d = -(P_1 \log_2 P_1 + P_2 \log_2 P_2) \tag{5.51}$$

which comes from a binary information source. For every cycle of the demon's operation, this much information is needed in order for the system's entropy to be reduced. The largest amount of information needed must correspond to the equiprobable case:

$$P_1 = P_2 = \frac{1}{2} \tag{5.52}$$

which also implies that

$$V_1 = V_2 = \frac{1}{2}V \tag{5.53}$$

Then the largest amount of information required by the demon is

$$I_{max} = 1 \text{ bit/cycle} \tag{5.54}$$

The demon can use this bit of information to reduce the system's entropy:

$$\Delta S_d = -k \ln 2 \tag{5.55}$$

By comparing Eqs. (5.47) and (5.55), we see that the net entropy change per cycle of operation is

$$\Delta S = \Delta S_c + AS_d \geq 0 \tag{5.56}$$

Again, the result is within the constraints of the second law of thermodynamics.

Now let us apply the entropy equation for an ideal monatomic gas (Eq. (4.26)). This equation can be simplified for our present purpose to the case of one molecule, $N = 1$:

$$S = K + k \ln V \tag{5.57}$$

where

$$K = \frac{5k}{2} + k \ln \left[g \left(\frac{4\pi mE}{3h^2} \right)^{3/2} \right]$$

is a positive constant. At the moment the molecule is located, the entropy decrease can be computed. For example, if the molecule is in chamber 1,

$$\Delta S_1 = k \ln \left(\frac{V_1}{V} \right) = k \ln P_1 \tag{5.58}$$

and if the molecule is in chamber 2,

$$\Delta S_2 = k \ln \left(\frac{V_2}{V} \right) = k \ln P_2 \tag{5.59}$$

Hence, the average entropy decrease per cycle is

$$\Delta S_d = P_1 \Delta S_1 + P_2 \Delta S_2 = -I_d k \ln 2 \tag{5.60}$$

where $I_d \leq 1$; the equality holds for $V_1 = V_2$. Therefore, for $V_1 \neq V_2$, the net entropy change per cycle of the demon's intervention is

$$\Delta S = \Delta S_c + \Delta S_d = k \left(\frac{hv}{kT} - I_d \ln 2 \right) > 0 \tag{5.61}$$

In concluding this section, we have demonstrated examples of Maxwell's demon in the application of modern information theory. We have shown that the receiving of information by the demon is inevitably accompanied by a certain compensatory (with some excess) increase in the entropy of some external source or system. This increase in the entropy traded for information is one of the basic relationships between information theory and physical science.

5.6 DIFFRACTION-LIMITED DEMON

The sorting demon of Maxwell has intrigued scientists for some time. Maxwell's demon is an excellent example in the entropy theory of information. Since we are in the computer epoch, we will extend Maxwell's demon to a diffraction-limited demon, for which we assume that the demon has a diffraction-limited eye, as shown in Fig. 5.4. Since the demon is located in a thermally isolated chamber, in order for him to see the molecules, we have provided him with a laser. By using the negentropy provided by the laser, the demon is able to let the molecules go through the trapdoor. In this manner, he is eventually able to decrease the entropy of the chamber from a higher entropy state to a lower state. However, a question is raised: Is the demon actually able to see the individual molecule since his eye is diffraction limited? If he is not able to resolve the individual molecule, then how can he be so sure that he is letting only the right molecule pass through the trapdoor? In order for him to do so, we equip the demon with a powerful computer. To perceive the arriving molecules, the demon turns on the laser, which is required to emit at least a quantum of light for him to make the observation, that is,

$$h\nu = kT \tag{5.62}$$

We assume that the quanta of light reflected by the approaching molecules is totally absorbed by the eye of the demon, which corresponds to an increase of entropy in the demon, that is,

$$\Delta S_d = \frac{h\nu}{T} \tag{5.63}$$

which is equivalent to the amount of information provided to the demon, that is,

$$I_d = \frac{\Delta S_d}{k \ln 2} \tag{5.64}$$

FIG. 5.4   Diffraction-limited demon exorcist.

Because of the diffraction-limited nature of the demon's eye, the demon needs to process the absorbed quanta to a higher resolution, so that he is able to resolve the molecules and allow the passage of the high- or low-energy molecules through the trapdoor. The amount of information gain, through the processing by the equipped computer, constitutes an equivalent amount of entropy increase, as given by

$$\Delta S_p = k\Delta I_d \ln 2 \tag{5.65}$$

where ΔI_d is the incremental amount of information provided by the computer. With this amount of information gain, the demon is able to reduce the entropy of the chamber to a lower state.

To compute the amount of entropy reduction by the demon's intervention, we let the initial entropy of the chamber be

$$S_0 = k \ln N_0 \tag{5.66}$$

where N_0 is the initial microscopic complexity of the chamber. After receiving I_d from the laser and ΔI_d from the computer, the demon is able to reduce the initial complexity from N_0 to N_1,

$$N_1 = N_0 - \Delta N \tag{5.67}$$

where ΔN is the net changed in complexity. The final entropy of the chamber is therefore

$$S_1 = k \ln N_1 \tag{5.68}$$

for which the net entropy reduction of the chamber is

$$\Delta S_1 = S_1 - S_0 = k \ln\left(1 - \frac{\Delta N}{N_0}\right) \approx -k - \frac{\Delta N}{N_0} \tag{5.69}$$

where $\Delta N \ll N_0$.

However, we see that the overall net entropy changed in the chamber, per trapdoor operation by the demon, would be

$$\Delta S = \Delta S_d + \Delta S_P + \Delta S_1 = k\left(\frac{h\nu}{kT} + \Delta I_d \ln 2 - \frac{\Delta N}{N_0}\right) > 0 \tag{5.70}$$

in which we see that the diffraction limit demon's exorcist is still within the limit of the second law of thermodynamics, where $h\nu > kT$, $\Delta I_d > 0$, and $\Delta N \ll N_0$.

5.7 MINIMUM COST OF ENTROPY

One question still remains unanswered: What would be the minimum cost of entropy required for the demon to operate the trapdoor? Let the arrival molecules at the trapdoor at an instant be one, two, or more molecules. Then the computer is required to provide the demon with a "yes" or a "no" information. For example, if a single molecule is approaching the trapdoor (say, a high-velocity one), the demon will open the trapdoor to allow the molecule to go through it. Otherwise, he will stand still.

For simplicity, if we assume a 50% probability of one molecule arriving at the trapdoor, then the demon needs one additional bit of information, from the computer, to open the trapdoor. This additional bit of information, however, corresponds to the amount of entropy increase provided by the computer, that is,

$$\Delta S_p = k \ln 2 \approx 0.7k \qquad (5.71)$$

which is the minimum cost of entropy required for the demon to operate the trapdoor. This minimum cost of entropy is in fact the result obtained by Szilard's demon in Section 5.5. From Eq. (5.70), we see that the overall net entropy increase in the chamber is

$$\Delta S = k \left(\frac{h\nu}{kT} + 0.7 - \frac{\Delta N}{N_0} \right) > 0 \qquad (5.72)$$

If one takes into account the other bit of "no" information, provided by the computer, then the average net entropy increase in the chamber per operation would be

$$\Delta S_{\text{ave}} = k \left(\frac{2h\nu}{kT} + 1.4 - \frac{\Delta N}{N_0} \right) \gg 0 \qquad (5.73)$$

in which two quanta of light radiation are required. It is trivial, but if one includes the possibility of a slower molecule approaching the trapdoor, then the average cost of entropy per operation is even higher.

Even though we omit the two quanta of light in the calculation, the overall net entropy change in the chamber is still increasing, that is,

$$\Delta S_{\text{ave}} = k \left(1.4 - \frac{\Delta N}{N_0} \right) > 0 \qquad (5.74)$$

where $\Delta N \ll N_0$. In other words, the entropy compensated by the computer is still higher than the entropy reduced by the demon. In view of this argument, we see that the computer we provided for the demon is also operated within the limit of the second law of thermodynamics.

We should now discuss the cost of entropy required to increase the resolution beyond the diffraction limit. Let a classical imaging system have the following resolution limit:

$$r_m = \frac{1.22\lambda f}{D} \qquad (5.75)$$

where r_m is the minimum separation, λ is the wavelength of the light source, and f and D are the focal length and diameter of the imaging aperture. To increase the resolution of the system, one could either reduce the wavelength or enlarge the aperture of the system, or both. However, if the wavelength reduction and the aperture enlargement are not the subjects to be considered, then the observed image can be processed by the computer beyond the resolution limit. However, the amount of resolution gain would be compensated by the amount of entropy increased. Since the minimum cost of entropy to resolve a point object is $0.7k$, for n object points' resolution, the minimum cost of entropy would be $0.7nk$. Nevertheless, in practice, the actual cost of entropy is very excessive.

For example, let A be the field of view of an imaging system and ΔA_0 the observation error, as shown in Fig. 5.5; then the amount of information obtained by this observation would be

$$I = \log_2 \frac{A}{\Delta A_0} \qquad (5.76)$$

where $\Delta A_0 = \pi(r_m)^2$ and r_m is the minimum resolvable distance of the optical system. If the observed image is then processed to a higher resolution, then a net amount of information gain is anticipated. In other words, the observation error of the imaging system can be reduced, by processing the observed image from ΔA_0 and ΔA_1. The net information gain provided by the computer is

$$\Delta I = \log_2 \frac{\Delta A_0}{\Delta A_1} \quad \Delta A_1 \ll \Delta A_0 \qquad (5.77)$$

Thus, we see that the minimum cost of entropy is

$$\Delta S = \Delta I k \ln 2 = k \ln 2 \log_2 \frac{\Delta A_0}{\Delta A_1} \qquad (5.78)$$

where ΔA_0 and ΔA_1 are the initial and final observation errors.

In short, we have achieved a very fundamental conclusion: one cannot get something from nothing; there is always a price to pay. Resolution beyond the resolution limit is possible, but only through the increase of entropy from a certain physical system (in our example, the computer). However, in practice the cost of entropy is very excessive. In fact, the additional

FIG. 5.5 Observation error: A is the field of view and ΔA_0 is the observation error.

information gain at best can only approach the equivalent amount of entropy traded off:

$$\Delta I \leq \frac{\Delta S}{k \ln 2} \tag{5.79}$$

5.8 GABOR'S PERPETUUM MOBILE OF THE SECOND KIND

We now discuss Gabor's [14] perpetual motion machine of the second kind. Gabor's problem essentially concerns a modified version of Szilard's machine, but without an intelligent demon's intervention. However, it relies on a certain relay activation; thus, the major difference in Gabor's problem is that the operation is accomplished strictly by physical devices.

In order to show that Szilard's machine operates within the limits of the second law of thermodynamics, we had to assume that such an observation by the demon cannot be obtained without a certain increase in entropy of some other source. In fact, this is one of the basic relationships between observation and information, which we elaborate on in Chapter 6.

Now we come to an illustration of Gabor's perpetuum mobile, as pictured in Fig. 5.6. Let us assume that a single

thermally agitated molecule is wandering within a cylinder of volume V at a constant temperature T. It is also assumed that the lower part of the cylinder is transparent and that a beam of light illuminates this lower chamber, with a volume of V_1, as shown in the figure. The light beam is intended to provide information on the location of the wandering molecule, and is derived from a heated filament. For simplicity, we assume a monochromatic source of frequency v. As shown in Fig. 5.6, a set of ideal reflectors (mirrors) is used to keep the light beam circulating through the lower chamber without any loss.

Now if the molecule wanders within the lower chamber of the cylinder, certain scattered light from it is expected. As in the previous examples, this scattered light can be detected by means of photodetectors. A relay is then activated in such a way that a frictionless piston is immediately inserted into the cylinder, and two sliding mirrors then cover the transparent portion. As a result of the random bombardment of the piston wall by the molecule, work can be performed by the slow expansion of the lower chamber. The corresponding entropy decrease after the molecule has been trapped in the lower chamber is therefore

$$\Delta S_d = k \ln \frac{V_1}{V} \tag{5.80}$$

where $V_1 < V$.

When the piston has moved to the extreme end of the cylinder ($V_1 = V$), it again slides out of the cylinder and returns to its original location. In the same manner, another cycle of operation can take place. If the work done by the machine is greater than the overall energy lost from the light source, then a perpetual motion machine of the second kind has been achieved—this time without help from a demon. Here we see again that in order to activate the relay, scattered light from the molecule is required. So we again come to the basic question of how much scattered light is needed for such an operation. Or, equivalently, how small is the amount of information required for such motion?

In order to answer this basic question, let t_0 be the average total amount of time consumed during every cycle of detection; Δt is the duration of the scattered light, and $\Delta t < t_0$. It can be seen that t_0 and Δt are directly proportional to the volumes of the cylinder V and V_1, respectively. Let us recall the information measure

$$I = \log_2 \frac{N_0}{N_1} \tag{5.81}$$

where N_0 and N_1 are the respective initial and final number of equiprobable stages. We also noted in our problem that

$$\frac{N_0}{N_1} = \left(\frac{t_0}{\Delta t} \right) \left(\frac{V}{V_1} \right) = \left(\frac{V}{V_1} \right)^2 \tag{5.82}$$

FIG. 5.6 Gabor's perpetuum mobile of the second kind.

Therefore, the smallest amount of information required to activate the relay is

$$I = 2 \log_2 \frac{V}{V_1} \qquad (5.83)$$

From Eq. (5.65), it can be seen that

$$Ik \ln 2 + \Delta S_d = k \ln \frac{V}{V_1} > 0 \qquad (5.84)$$

which is about twice the entropy decrease.

Since the information provided for the detection is derived from the increase in entropy from an external source—in our case, the light source—the expenditure of the entropy is excessive. It is a simple matter to show that the overall entropy increase is far greater than the amount of entropy the system can reduce:

$$\Delta S - Ik \ln 2 > 0 \qquad (5.85)$$

Thus,

$$\Delta S + \Delta S_d > 0 \qquad (5.86)$$

From the above discussion, it is clear that Gabor's perpetuum mobile still operates within the limits of the second law of thermodynamics.

REFERENCES

1. L. Szilard, Über die Entropieverminderung in Einem Thermodynamischen System bei Eingriffen Intelligenter Wesen, *Z. Phys.*, vol. 53, 840 (1929).
2. L. Szilard, On the Decrease of Entropy in a Thermodynamic System by the Intervention of Intelligent Beings (translated by A. Rapoport and M. Knoller), *Behav. Sci.*, vol. 9, 301 (1964).
3. L. Brillouin, Maxwell's Demon Cannot Operate. Information Theory and Entropy I, *J. Appl. Phys.*, vol. 22, 334 (1951).
4. L. Brillouin, *Science and Information Theory*, 2nd ed., Academic, New York, 1962.
5. F. W. Sear, *Thermodynamics, the Kinetic Theory of Gases, and Statistical Mechanics*, Addison-Wesley, Reading, MA, 1953.
6. C. E. Shannon, A Mathematical Theory of Communication, *Bell Syst. Tech. J.*, vol. 27, 379–423, 623–656 (1948).
7. C. E. Shannon, Communication in the Presence of Noise, *Proc. IRE*, vol. 37, 10 (1949).
8. C. E. Shannon and W. Weaver, *The Mathematical Theory of Communication*, University of Illinois Press, Urbana, 1949.
9. J. H. Jeans, *Dynamical Theory of Gases*, 3rd ed., Cambridge University Press, New York, 1921, p. 183.
10. M. von Smoluchowski, Experimentell Nachweisbare, der Üblichen Thermodynamik Widersprechende Molekular-Phänomene, *Phys. Z.*, vol. 13, 1069, (1912).
11. P. M. Demers, Les Démons de Maxwell et le Second Principle de la Thermodynamique, *Can. J. Res.*, vol. 22, 27 (1944).
12. P. M. Demers, Le Second Principle et la Théorie des Quanta, *Can. J. Res.*, vol. 23, 47 (1945).
13. L. Brillouin, The Negentropy Principle of Information, *J. Appl. Phys.*, vol. 24, 1152 (1953).
14. D. Gabor, Light and Information, in E. Wolf, ed., *Progress in Optics*, vol. I, North-Holland, Amsterdam, 1961.
15. G. N. Lewis, The Symmetry of Time in Physics, *Science*, vol. 71, 569 (1930).
16. J. C. Slater, *Introduction to Chemical Physics*, McGraw-Hill, New York, 1939.

6 Observation and Information

In the classical theory of light, an observation can be made as small as we please. However, for some very small objects this assumption does not hold true. For example, let us take a Mach–Zehnder interferometer [1], as shown in Fig. 6.1. It can be seen that only a very small fraction of light is required to illuminate the object transparency $s(x, y)$. This small fraction of light, however, carries all the information that we intended. However, the other path of light carries almost all the energy of the light source. The corresponding irradiance distributed on the photographic emulsion is [2]

$$I(p,q) = R^2 + |S(p,q)|^2 + 2RS(p,q)\cos\left[\alpha_0 p + \phi(p,q)\right] \quad (6.1)$$

where R is the background (reference) beam, $S(p, q)$ is the corresponding Fourier spectrum of the object transparency, (p, q) are the corresponding spatial frequency coordinates, α_0 is an arbitrary constant, and

$$S(p,q) = |S(p,q)|\exp\left[i\phi(p,q)\right] \quad (6.2)$$

In our example, it is assumed that $R \gg |S(p, q)|$. Thus, the interference term (the last term) is considered very small in comparison with the background irradiance. In principle, it is possible to enhance the weak interference term by means of a coherent optical processor [2], so it is simply a matter of making $s(x, y)$ observable. In the classical theory, it is possible to observe with as weak an object beam as we wish. However, we know instinctively that this is not practical, since an increase in background light would at the same time increase background fluctuation. Thus, the informational term (the interference term) could be completely buried in the background fluctuation. So in practice, there exists a practical lower limit. Beyond this limit, it is not possible to retrieve the information (the object).

In our case, the informational term must be at least equal to or greater than the mean square fluctuation of the background irradiance:

$$R^2\left\langle S^2(p,q)\right\rangle \geq \left\langle (\delta R^2)^2\right\rangle \quad (6.3)$$

where $\langle\ \rangle$ denotes the spatial ensemble, which implies that

$$\left\langle S^2(p,q)\right\rangle \geq \varepsilon_0 \quad (6.4)$$

with ε_0 the lower limit of irradiance.

Substituting the equality in the above equation in Eq. (6.3) yields

$$\frac{R^2}{\varepsilon_0} = \frac{\left\langle (\delta R^2)^2\right\rangle}{\varepsilon_0^2} \quad (6.5)$$

This is the minimum background irradiance, in terms of ε_0, that allows the observation to be made. As Gabor [3,4] noted, Eq. (6.5) is Poisson's *law of rare events*. This was what was accounted for in the hypothesis that monochromatic light arrives in quanta. In our case, these quanta of light arrive in a random fashion and are subjected to constraints; that is, on the average, the background irradiance R^2/ε_0 arrives during the observation (the recording).

Thus, no observation is possible if less than *one quantum* of light arrives from the object transparency. This simple statement essentially leads us to the concept of the quantum theory of light; namely, the monochromatic light field is perceived in *discrete quanta*.

In Chapter 5, we reached a significant conclusion: an observation can take place only when there is a certain compensatory increase in the entropy of some external source. This result is regarded as the basic relationship between observation and information theory, which we discuss here. We utilize several important results, which were demonstrated by Brillouin [5–7].

6.1 OBSERVATION WITH RADIATION

Since light is electromagnetic in nature, it is the purpose of this section to discuss the probabilistic analysis of observations made by radiation. We note that as the radiation frequency increases, the quantum effect takes place. When the quantum $h\nu$ is no longer a small quantity compared with kT, a photodetector takes the quantum effect into account:

$$E_n = nh\nu \quad n = 1, 2, 3, \ldots$$

where the E_n's are the quantized energy levels.

These energy levels at a temperature in Kelvin have the following probability distribution (see Appendix C), known as *Gibb's distribution*:

$$P_n = Ke^{-nh\nu/kT} \quad (6.6)$$

with

$$\sum_{n=0}^{\infty} P_n = 1 \quad (6.7)$$

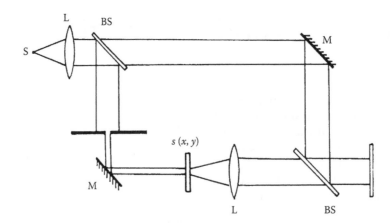

FIG. 6.1 Paradox of observation without illumination. S, monochromatic point source; M, mirror; $s(x, y)$, object transparency; P, photographic plate; BS, beam splitter; L, lens.

where

$$K = 1 - e^{-hv/kT}$$

Thus, Eq. (6.6) can be conveniently written as

$$P_n = e^{-hv/kT} - e^{-(n+1)hv/kT} \qquad (6.8)$$

The mean quantum state can be determined by

$$\bar{n} = \sum_{n=0}^{\infty} nP_n = \frac{1}{e^{hv/kT} - 1} \qquad (6.9)$$

It is noted that if $hv \ll kT$, then

$$\bar{n} \simeq \frac{kT}{hv} \quad \text{for } hv \ll kT \qquad (6.10)$$

which is the average energy of the photodetector:

$$\bar{E} = \bar{n}hv = kT \qquad (6.11)$$

In the observation analysis, we denote the probabilities

$$P(0 \le n < g) = \sum_{n=0}^{g-1} P_n = 1 - e^{-ghv/kT} \qquad (6.12)$$

and

$$P(n \ge g) = \sum_{n=g}^{\infty} P_n = 1 - P(0 \le n < g) = e^{-ghv/kT} \qquad (6.13)$$

where g is an arbitrary quantum state.

Now let us observe the photodetector. Let the quantum state g be used as a threshold decision level for the observation. On the photodetector, at time t, energy corresponding to $n \ge g$ quanta is called a *positive observation* (reading). It can be seen that the probability of making an observation error is $P(n \ge g)$, primarily because of the thermal fluctuation of the photodetector. However, the probability of making a correct observation is $P(0 \le n \le g)$, since it is basically due to the absorption of additional quanta from some external source that are carried over the g threshold level.

We can now compute an optimum decision level for the observation. In order to do so, we must first compute the corresponding *median quantum state*. By definition, let us equate either Eq. (6.12) or (6.13) to $\frac{1}{2}$, which yields

$$m = \frac{kT}{hv} \ln 2 \qquad (6.14)$$

where m denotes the median quantum state. Since $hv \ll kT$, we see from Eq. (6.10) that

$$m = \bar{n} \ln 2 < \bar{n} \qquad (6.15)$$

which is smaller than the average quantum state $\bar{n}$. The corresponding *median energy* of the photodetector is

$$E_m = mhv = kT \ln 2 \qquad (6.16)$$

From Eq. (6.11), we see that

$$E_m < \bar{E} \qquad (6.17)$$

As we see later, the situation is entirely reversed for high-frequency observation.

We will now calculate the average quantum stages for the cases of $0 \leq n < g$, and $n \geq g$. Let us define the following reduced sample space probabilities:

$$P(n/0 \leq n < g) = \frac{P_n}{P(0 \leq n < g)} \qquad (6.18)$$

and

$$P(n/n \geq g) = \frac{P_n}{P(n \geq g)} \qquad (6.19)$$

Then the corresponding average quantum states are

$$\bar{n}_1 = \sum_{n=0}^{g-1} nP(n/0 \leq n < g) = \bar{n} - \frac{g}{e^{ghv/kT} - 1} \qquad (6.20)$$

and

$$\bar{n}_2 = \sum_{n=g}^{\infty} nP(n/n \geq g) = \bar{n} + g \qquad (6.21)$$

It is also noted that

$$\bar{n}_1 = \bar{n}_1 P(0 \leq n < g) + \bar{n}_2 P(n \geq g) \qquad (6.22)$$

For an optimum decision observation, we adopt the median m for the threshold level, $g = m$, and then Eqs. (6.20) and (6.21) yield, respectively,

$$\bar{n}_1 = \bar{n} - \frac{m}{e^{mhv/kT} - 1} \qquad (6.23)$$

and

$$\bar{n}_2 = \bar{n} + m \qquad (6.24)$$

It is noted that $hv \ll kT$. Then Eqs. (6.23) and (6.24) can be written as

$$\bar{n}_1 \simeq \bar{n}(1 - \ln 2) = 0.3\bar{n} \qquad (6.25)$$

and

$$\bar{n}_2 = \bar{n}(1 + \ln 2) = 1.7\bar{n} \qquad (6.26)$$

If one accepts that $g = m$ is the threshold decision level for the observation, then we have an error probability of 50% per observation. However, if we choose a higher quantum state for the decision level ($g > m$), then a more reliable

observation can be obtained. For example, if $g = 4m$ is chosen, then the error probability of the observation will be about 2%.

It is interesting to note from Eq. (6.21) that the average energy above the g quantum state is

$$\bar{E}_2 = \bar{n}_2 hv = \bar{n}hv + ghv \qquad (6.27)$$

where ghv can be regarded as the *excess* energy. This excess energy can come from two possible sources, either from an internal thermal fluctuation or from an external source. If it emanates purely from an external source, then g quanta were actually absorbed in the photodetector and added to the normal average energy of $\bar{n}hv$. But if this excess energy is purely from an internal thermal fluctuation, then it will be compensated for by a loss in energy at a later instant in time. Thus, the photodetector has an average energy of kT. However, if it is from an external source, then the excess energy of ghv will eventually be dissipated into the surrounding environment of T in the photodetector. This corresponds to an entropy increase in the photodetector:

$$\Delta S = \frac{ghv}{T} \qquad (6.28)$$

We see that if $g = m$ is chosen for the decision level, then Eq. (6.28) becomes

$$\Delta S = k \ln 2 \qquad (6.29)$$

since $m = kT \ln 2/hv$. Equation (6.29) can be regarded as the *minimum cost* (the optimum) in entropy needed for an observation, with a 50% chance of error. In terms of the amount of information, Eq. (6.29) is essentially equal to 1 bit:

$$I = \frac{\Delta S}{k \ln 2} = 1 \, \text{bit} \qquad (6.30)$$

This is the amount of information obtained per observation by the photodetector. In fact, Eq. (6.29) is a consequence of Szilard's [8,9] results mentioned in Chapter 5. We emphasize that the result of Eq. (6.29) is not restricted only to the low-frequency case; it can also be applied to high-frequency observation.

We can now discuss high-frequency observation, and whether the quantity of hv is higher or lower in the order of kT. In this case, hv/kT is no longer a small quantity. Let us first state the quantity:

$$hv \gg kT \qquad (6.31)$$

We know that the smallest amount of energy the photodetector can absorb is one quantum. In view of $hv \gg kT$ and the

50% chance of error criterion ($g = m$), $m = 1$. From Eq. (6.15), we see that

$$m = \bar{n} = 1 \qquad (6.32)$$

and with Eq. (6.14), we now have

$$hv = kT \ln 2 \qquad (6.33)$$

Since the total probability for the higher quantum states ($n > m$) is 50%, the probability of the ground state is also 50%. Hence, if we observe one quantum or more in the photodetector, the chance that the absorption is from an external source is 50%, and the chance that it is caused by thermal fluctuation is also 50%. Now if one quantum of absorption (the minimum) is due to an external source, then this corresponds to an entropy increase in the photodetector:

$$\Delta S = k \ln 2 \qquad (6.34)$$

which is identical to Eq. (6.29) for low-frequency observation and corresponds to a minimum cost in entropy in high-frequency observation.

In view of $hv \gg kT$, the minimum cost we pay for high-frequency observation is even greater:

$$\Delta S > k \ln 2 \qquad (6.35)$$

However, since $m < 1$, we use $g = 1$, an integer, which provides us with a more reliable observation. Of course, this is better than a 50% chance of error, and the amount of information obtained is still the same, exactly 1 bit. This is the minimum information required for the functioning of Szilard's demon, but the cost is much higher when this method is used.

We emphasize that this information is available only for a very short instant in time. The excess energy (or rather, the entropy) provided by the external source is eventually quietly dissipated in the photodetector. This excess energy dissipation provides the observation within the limits of the second law of thermodynamics.

Thus far, our discussion has been restricted to only the ensemble point of view. That is, for a special observation, the minimum cost may be lower than $k \ln 2$, but under no circumstances have we been able to predict when this will occur. Thus, only in terms of statistical averages will we be provided with a logical answer. However, our discussion has revealed the basic relationship between information theory and observation; we use this basic relationship to further explore observation and information.

6.2 SIMULTANEOUS OBSERVATIONS

In this section, we consider an interesting example of many simultaneous observations. We see that the minimum cost in entropy for simultaneous observations is even higher than $k \ln 2$.

Let us assume that we are simultaneously observing α photodetectors. The observation can be performed with one or with more than one of these α detectors, so that the energy is seen to be greater than an arbitrary threshold level of the detector. For simplicity, let us assume that all these photodetectors are identical and are maintained at the same Kelvin temperature T. First, we consider the low-frequency observation, $hv \ll kT$, and follow this immediately with the high-frequency observation, $hv \gg kT$. Let us recall the probabilities of Eqs. (6.12) and (6.13) and denote them, respectively, as P_1 and P_2:

$$P_1 = P(0 \le n < g) = 1 - e^{-E_0/kT} \qquad (6.36)$$

and

$$P_2 = P(n \ge g) = e^{-E_0/kT} \qquad (6.37)$$

where $E_0 = ghv$, a threshold energy level.

Since there are α photodetectors, the probability that β detectors deflect energy below E_0, at an instant in time, is

$$P_{\alpha,\beta} = \frac{\alpha!}{\beta!(\alpha - \beta)!} P_1^\beta P_2^{\alpha-\beta} \qquad (6.38)$$

which is a *binomial* probability distribution.

We first assume that an observation gives a positive result if any one of the α photodetectors gives rise to an energy level equal to or above E_0. Therefore, the probability of a correct observation is P_1^α, and $1 - P_1^\alpha$ corresponds to the observation error. If we impose a 50% chance of error constraint on the α simultaneously observed detectors, we will then have

$$P_1^\alpha = (1 - e^{-E_0/kT})^\alpha = \frac{1}{2} \qquad (6.39)$$

It can be seen that if $\alpha = 1$, Eq. (6.39) is reduced to the previous result:

$$E_0 = kT \ln 2 \quad \text{for } \alpha = 1 \qquad (6.40)$$

From Eq. (6.39), one can determine that E_0 is a function of α:

$$E_0 = -kT \ln\left[1 - \left(\frac{1}{2}\right)^{1/\alpha}\right] > 0 \qquad (6.41)$$

where α is the number of photodetectors. It can be seen that E_0 increases as the number of detectors α increases. The lower limit of E_0 corresponds to $\alpha = 1$, as in Eq. (6.40).

If we take the natural logarithm of Eq. (6.39), we will have

$$\alpha \ln(1 - e^{-E_0/kT}) = -\ln 2 \qquad (6.42)$$

We see that E_0 increases as α increases. By the logarithmic expansion

$$\ln(1-x) = -\left(x + \frac{x^2}{2} + \frac{x^3}{3} + \cdots\right) \quad \text{for } x^2 < 1 \qquad (6.43)$$

Eq. (6.42) can be approximated:

$$\alpha e^{-E_0/kT} \simeq \ln 2 \qquad (6.44)$$

which corresponds to large values of α and E_0. Thus, Eq. (6.41) can be further reduced to

$$E_0 \simeq kT\left[\ln\alpha - \ln(\ln 2)\right] \quad \text{for } \alpha \gg 1 \qquad (6.45)$$

Since the additional energy from the external source eventually dissipates in the photodetectors, it corresponds to an increase in entropy:

$$\Delta S = k(\ln\alpha + 0.367) > k\ln 2 \qquad (6.46)$$

where $\ln(\ln 2) = -0.367$.

Equation (6.46) indicates the minimum cost in entropy for *one* of the positive readings from the α photodetectors. It is also clear that additional energy from the external source is absorbed by *all* the other photodetectors. Since this energy is assumed to be less than the threshold energy level E_0, the actual cost in entropy is even higher:

$$\Delta S > k(\ln\alpha + 0.367) \qquad (6.47)$$

By a similar analysis, one can show that for γ simultaneous positive readings (correct observations) from α photodetectors, the corresponding net entropy increase is

$$\Delta S = \gamma k(\ln\alpha + 0.367) \qquad (6.48)$$

where $1 < \gamma < \alpha$. From the same account of low-energy absorption for the remaining $(\alpha - \gamma)$ detectors, we conclude that

$$\Delta S > \gamma k(\ln\alpha + 0.367) \qquad (6.49)$$

The net entropy increase in all the detectors is even higher than the net entropy increase shown in Eq. (6.48).

With this basic analysis for low-frequency observations, we can now carry the analysis to the high-frequency case, $h\nu \gg kT$.

As noted earlier, a single quantum of $h\nu$ is sufficient for a reliable observation (reading), since in practice high-frequency oscillation does not exist in the photodetector (a blackbody radiator). Hence, we simply take the threshold energy level

$$E_0 = h\nu > kT\ln 2 \quad \text{for } h\nu \gg kT$$

for the case of only *one* reading from the α photodetectors. By the same reasoning as in the low-frequency observations, the net entropy increase is

$$\Delta S > \frac{h\nu}{T} > k\ln 2 \qquad (6.50)$$

Furthermore, for γ positive observations out of the α simultaneous observations, the net entropy increase is

$$\Delta S > \frac{\gamma h\nu}{T} > \gamma k\ln 2 \qquad (6.51)$$

In concluding this section, we point out that for many simultaneous observations, the minimum cost in entropy is higher than that for a single observation. The cost in entropy is generally proportional to the number of simultaneous positive observations. As for high-frequency observation, the cost in entropy is even higher, since $h\nu \gg kT$. But a more reliable observation in the high-frequency case can generally be achieved. Furthermore, it is also emphasized that the case of $h\nu \simeq kT$ should be analyzed separately and under different special conditions.

6.3 OBSERVATION AND INFORMATION

We now come to the basic problem of observation and information. As we have stressed, the amount of information obtained from an observation is ultimately derived from some compensatory increase in the entropy of other sources. We have also demonstrated, with physical observations, that no one has been able to claim a perfect observation. Thus, certain observation errors are bound to occur.

In order to emphasize the notion of information as applied to an optical observation, let us define the information measure for the observation [10]:

$$I = \log_2 \frac{N_0}{N_1} \text{ bits} \qquad (6.52)$$

where N_0 is the number of equiprobable states *before* the observation, and N_1 is the number of equiprobable states *after* the observation.

Let us designate the total field of view A, an optical system. For simplicity, the optical system is composed of compound lenses. We observe a position at a certain point in A, say in a (x, y) spatial domain, with an observation error of ΔA, as shown in Fig. 6.2. ΔA is the spatial resolution limit of the optical system. The amount of information in this observation can be obtained from the ratio of the initial uncertainty of A to the final uncertainty of ΔA:

$$I = \log_2 \frac{A}{\Delta A} \text{ bits} \qquad (6.53)$$

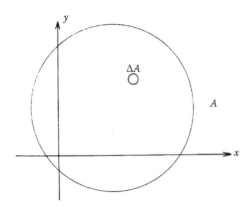

FIG. 6.2 Experimental error observation: A is the total field of view, and ΔA is the observation error at position (x, y).

Accordingly, the accuracy of the observation can be defined as [6]

$$A \triangleq \frac{A}{\Delta A} \qquad (6.54)$$

Thus, the amount of information obtained by the observation can also be written as

$$I = \log_2 A \quad \text{bits} \qquad (6.55)$$

In application, let the resolving power of the optical system be [2,7,10]

$$\Delta A_i = \pi \left(\frac{1.22\lambda f}{D} \right)^2 \qquad (6.56)$$

where ΔA_i is the remaining uncertainty of the observation. The subscript i refers to the image plane, λ is the wavelength of the light source, f is the focal distance, and D is the diameter of the optical aperture.

By substituting Eq. (6.56) into Eq. (6.53), the amount of information obtained is

$$I = \log_2 \frac{A_i}{\pi} \left(\frac{D}{1.22\lambda f} \right)^2 \quad \text{bits} \qquad (6.57)$$

where A_i is the total field of view, which refers to the image plane. From the above equation, we see that the amount of information increases as the size of the optical aperture increases.

It is interesting to note that if the optical system makes several completely different observations, then the total amount of information obtained is

$$I_0 = \sum_{n=1}^{N} I_n \quad \text{bits} \qquad (6.58)$$

where

$$I_n = \log_2 \frac{A_{in}}{\pi} \left(\frac{D}{1.22\lambda f} \right)^2 \qquad (6.59)$$

and N denotes the total number of different observations. However, if the N observations are completely repetitive (redundant), then the total information obtained is no more than that of a single observation:

$$I_0 = I_n \qquad (6.60)$$

Thus, with N sequential observations, one expects that

$$I_n \leq I_0 \leq \sum_{n=1}^{N} I_n \qquad (6.61)$$

We now discuss several simultaneous observations made by several optical systems. We see that if these optical systems see different fields of view, then the amount of information provided is equal to the sum. However, if the fields are not completely different, then the amount of information provided is somewhat smaller than the sum. If the fields of view are identical, then the amount of information provided is equal to that of any one of the channels.

Now let us discuss the optical spatial channels. Let N be the number of independent spatial channels with channel capacities of $C_1, C_2, ..., C_N$, as shown in Fig. 6.3. The overall spatial capacity is then

$$C = \sum_{n=1}^{N} C_n \qquad (6.62)$$

where the C_n are defined as the upper limit of the spatial mutual information [11]:

$$C_n \triangleq \max I_n(f, g) \qquad (6.63)$$

where $f(x, y)$ and $g(x, y)$ are the input–output spatial signal ensemble, and n denotes the nth spatial channel. Thus, the

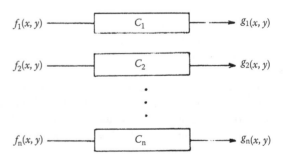

FIG. 6.3 Block diagram of the input–output spatial channels.

overall spatial capacity is the sum of the maximum mutual information.

6.4 ACCURACY AND RELIABILITY IN OBSERVATIONS

We discuss problems of observation within a space and time domain. This involves observing a position on a spatial plane, at an instant in time, with a certain observation error. It is emphasized that the spatial domain of observation must be bounded; otherwise, the information cannot be defined, since the information provided would lead to an infinite amount.

In order to illustrate the accuracy of observation, we assume a spatial domain A, which corresponds to the total field of view of an optical system. The spatial domain A is then subdivided into small areas ΔA, which are limited by the resolvable power of the optical system. Of course, light is necessary for the observation, in which a particle or object is assumed to be wandering within the spatial domain A. In practice, we look at each subdivision ΔA until we locate the particle. The accuracy of observation, as defined in Eq. (6.54), is

$$A \triangleq \frac{A}{\Delta A} = \alpha \qquad (6.64)$$

where α is the total number of ΔA's within A. To look for the particle, we simply illuminate each ΔA by a beam of light, and each ΔA is assumed to be equipped with a photodetector able to detect scattered light, if any, from the particle. As in previous examples, we assume that each of the photodetectors is maintained at a constant Kelvin temperature T. Later, we discuss low-accuracy observation that corresponds to $hv \ll kT$. We also discuss separately sequential and simultaneous observations.

Let us start with sequential observations. We investigate each of the succeeding photodetectors until a positive reading is obtained, say from the qth ΔA, where $q \leq \alpha$. The reading may be caused by thermal fluctuation in the detector, but if it happens to be a positive reading, then the particle has been found in one of the ΔA's out of the q possibilities. Hence, the amount of information obtained by this sequential observation is

$$I = \log_2 q \quad \text{bits} \qquad (6.65)$$

Since the positive reading was obtained from the absorption of scattered light by the qth photodetector, then according to Eq. (6.41), the accompanying entropy increase in the qth detector is

$$\Delta S \geq -k \ln\left[1 - \left(\frac{1}{2}\right)^{1/q}\right] \quad q \geq 1 \qquad (6.66)$$

For a large value of q, the right-hand side of Eq. (6.66) can be approximated (see Eq. (6.47)):

$$\Delta S \geq -k(\ln q + 0.367) > k \ln 2 \qquad (6.67)$$

Thus,

$$\Delta S - Ik \ln 2 \geq 0.367k > 0 \qquad (6.68)$$

which is a positive quantity.

For the case of simultaneous observations, we assume that we have observed γ positive readings simultaneously from the α photodetectors. The amount of information obtained is therefore

$$I = \log_2 \frac{\alpha}{\gamma} \quad \text{bits} \qquad (6.69)$$

where $\gamma \leq \alpha$. Again, we see that any reading could be due to thermal fluctuation. Since there are γ detectors absorbing the scattered light, and the observations are made on all the α photodetectors, the overall amount of entropy increase in the γ photodetectors is

$$\Delta S \geq -yk \ln\left[1 - \left(\frac{1}{2}\right)^{1/\alpha}\right] \qquad (6.70)$$

For a large value of α, Eq. (6.70) can be approximated as

$$\Delta S \geq \gamma k(\ln \alpha + 0.367) \qquad (6.71)$$

which increases with respect to γ and α. Thus,

$$\Delta S - Ik \ln 2 \geq k\left[\ln \gamma + (\gamma - 1)\ln \alpha + 0.367\gamma\right] > 0 \quad (6.72)$$

It is interesting to note that if it takes only one of the α photodetectors to provide a positive reading ($\gamma = 1$), then Eq. (6.72) is essentially identical to Eq. (6.68). However, if $\gamma \gg 1$, then the amount of information obtained from the simultaneous observations (Eq. (6.69)) is somewhat less than that obtained from the sequential observations of Eq. (6.65), for $q \gg 1$, and the amount of entropy increase for the simultaneous observations is also greater.

Since it is assumed that only one particle is wandering in the spatial domain A, for $\gamma = 1$, Eq. (6.72) yields the smallest trade-off of entropy and information. At the same time, for a large number of photodetectors ($\alpha \gg 1$), Eq. (6.72) is the asymptotic approximation used in high-accuracy observation ($hv \gg kT$). It is also emphasized that any other arrangement of the photodetectors may result in higher entropy. For example, if all the photodetectors are arranged to receive light directly, rather than from scattered light, then it can be seen that a high entropy cost will result.

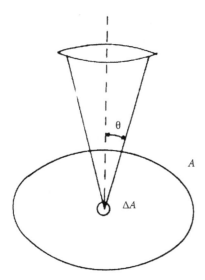

FIG. 6.4 High-accuracy observation.

We now discuss high-accuracy observation. It is noted that if ΔA becomes very small, then higher-frequency illumination (a shorter wavelength) is necessary for the observation. As illustrated in Fig. 6.4, this observation cannot be efficient unless the wavelength of the light source is shorter than $1.64d \sin \theta$ [2]:

$$\gamma \leq 1.64d \sin \theta \qquad (6.73)$$

where d is the diameter of ΔA, and θ is the subtended half-angle of the lens aperture. Accordingly,

$$d = \frac{1.22\lambda}{2\sin\theta} \qquad (6.74)$$

is the well-known formula for resolving power, where $2 \sin \theta$ is the *numerical aperture*. Thus, the frequency required for the observation must satisfy the inequality

$$v = \frac{c}{\lambda} \geq \frac{c}{1.64d \sin \theta} \qquad (6.75)$$

where c is the speed of light.

Now let us use the lower limit of Eq. (6.75) for a definition of the characteristic diameter or distance of the detector, which is assumed to maintain a constant temperature T:

$$\frac{hv}{kT} = \frac{hc}{1.64kTd \sin d} = \frac{d_0}{1.64d \sin \theta} \qquad (6.76)$$

The characteristic diameter d_0 is therefore

$$d_0 = \frac{hc}{kT} \simeq \frac{1.44}{T} \qquad (6.77)$$

Thus, for high-frequency observation, such as $hv \gg kT$, we see that

$$d \ll d_0 \qquad (6.78)$$

However, for low-frequency observation, such as $hv \ll kT$, we see that

$$d \gg d_0 \qquad (6.79)$$

It is emphasized that d_0 possesses no physical significance except that, at a given temperature T, it indicates the boundary between low- and high-frequency observations.

Let us recall Eq. (6.41) for high-frequency observation:

$$hv > E_0 = -kT \ln\left[1 - \left(\frac{1}{2}\right)^{1/\alpha}\right] \qquad (6.80)$$

where E_0 is the threshold energy level for the photodetectors. For $\alpha \gg 1$, Eq. (6.80) can be approximated by

$$hv > kT(\ln \alpha + 0.367) \qquad (6.81)$$

Since the absorption of one quantum is adequate for a positive reading, the corresponding entropy increase is

$$\Delta S = \frac{hv}{T} > k(\ln \alpha + 0.367) \qquad (6.82)$$

From Eq. (6.69), we obtain the information

$$I = \log_2 \alpha \quad \text{bits} \qquad (6.83)$$

Therefore, we conclude that

$$\Delta S - Ik \ln 2 > 0.367k > 0 \qquad (6.84)$$

Except for the equality, this ΔS is identical to that of the low-frequency observation of Eq. (6.68). However, the entropy increase is much higher, since v is very high. We emphasize that the higher the frequency used, the finer the observation obtained. This is the price we pay in entropy for greater accuracy of observation.

One must discuss the case of $hv \simeq kT$ separately and with special attention. For instance, if Eq. (6.70) holds, it does not necessarily mean that its asymptotic approximation (Eq. (6.71)) also holds.

We now come to the reliable observation. One must distinguish the basic difference between accuracy and reliability in observations. From the viewpoint of statistical communication [12–14], a reliable observation is directly related to the chosen decision threshold level E_0. That is, the higher the threshold level, the greater the reliability. However, accuracy in observation is inversly related to the spread of the pulse signal: the narrower the spread, the greater the accuracy. A simple example illustrating the difference between these two concepts is shown in Fig. 6.5. It is evident that the higher the threshold energy level E_0 chosen, the greater the reliability of observation. However, higher reliability also corresponds to a higher probability of a miss. If the decision level E_0 is set at a lower level, then a less reliable observation is expected. Thus, a higher probability of error (a false alarm) results, because of thermal fluctuation (noise). It is noted that in *decision theory* [13,14], given an *a priori* probability of noise, the cost of a miss, and the cost of a false alarm, an optimum decision level can be found.

Now instead of considering only a 50% chance of error in observation, we equate the error probability to the inverse of the reliability [6]:

$$P(n \ge g) = \frac{1}{R} \qquad (6.85)$$

where R is the reliability. We consider first the reliability in low-frequency observation, $hv \ll kT$. It is also noted that low-frequency observation corresponds to low-accuracy observation.

Let us use the error probability of Eq. (6.13) to determine the reliability:

$$P(n \ge g) = e^{-ghv/kT} = \frac{1}{R} \qquad (6.86)$$

which can be written as

$$\ln R = \frac{ghv}{kT} \qquad (6.87)$$

From Eq. (6.27), we see that after the observation, the excess energy ghv is eventually dissipated in the photodetector; thus, it corresponds to an increase in entropy:

$$\Delta S \ge \frac{ghv}{T} = k \ln R \qquad (6.88)$$

It is also interesting to consider the reliability in the problem of many simultaneous observations. Let us recall Eq. (6.39),

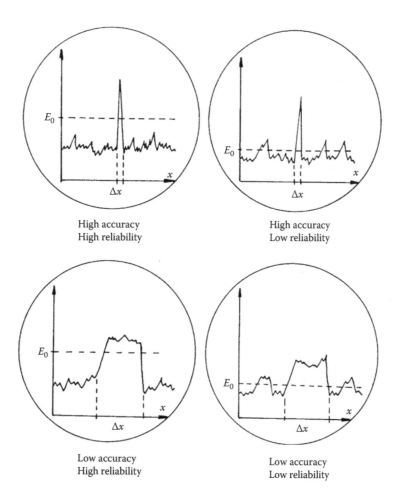

High accuracy
High reliability

High accuracy
Low reliability

Low accuracy
High reliability

Low accuracy
Low reliability

FIG. 6.5 Examples of accuracy and reliability of observation.

the probability of a correct observation when simultaneously observing α photodetectors. We have

$$(1 - e^{-gh\nu/kT})^\alpha = 1 - \frac{1}{R} \tag{6.89}$$

It can be seen that for $\alpha = 1$ (the case of a single photodetector), Eq. (6.89) is identical to Eq. (6.87). Alternatively, Eq. (6.89) can be written as

$$\alpha \ln(1 - e^{-gh\nu/kT}) = \ln\left(1 - \frac{1}{R}\right) \tag{6.90}$$

From Eq. (6.43), we determine that for large values of α and R, the left- and right-hand sides of Eq. (6.90) can be approximated by

$$\alpha \ln(1 - e^{-gh\nu/kT}) \simeq -\alpha e^{-gh\nu/kT} \tag{6.91}$$

and

$$\ln\left(1 - \frac{1}{R}\right) \simeq -\frac{1}{R} \tag{6.92}$$

Thus, we have

$$\alpha e^{-gh\nu/kT} \simeq \frac{1}{R} \tag{6.93}$$

or

$$\ln(\alpha R) \simeq \frac{gh\nu}{kT} \quad \text{for } \alpha R \gg 1 \tag{6.94}$$

Since Eq. (6.94) is identical to Eq. (6.87), for $\alpha = 1$, we see that Eq. (6.94) is a good approximation, even for a low value of α. Since it is assumed that there are γ correct observations from the α detectors, the entropy increase is

$$\Delta S \geq \gamma \frac{gh\nu}{T} = \gamma k \ln(\alpha R) \tag{6.95}$$

We now discuss high-frequency observation, $h\nu \gg kT$. As noted previously, one quantum $h\nu$ per photodetector is adequate for an observation, since $h\nu \gg kT$. Thus, the entropy increase is (see Eq. (6.51))

$$\Delta S \geq \gamma \frac{h\nu}{T} \tag{6.96}$$

Again, for intermediate-frequency observation, $h\nu \simeq kT$, the reliability analysis must be treated separately in each case.

We now consider the efficiency of observation. As defined in Chapter 5, the efficiency of observation is

$$\eta = \frac{Ik \ln 2}{\Delta S} \tag{6.97}$$

In the low-frequency case ($h\nu \ll kT$), the efficiency of many simultaneous observations, for $\gamma = 1$ (from Eqs. (6.69) and (6.71)), is

$$\eta = \frac{\ln \alpha}{\ln \alpha + 0.367} = \frac{1}{1 + 0.367/\ln \alpha} \simeq 1 - \frac{0.367}{\ln \alpha} \tag{6.98}$$

It can be seen that the higher α, and the higher the accuracy A, the higher the efficiency of observation. Thus, the efficiency approaches unity as $\alpha \to \infty$.

It would be interesting to determine the efficiency of observation as a function of reliability. Again, for simplicity, we let $\gamma = 1$; therefore, the efficiency is (from Eqs. (6.83) and (6.95))

$$\eta = \frac{Ik \ln 2}{\Delta S} = \frac{\ln \alpha}{\ln(\alpha R)} = \frac{1}{1 + \ln R/\ln \alpha} \tag{6.99}$$

Thus, we see that the observation efficiency depends on the ratio of logarithmic reliability to logarithmic simultaneous observations, α. The higher this ratio is, the lower the observation efficiency.

In the case of high frequency, $h\nu \gg kT$, the efficiency of observation can be obtained from Eqs. (6.83) and (6.96). Thus, we have

$$\eta = kT \frac{\ln \alpha}{h\nu} \tag{6.100}$$

Since $h\nu > kT(\ln \alpha + 0.367)$, the efficiency is lower than that in the low-frequency case.

In concluding this section, we point out that high-frequency observation corresponds to high accuracy and high reliability. However, the observation efficiency is somewhat lower than that in the low-frequency case.

6.5 OBSERVATION BY INTERFERENCE AND BY MICROSCOPE

In Section 6.4, we discussed accuracy and reliability in observation. We illustrated with examples the amount of entropy traded for the amount of information obtained and found it to be relatively small. We saw that from Eq. (6.98)—at least for relatively low reliability—one is able to push the observation efficiency close to unity. From Eq. (6.99), we concluded that observation efficiency can be high for relatively high reliability and high accuracy. However, efficiency is generally somewhat less in high-frequency observation than in the low-frequency case. With a relatively high cost in entropy in high-frequency observation, both high accuracy and high reliability can be achieved.

In this section, we discuss and illustrate two other methods of observation, namely, interference and the microscope. We use these two methods to show that the amounts of entropy involved are indeed great, and that the observation efficiencies are even lower than in the previous examples.

It is common knowledge in the field of interferometry that the separation between two reflecting walls or two particles can be obtained by means of an interferometric technique. For example, if one wishes to determine the distance between two particles, one observes or measures the interference fringes between the two particles, as shown in Fig. 6.6. It is clear that these interference fringes, resulting from standing waves between them, can be obtained simply by illuminating the particles with a monochromatic light source, as shown in the figure. Here we see that the half-wavelength of the light source is smaller than the separation of the two particles:

$$\frac{\lambda}{2} \leq x_1 \qquad (6.101)$$

where x_1 is the distance between the two particles. By counting the number of interference fringes, one can determine the separation x_1 between the particles:

$$x_1 = \frac{\alpha \lambda}{2} \qquad (6.102)$$

where α is the number of interference fringes.

If it is assumed that the observation is accomplished by scanning a tiny photodetecting probe between the particles (providing the perturbation of the field is negligible), then the detector requires g quanta $h\nu$ in order for it to respond properly at the location of maximum interference. Since there are α fringes in each scanning, the total amount of energy required is

$$\Delta E \geq \alpha g h\nu \qquad (6.103)$$

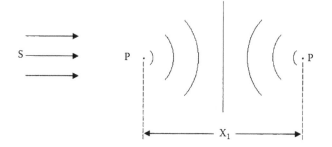

FIG. 6.6 Observation by interference. S, monochromatic plane wave; P, particle; X_1, separation.

The corresponding amount of entropy increase in the photodetector per scanning is

$$\Delta S \geq \frac{\Delta E}{T} = \frac{\alpha g h\nu}{T} \qquad (6.104)$$

Let us define the characteristic distance (see Eq. (6.77)):

$$\Delta x_0 = \frac{hc}{kT} \simeq \frac{1.44}{T} \qquad$$

where c is the speed of light. Then, according to Eq. (6.102), the frequency of the light source can be written as

$$\nu = \frac{c}{\lambda} = \frac{\alpha c}{2x_1} \qquad (6.105)$$

Thus, the entropy increase can be written as

$$\Delta S \geq \frac{\alpha k g \Delta x_0}{2\Delta x} \qquad (6.106)$$

where $\Delta x = x_1/\alpha$, the separation between fringes. From Eq. (6.106), we can determine the minimum number of quanta g required for the interference observation. We first discuss low-frequency interference observation. This corresponds to a larger separation between the two particles, compared with the wavelength λ.

Now for the low-frequency case, $h\nu \ll kT$, we have

$$\frac{h\nu}{kT} = \frac{\alpha hc}{2kTx_1} = \frac{\Delta x_0}{2\Delta x} \ll 1 \qquad (6.107)$$

In interference observation, it is quite analogous to the problem of many simultaneous observations, discussed in Section 6.2. Thus, for a 50% error probability (reliability $R = 2$), we have the decision threshold level (Eq. (6.41)):

$$E_0 = g h\nu = -kT \ln\left[1 - \left(\frac{1}{2}\right)^{1/\alpha}\right] \qquad (6.108)$$

Therefore, the minimum number of quanta required for the observation per scanning is

$$g = -\frac{kT \ln\left[1 - \left(\frac{1}{2}\right)^{1/\alpha}\right]}{h\nu} \qquad (6.109)$$

By substituting Eq. (6.109) into Eq. (6.104), the entropy increase is found:

$$\Delta S \geq -\alpha k \ln\left[1 - \left(\frac{1}{2}\right)^{1/\alpha}\right] \qquad (6.110)$$

It is also clear that the amount of information per scanning, using the interference observation method, is

$$I = \log_2 \frac{x_1}{\Delta x} = \log_2 \alpha \qquad (6.111)$$

which is essentially the same as Eq. (6.83). Thus, the efficiency of the interference observation can be determined:

$$\eta = \frac{Ik \ln 2}{\Delta S} = \frac{\ln \alpha}{-\alpha \ln \left[1 - \left(\frac{1}{2} \right)^{1/\alpha} \right]} \qquad (6.112)$$

If one compares Eq. (6.112) with Eq. (6.98), it is evident that the efficiency is much smaller here than in the previous low-frequency cases. Particularly when α is large, Eq. (6.112) is approximated by

$$\eta \simeq \frac{\ln \alpha}{\alpha(\ln \alpha + 0.367)} < 1 \qquad (6.113)$$

Again, we see that

$$\Delta S - Ik \ln 2 \geq k \left\{ -\alpha \ln \left[1 - \left(\frac{1}{2} \right)^{1/\alpha} \right] - \ln \alpha \right\} > 0 \qquad (6.114)$$

Now let us discuss the high-frequency case, $h v \gg kT$. As noted, high-frequency observation corresponds to high-accuracy observation. In interference observation, it also corresponds to shorter-distance observation. Thus, we have

$$\frac{h v}{kT} = \frac{\Delta x_0}{\Delta x} \gg 1 \qquad (6.115)$$

Here the high-frequency interference observation is well beyond the limit of blackbody radiation of the photodetector, which is maintained at Kelvin temperature T. As noted earlier, in high-frequency observation one quantum per fringe detection is sufficient; thus, the total amount of energy required per scanning observation is

$$\Delta E \geq \alpha h v = \frac{\alpha hc}{2\Delta x} \qquad (6.116)$$

The corresponding entropy increase in the photodetector is

$$\Delta S \geq \frac{\alpha h v}{T} = \frac{\alpha hc}{T 2 \Delta x} \qquad (6.117)$$

Moreover, from Eq. (6.117) we see that

$$E\ x \geq \frac{\alpha hc}{2} \qquad (6.118)$$

which bears a resemblance to the uncertainty relation. However, the physical significance is quite different from that of the uncertainty relation. That is, in the uncertainty relation (discussed in Section 6.5) ΔE corresponds to the error in energy, while in the interference observation ΔE is the minimum energy required for the observation. It is noted that ΔE of the interference observation is at least partially dissipated into heat during the observation.

Furthermore, from Eq. (6.118) we see that as Δx becomes infinitesimally small, the minimum required energy for observation increases to an infinite amount. However, in practice, there exists a physical limit. As Δx approaches a certain limit, the Heisenberg uncertainty principle applies. The observation cannot be obtained without introducing errors. This subject of uncertainty and observation is discussed in greater detail in Section 6.6.

Now let us discuss observation under a microscope. In this problem, a large number of photons are required. The amount of entropy required in microscopic observation is great in comparison with the previous cases.

For simplicity, let us assume that a circular waveguide of radius r_0 is used to focus the range of observation, as shown in Fig. 6.7. The light propagated through the waveguide has discrete wavelengths (quanta) with which to satisfy the boundary conditions. Superimposing these discrete quanta of the light field results in a focus spot, say at $Z = Z_0$. Let us then suppose a circularly polarized transverse electric wave exists in the waveguide, for which the wave equation is

$$\nabla^2 H_z = \frac{1}{c^2} \frac{\partial^2 H_z}{\partial t^2} \qquad (6.119)$$

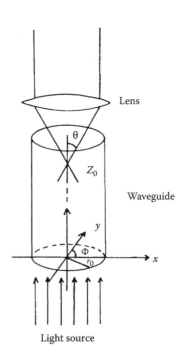

FIG. 6.7 Observation under a microscope.

where ∇^2 is the Laplacian operator, and c is the speed of light. The corresponding magnetic vector is [15,16]

$$H_z = \sum_{n=1}^{\infty} H_{0n}(\cos n\phi + i \sin n\phi)J_n(ur)\cos(\omega t - \gamma z) \qquad (6.120)$$

where the H_{0n} are arbitrary constants, the $J_n(ur)$ are Bessel functions of the first kind, γ is a propagation constant, $\omega = 2\pi v$, and v is the frequency of the light source. The corresponding cutoff wavelength is

$$\lambda_c = \frac{2\pi r_0}{u'_{nr}} \qquad (6.121)$$

where u'_{nr} is the rth root of $\partial J_n(ur)/\partial r = 0$.

It can be seen from Eq. (6.121) that the size of the focus spot decreases as the terms of the summation increase. This in turn requires a larger lens aperture and a higher light source frequency. However, in practice only a finite limited focus size can be achieved. Thus, let there be a finite summation N for Eq. (6.120) and let the source wavelength be smaller than the cutoff wavelength of Eq. (6.121). Then a finite radius of the focus spot can be defined:

$$\Delta r \geq \frac{1.22 r_0}{N} \qquad (6.122)$$

The corresponding focal area is

$$\Delta A = 2\pi(\Delta r)^2 \geq \pi \frac{(1.22 r_0)^2}{N^2} \qquad (6.123)$$

To avoid any possible reflection of the waveguide, we assume that the waveguide is properly terminated by a matched impedance.

To compute the amount of entropy required for this observation, it is necessary for us to assume that short light pulses of duration Δt are used. To determine the finite number of degrees of freedom, we let the light pulses be repeated periodically at every t_0. If we take the effective bandwidth of the pulse (i.e., the spectrum extended from $v = 0$ up to $1/2 \Delta t$, instead of $1/\Delta t$), the number of degrees of freedom per period of t_0 is

$$M \geq \frac{t_0}{2\Delta t} \qquad (6.124)$$

Thus, the light beam has a total of $N^2 M$ degrees of freedom per period t_0.

Let us first consider $hv \ll kT$, a low-frequency observation. Since at $hv \ll kT$ each degree of freedom has an energy of kT, the total thermal energy per period of t_0 is approximately

$$E_T \simeq N^2 MkT \qquad (6.125)$$

To overcome this thermal background, we use energy from a light source greater than this amount (see Eq. (6.45)):

$$\Delta E \geq N^2 Mk\,T[\ln \alpha + 0.367] \quad \alpha \gg 1 \qquad (6.126)$$

The corresponding entropy increase is

$$\Delta S = \frac{\Delta E}{T} \geq N^2 Mk[\ln \alpha + 0.367] \qquad (6.127)$$

Let us now determine the amount of information per period of t_0. But in order to do so, we must first determine the accuracy of the focal spot:

$$A = \frac{A}{\Delta A} \cdot \frac{t_0}{\Delta t} \qquad (6.128)$$

where $A = \pi r_0^2$, the total field of view. Thus, the amount of information obtained is

$$I = \log_2 A = \log_2 \left[2M \left(\frac{N}{1.22} \right)^2 \right] \qquad (6.129)$$

It is also noted from Eq. (6.128) that the entropy of Eq. (6.127) can be written as

$$\Delta S \geq \frac{(1.22)^2}{2} Ak[\ln \alpha + 0.367] \quad \alpha \gg 1 \qquad (6.130)$$

Thus, we see that

$$\Delta S - Ik \ln 2 > 0 \qquad (6.131)$$

From Eq. (6.128), we see that for a larger field of view, A corresponds to higher accuracy; hence, a larger amount of information is obtained. However, from Eq. (6.130) we see that the entropy is even greater than the information obtained. This result is one of the basic facts in practical problems: observation of a focus without field and time limitations eventually leads to an infinite cost in entropy (energy). This is one of the results that Toraldo di Francia [17] obtained. He has pointed out that a two-point resolution is impossible unless the observer has *a priori* an infinite amount of information about an object. Thus, infinite entropy (energy) is required. In all practical problems, observations are made under finite limitations of field and time.

Finally, in the high-frequency case, $hv \gg kT$, although high accuracy and high reliability can be obtained, the entropy required is even greater than for a low-frequency observation. This increase in entropy is essentially what we trade for high accuracy and reliability.

6.6   UNCERTAINTY AND OBSERVATION

We now examine the fundamental constraints of physical observations. As we are all aware, all physical problems are ultimately restricted by certain limitations. When quantum conditions are in use, all limitations are essentially imposed by the basic Heisenberg uncertainty principle and by wave mechanics. We discuss the basic limitation in observation imposed by the uncertainty principle. Time and energy smaller than Planck's constant cannot be observed or measured simultaneously [18]:

$$\Delta E \Delta t \geq h \qquad (6.132)$$

where ΔE can be regarded as the unknown energy perturbation of the observation system and specimen, and Δt is the observation time accuracy.

In applying the Heisenberg uncertainty principle to our observation problems, let us refer to Section 6.1, the statistical analysis of observations made by radiation. That is, we compare the energy ΔE required for an observation with the mean square thermal fluctuation of the photodetector γkT, where γ is the number of degrees of freedom. Essentially, this is the number of low-frequency vibrations ($h\nu \ll kT$). Thus, if $\Delta E < \gamma kT$, then according to the uncertainty relation of Eq. (6.132) we have

$$\Delta t \gg \frac{h}{\gamma kT} \qquad (6.133)$$

From the above inequality, we see that a larger time resolution Δt can be obtained with low-frequency observation. Since ΔE is small, the perturbation due to the time resolution is very small and can by comparison be ignored.

However, if the radiation frequency ν becomes greater, such that $\Delta E = h\nu > \gamma kT$, then using the uncertainty relation of Eq. (6.132), we have

$$\Delta t \leq \frac{h}{\gamma kT} \qquad (6.134)$$

In this case, we see that as the radiant energy required for the observation increases, the more accurately the time resolution can be measured and observed. But the perturbation of the observation is also greater. Thus, the time resolution Δt obtained with the observation may not even be correct, since ΔE is large.

The observation system is generally assumed to be nonperturbable, as in the classical theory of light, in which case precise observation can always be obtained and can be repeated many times with consistent results. In other words, the observations provide reproducible results, and as we have noted, this assumption is generally true for the many-particles problem and a large number of quanta in observation. With the classical theory, the accuracy of observation is not expected

to be too great, since this condition is imposed far away from the limitation of the uncertainty principle:

$$\Delta E \Delta t \gg h \qquad (6.135)$$

or, equivalently,

$$\Delta p \Delta x \gg h \qquad (6.136)$$

where Δp and Δx are the respective momentum and position errors. From Eqs. (6.135) and (6.136), it can be seen that in order to observe the quantities E, t, p, and x, the errors of ΔE, Δt, Δp, and Δx must not be too small.

However, as pointed out earlier, when quantum conditions occur, a nonperturbing system simply does not exist. When the higher-quantum $h\nu$ is used, a certain perturbation within the system is bound to occur; hence, high accuracy in high-frequency observation is limited by the uncertainty principle.

Let us now investigate the problem of observing extremely small distances. In Section 6.5 (Eq. (6.101)), we stated that for the observation of a small distance Δx between two particles, one must use a light source having a wavelength λ that satisfies the condition

$$\lambda \leq 2\Delta x \qquad (6.137)$$

Since Δx is assumed to be extremely small, a high-frequency light source is required for the observation. Thus, we see that high-frequency observation corresponds to higher momentum:

$$p = \frac{h}{\lambda} \geq \frac{h}{2\Delta x} \qquad (6.138)$$

In turn, this high-frequency source of radiation corresponds to a higher-quantum $h\nu$, in which it interacts with the observed specimen as well as with the observing equipment, and ultimately affects the whole observing system, causing it to change its momentum, which may vary from $-p$ to p. Thus, the change in momentum of the observing system is

$$\Delta p = 2p \gg \frac{h}{\Delta x} \qquad (6.139)$$

Since the radiant energy provided can be written as

$$\Delta E = h\nu = \frac{hc}{\lambda} \geq \frac{hc}{2\Delta x} \qquad (6.140)$$

we can conclude that

$$\Delta E \Delta x \geq \frac{hc}{2} \qquad (6.141)$$

It can be seen from the above equation that, mathematically speaking, there is no lower limit to Δx as long as ΔE *is* able to increase. But Eq. (6.139) tells us that as ΔE increases, the perturbation of the system under observation cannot be ignored. Thus, in practice, when ΔE reaches a certain quantity, the precise observation of Δx is obstructed and the observation of smaller and smaller objects presents ever-increasing difficulty. If certain sophisticated observation equipment were available for use in counteracting the perturbations, the cost of observation would be even greater. Eventually, the cost of observation would become so great that no one would be able to afford it.

Finally, let it be emphasized that Heisenberg's principle of uncertain observation is restricted to the ensemble point of view. That is, for a special observation, the uncertainty may be violated. However, we have never been able to predict when this observation will occur. Therefore, a meaningful answer to the Heisenberg uncertainty principle is present only in the sense of statistical ensemble average.

6.7 REMARKS

It has been shown that the amount of information obtained from a physical observation depends on the logarithm of the accuracy:

$$I = \log_2 A \tag{6.142}$$

but the cost in entropy is somewhat higher:

$$\Delta S > Ik \ln 2 \tag{6.143}$$

Thus, the overall observation system satisfies the second law of thermodynamics, which says

$$\Delta S - Ik \ln 2 > 0 \tag{6.144}$$

The efficiency of observation has been defined as

$$\eta = \frac{Ik \ln 2}{S} \leq 1 \tag{6.145}$$

We have seen that for low-frequency observation, the efficiency is somewhat greater than for high-frequency observation. However, for very small distances or small objects, high-frequency observation is better. And although the efficiency is somewhat less, high-frequency observation results in greater accuracy and reliability. However, it is emphasized that aside from the energy constraints, in high-frequency observation the ultimate limitation is imposed by Heisenberg's uncertainty principle; that is, the smaller the distances or the objects to be observed, the greater the amount of energy required for

the observation. Thus, perturbation in the observation system cannot be avoided up to certain energy limits. Furthermore, it is clear that direct observation usually provides greater efficiency. However, in practice, it cannot always be obtained, for example, where very small distances are involved.

We also see that every observation requires a certain compensatory increase in entropy. However, in some practical cases the amount of entropy compensation may be negligible, since it is very small when compared with the total entropy of the observation system.

Furthermore, from Eq. (6.144) we see that observation is essentially an irreversible process from the thermodynamic point of view [19].

REFERENCES

1. M. Born and E. Wolf, *Principles of Optics*, 2nd rev. ed., Pergamon, New York, 1964.
2. F. T. S. Yu, *Introduction to Diffraction, Information Processing, and Holography*, MIT Press, Cambridge, MA, 1973.
3. D. Gabor, Light and Information, in E. Wolf, ed., *Progress in Optics*, vol. I, North-Holland, Amsterdam, 1961.
4. D. Gabor, Informationstheorie in der Optik, *Optik*, vol. 39, 86 (1973).
5. L. Brillouin, The Negentropy Principle of Information, *J. Appl. Phys.*, vol. 24, 1152 (1953).
6. L. Brillouin, *Science and Information Theory*, Academic, New York, 1956.
7. L. Brillouin, *Scientific Uncertainty and Information*, Academic, New York, 1964.
8. L. Szilard, Über die Entropieverminderung in Einem Thermodynamischen System bei Eingriffen Intelligenter Wesen, *Z. Phys.*, vol. 53, 840 (1929).
9. L. Szilard, On the Decrease of Entropy in a Thermodynamic System by the Intervention of Intelligent Beings (translated by A. Rapaport and M. Knoller), *Behav. Sci.*, vol. 9, 301 (1964).
10. F. T. S. Yu, Observation, Information, and Optical Synthetic Aperture of Spherical Lenses, *Optik*, vol. 38, 425 (1973).
11. R. Fano, *Transmission of Information*, MIT Press, Cambridge, MA, 1961.
12. M. Schwartz, *Information Transmission, Modulation, and Noise*, 2nd ed., McGraw-Hill, New York, 1970.
13. W. W. Harman, *Principles of the Statistical Theory of Communication*, McGraw-Hill, New York, 1963.
14. I. Selin, *Detection Theory*, Princeton University Press, Princeton, NJ, 1965.
15. J. D. Kraus and K. P. Carver, *Electromagnetics*, 2nd ed., McGraw-Hill, New York, 1973.
16. S. Ramo and J. R. Whinnery, *Fields and Waves in Modern Radio*, John Wiley, New York, 1953.
17. G. Toraldo di Francia, Resolving Power and Information, *J. Opt. Soc. Am.*, vol. 45, 497 (1955).
18. J. L. Powell and B. Crasemann, *Quantum Mechanics*, Addison-Wesley, Reading, MA, 1961.
19. F. W. Sears, *Thermodynamics, the Kinetic Theory of Gases, and Statistical Mechanics*, Addison-Wesley, Reading, MA, 1953.

7 Image Restoration and Information

In Chapter 6, we presented in detail the relationship between observation and information, arriving at a very important result, namely, that for every observation there is a compensatory increase in entropy of other sources. We have shown that every physical observation, as well as every physical device, is ultimately limited by Heisenberg's uncertainty principle from quantum mechanics. In observing the most minute object or a small distance, there is a practical limit at which great accuracy cannot be achieved. Also, the smaller the distances or objects to be observed are, the more costly the energy (or entropy) that is required. In some instances, the amount of energy required may even approach infinity.

Now we come to the problem of image restoration and information. We first discuss image restoration techniques and certain limitations that are encountered. We illustrate that it is not possible to achieve infinite precision even with finite *a priori* knowledge of the image. We also demonstrate that the restoration of lost information may lead to an infinite amount of energy (or entropy) compensation. Finally, an example of the restoration of a smeared photographic image is given.

7.1 IMAGE RESTORATION

A smeared or out-of-focus photographic image can be restored by means of a coherent optical processor [1–3]. However, as will be shown, the restored image will not be better than that in a photograph taken with the *maximum allowable exposure time*, that is, the longest possible exposure time during which the recorded image will not smear significantly. The allowable time depends on the size and details of the object being photographed. For example, the smaller the size or the finer the detail, the shorter the allowable time required for exposure. In other words, the restored image theoretically can only approach the case in which the image is recorded without being smeared or becoming out of focus.

In general, the problem of image restoration can be divided into two categories:

1. Restoration from the distorted image
2. Restoration by superimposing the images that were distorted by smearing or by being out of focus

In the first category, it is possible for the image to be restored, but the process does not use the smeared image imposed on the film. In the second category, it is possible not only to restore the image, but also to utilize the excessive recording of the smeared image.

We see that if spatial random noise is ignored, the results of the above two restorations are essentially identical, since they provide the same amount of information. However, if the random noise of the recorded image is taken into account, then the second restoration category will have a higher information content. We show that the second restoration category is impossible to achieve in practice [4,5].

For simplicity, a few techniques of image restoration in which the smear is due to linear image motion are discussed. Certain physical constraints of the restoration are also included in the discussion.

Let an image be distorted by some physical means; then the distorted image may be described by its Fourier transform:

$$G(p) = S(p)D(p) \tag{7.1}$$

where $G(p)$ is the distorted image function, $D(p)$ is the distorting function, $S(p)$ is the distortionless image function, and p is the spatial frequency.

We know that the distortionless image functions $S(p)$ can be recovered if we know the distorting function $D(p)$:

$$S(p) = G(p) \cdot \frac{1}{D(p)} \tag{7.2}$$

First, we momentarily disregard the physical realizability of the inverse filter $1/D(p)$. Then we ask whether this *inverse filtering process* increases the information content of the distorted signal $G(p)$. The answer is no, it does not, because we know exactly how it is distorted. The additional amount of information we gained from inverse filtering was essentially derived from the *a priori* knowledge of $D(p)$. One might argue that if this method of recovering the image does not increase the information content, why do we go to the trouble of restoring it? The answer apparently involves an image recognition problem. Since we recognize $G(p)$, as well as $S(p)$, through the knowledge of $D(p)$, these two images are essentially no different from each other—at least from the information standpoint. But for those who do not know how the image was distorted *a priori*, certainly a more realistic image $S(p)$ contains more information than a distorted one. This is the major reason we go to the trouble of restoring the image. However, it is noted that the information gain apparently comes from the compensation of certain energy (entropy) in converting $G(p)$ to $S(p)$.

Let us now assume that a resolvable point image of constant irradiance I is projected on the recording medium of a camera. If the point object moves at a constant velocity v, and the exposure time is t, then the moving point image recorded on the photographic film will smear into a straight line of length $\Delta x = mvT$, with m a proportionality constant. The corresponding transmittance may be described by the expression

$$f(x) = \begin{cases} A & \text{for } -\dfrac{\Delta x}{2} \le x \le \dfrac{\Delta x}{2} \\ 0 & \text{otherwise} \end{cases} \tag{7.3}$$

where A is a positive constant proportional to I. The corresponding Fourier transform is

$$F(p) = A\Delta x \frac{\sin(p\Delta x/2)}{p\Delta x/2} \tag{7.4}$$

In the image restoration, we apply the inverse filtering process, giving

$$F(p)\frac{A}{F(p)} = A\mathbf{F}[\delta(x)] \tag{7.5}$$

where $\mathbf{F}$ denotes the Fourier transform, and $\delta(x)$ is the Dirac delta function.

From Eq. (7.4), it can be seen that an inverse filter $A/F(p)$ cannot be realized in practice, since it contains poles at $p = 2n\pi/\Delta x$, $n = 0, 1, 2, \ldots$. However, if one assumes that the inverse filter can be completely or partially realized, then the restored image is at best equal to $A\delta(x)$, the minimum resolvable point image. In other words, the inverse filtering process restores the image but does not increase the overall image spectrum over that obtained with the maximum allowable exposure. But is it possible to have a physical filter $H(p)$ such that

$$F(p) \cdot H(p) = B\mathbf{F}[\delta(x)] \tag{7.6}$$

where B is a positive real constant. The question is, can B be *greater* than A?

The answer to this question is no, since a coherent optical information processing system is *passive*. In view of the practical complications of filter synthesis, we are led to a different approach to our solution.

Again, suppose a moving image $f(x)$ is recorded on film, and that the length is Δx. Image smearing can be controlled by exposure modulation, such as that composed of a finite sequence of identical functions. Then the transmittance of the recorded function can be described as

$$g(x) = f(x) + f(x - \Delta l) + f(x - 2\Delta l) + \cdots + f(x - N\Delta l)$$

$$= \sum_{n=0}^{N} f(x - n\Delta l) \tag{7.7}$$

where $N = \Delta x/\Delta l$, and Δl is the incremental translation of $f(x)$. The corresponding Fourier transform is

$$G(p) = \sum_{n=0}^{N} F(p)e^{-ipn\Delta l} \tag{7.8}$$

where $G(p)$ and $F(p)$ are the Fourier transforms of $g(x)$ and $f(x)$, respectively. From Eq. (7.8), the image restoration may be attempted by

$$G(p)H(p) = F(p) \tag{7.9}$$

where the filter is

$$H(p) = \frac{1}{\displaystyle\sum_{n=0}^{N} e^{-ipn\Delta l}} \tag{7.10}$$

Now if we properly translate the x axis of Eq. (7.7), then by the *Lagrangian identity* [6] the denominator of Eq. (7.10) can be written as

$$\sum_{-N/2}^{N/2} e^{-inp\Delta l} = \frac{\sin\left(N/2 + \dfrac{1}{2}\right)p\,\Delta l}{\sin\dfrac{1}{2}p\Delta l} \tag{7.11}$$

For a large N and a small Δl, Eq. (7.10) can be approximated by

$$H(p) \triangleq \frac{1}{2}\frac{p\Delta x/N}{\sin(p\Delta x/2)} \tag{7.12}$$

Practically speaking, $H(p)$ is not a physically realizable function. However, if we assume $H(p)$ is physically realizable, then by this same interpretation we can say that, at best, the restored image can only be equal to the image recorded with the maximum allowable exposure time.

But because of the physical unrealizability of the filter, we must modify the exposure-modulating process. From Eqs. (7.9) and (7.10), it appears that the filter $H(p)$ can be made physically realizable if unity is added to the denominator of Eq. (7.10):

$$H(p) = \frac{1}{1 + \displaystyle\sum_{n=0}^{N} e^{-ipn\Delta l}} \tag{7.13}$$

We see that if the filter of Eq. (7.13) is used, the physical realizability condition is

$$|H(p)| \le 1 \quad \text{for all } p \tag{7.14}$$

One of the obvious procedures in making $H(p)$ physically realizable is to control the exposure properly, so that the resulting transmittance is

$$g(x) = 2f(x) + f(x - \Delta l) + f(x - 2\Delta l) + \cdots + f(x - N\Delta l)$$

$$= f(x) + \sum_{n=0}^{N} f(x - n\Delta l) \qquad (7.15)$$

where $N = \Delta x/\Delta l$, and $\Delta x = mvT$. The corresponding Fourier transform is

$$G(p) = F(p)\left(1 + \sum_{n=0}^{N} e^{-ipn\Delta l}\right) \qquad (7.16)$$

Thus, the inverse filter function is

$$H(p) = \frac{+}{1 + \sum_{n=0}^{N} e^{-inp\Delta l}} \simeq \frac{1}{1 + N\dfrac{\sin(p\Delta x/2)}{p\Delta x/2}} \qquad (7.17)$$

and is a physically realizable filter. However, it can be seen that the restored image spectrum is equal to $F(p)$, which is only one-half of the first image function in Eq. (7.15).

Now we discuss an interesting image restoration technique. If the recorded image can be modulated such that each of the resolvable points can be imaged by itself, then it may be possible to reconstruct the image by means of coherent illumination. Not only will the smearing of the image be corrected, but in addition, the overall image spectrum may be improved.

Suppose a movable point image is recorded on the film by means of a certain shutter process, so that the amplitude transmittance is

$$T(x) = \frac{1}{2} - \frac{1}{2}\cos\frac{\pi}{\lambda f}x^2 = \frac{1}{2} - \frac{1}{4}e^{i(\pi/f\lambda)x^2} - \frac{1}{4}e^{-i(\pi/f\lambda)x^2} \qquad (7.18)$$

where λ is the wavelength of the coherent source, and f is the focal length of the one-dimensional *zone lens* [5]. It is noted that a zone lens is similar to a *zone plate* [5], except that the transmittance is a sinusoidal function of x^2.

If this modulated transparency of Eq. (7.18) is illuminated by a monochromatic plane wave front, as shown in Fig. 7.1, then the complex light field at the focal length f may be determined by the Fresnel–Kirchhoff theory or by Huygens' principle [7] (Section 2.2):

$$U(\alpha) = \int_{-\infty}^{\infty} T(x)u(\alpha - x; k)\,dx \qquad (7.19)$$

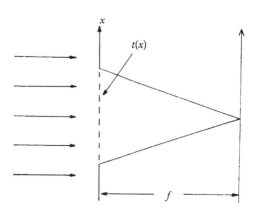

FIG. 7.1 Under coherent illumination.

where

$$u(x) = \frac{1}{f}\exp\left[i\frac{2\pi}{\lambda}\left(f + \frac{x^2}{2f}\right)\right]$$

is known as the *spatial impulse response*.

By substituting Eq. (7.18) into Eq. (7.19), the solution is

$$U(\alpha) = \left[\frac{1}{2}\left(\frac{1}{\lambda}\right)^{1/2}\exp\left(i\frac{\pi}{4}\right) - \frac{1}{4}\left(\frac{1}{2f\lambda}\right)^{1/2}\exp\left[i\frac{\pi}{2f\lambda}\alpha^2 + \frac{\pi}{4}\right]\right.$$
$$\left. - \frac{1}{4}\exp\left(-i\frac{\pi}{f\lambda}\alpha^2\right)\right]\delta(\alpha) \qquad (7.20)$$

where $\delta(\alpha)$ is a Dirac delta function.

It is clear that the first term of Eq. (7.20) is the zero-order diffraction. The second term is the divergent diffraction. The last term corresponds to the convergent diffraction. In other words, a zone lens such as the one in Eq. (7.18) is equivalent to three lenses (i.e., direct current [dc], divergent, and convergent). However, if we examine Eq. (7.20) a bit more carefully, we see that the convergent image term is reduced by a factor of 4 (i.e., a factor of 16 with respect to irradiance) compared with the one used for the maximum allowable exposure time. Obviously, the excessive recording (energy) on the film due to the modulation was not effectively used. On the contrary, this excess recording has been converted into useless diffraction in the zero and divergent terms. Thus, from this simple analysis we are led to doubt that there exists a realizable shutter modulation technique in which the excessive recording can be fully or partially converted into useful diffraction. The restoration of a smeared image may be seen as an energy and time problem for which a physically realizable solution involving the expenditure of a finite amount of energy does not exist.

In general, a smeared image can be represented in a three-dimensional orthogonal space with a closed time interval $(t_1 \leq t \leq t_2)$. The smeared image can therefore be regarded as a time and space problem. If we claim to be able to restore the image due to smearing by piling up the excessive recording on the film, this is equivalent to saying that it is possible to separate the times and spaces from the smeared image. Of course, from the basic theory of relativity [8], it is impossible for us to do so without expending an excessive amount of entropy.

7.2 UNCERTAINTY AND IMAGE RESTORATION

In order to demonstrate that restoration of a smeared image is a problem related to Heisenberg's uncertainty principle [9], we start with the basic inequality of a photographic film:

$$It \geq E_0 \qquad (7.21)$$

where E_0 is the minimum energy (exposure) per unit area of the film required for an object to be properly recorded, I is the radiance intensity of the minimum resolvable image, and t is the exposure time of the film. In other words, if the image is recorded below the minimum required exposure E_0, it is impossible for it to be resolved. This minimum exposure E_0 can be called the *threshold level* of the photographic film.

To demonstrate that the foregoing inequality is equivalent to Heisenberg's uncertainty relation, one can simply substitute $E_0 = hv$ in Eq. (7.21); thus,

$$Et \geq h \qquad (7.22)$$

where $E = I/v$, h is Planck's constant, and v is the frequency.

We see that these inequalities give the theoretical limit of the film. The inequality of Eq. (7.21) or (7.22) gives the *detectability* (or *recordability*) of the film. If the situation does not satisfy the detectability, it will be impossible for the recorded image to be resolved. In the following, we use this inequality to connect the theory of information.

For stationary objects and film, if the signal-to-noise ratio of the image is high, there is no essential difficulty in satisfying the inequality of Eq. (7.21) or (7.22). However, for moving objects, for film, or for both, there exists a maximum allowable exposure time with which the recorded image will not be distorted as a result of smearing:

$$t \leq t_{max} \qquad (7.23)$$

where t_{max} is the maximum allowable exposure time. In general, t_{max} is linearly related to the size of the recorded image.

From the inequality of Eq. (7.22), a trade of energy and exposure time is possible; with high irradiance, the film can record smaller objects. But if the irradiance from the object is low, it may be impossible to reduce the maximum allowable exposure time without violating the detectability conditions.

If we are tempted to satisfy the detectability conditions by letting the exposure time become greater than the allowable time, then the recorded image will be further degraded (as a result of smear). Some of the information recorded in the allowable time interval, $0 \leq t \leq t_{max}$, is partially or completely destroyed by this excessive recording. This amount of

information loss can also be regarded as an increase in the entropy of the film. Hence, a smeared image contains less information than an unsmeared one under the maximum allowable time conditions. This loss of information due to additional recording is also the physical basis of the unrealizability of the inverse filter discussed in Section 7.1. In order to bring back the information loss as a result of smearing, the expenditure of an infinite amount of energy is required.

In coherent optical processing, a great deal of energy available is from the source. Can it be used? The answer appears to be no, since the energy in a coherent system is used to convey information; it is therefore not convertible into information. For example, suppose a printed page contains a certain amount of information. If it is located in a dark room, the information content has no way of reaching us. We need a certain quantity of light in order to make the information observable. This quantity of light is the cost in energy paid for the information transmission. Moreover, if this page contains a number of misprinted words, and these misprinted words are independent of the correct ones, then a great amount of information (or energy) will be required for us to restore the intended words.

We conclude that the extra energy (or bits of information) we intended for the film to absorb in order to correct the smear actually further degrades the information content of the image. However, to say that it is possible to reconstruct completely the original image from film degraded by smearing or modulation is equivalent to saying that it is possible to make the recording instrument record an unrecordable object after a recording; the contradiction is apparent.

But one can still argue that if an object is embedded in noise (such as turbulence), why is it possible to restore the image by using complex spatial filtering? Before we attempt to answer this question, we should ask whether the source irradiance and the exposure time satisfy the detectability conditions. If they do not, can we restore the imaging object? We cannot. However, if they *do* satisfy the detectability conditions, can we restore the imaging object better than we could in the case without noise? Again, we cannot. At best, we can only approach the case without noise as a limit. Therefore, we conclude that the imaging object can be restored (with a probability of error) from the degraded random noise if and only if the intensity from the object and the exposure time of the film fulfill the basic detectability conditions:

$$It > E_0 \qquad (7.24)$$

Furthermore, the restored image, at best, can only be equal to the image obtained under the conditions without random noise (such as turbulence).

A few questions can still be raised: Is there a coding process (such as a modulating camera) that can improve the smearing correction? No, because to improve the information transmission through a communication channel, the coding process must take place at the transmitted end—not at the receiving end.

How about coherent detection techniques, such as correlation detection and sampling? Can these improve image restoration? No, since we have no prior knowledge about the object (the recorded image). There is no way to correlate the recorded smeared image. And so we summarize these considerations as follows:

1. A smeared image is basically correctable. However, the corrected results can only approach the case in which the image is recorded with the maximum allowable exposure time $(t - t_{max})$ of the film. In practice, the corrected result is far below the t_{max} criterion, because of the unrealizability of the filter.

2. The modulating camera technique, or any other modulating process, is unable to improve the image resolution over that of the maximum allowable time criterion, because the information content of the image is further degraded by this modulating process.

3. The smear correction problem is basically a time and space problem. It is physically impossible to restore the image by partially or completely piling up the excessive recording due to smearing.

4. If the irradiance from the object and the exposure time of the film do not satisfy the detectability conditions of the recording medium, it is physically impossible to restore the recorded image. To do so would violate the uncertainty principle. It is possible to restore the recorded image to a point where it equals the image obtained with the maximum allowable exposure time if and only if the detectability conditions are satisfied.

5. The unrealizability of an inverse filter can be explained from the standpoint of entropy and information; the amount of information loss as a result of degradation of the recorded image (such as smearing) can be recovered only with the expenditure of an infinite amount of energy (i.e., infinite entropy compensation).

6. None of the existing coding processes and coherent detection techniques can improve image restoration, because efficient coding must be done at the transmitting end, not at the receiving end. In coherent detection, there is a lack of prior knowledge of the image recorded.

However, some image restoration after smearing is still possible. The optimum image restoration obtainable can be found by means of a mean square error approximation, such as

$$\overline{\varepsilon^2(x, y)} = \lim_{\substack{x_1 \to \infty \\ y_1 \to \infty}} \frac{1}{x_1 y_1} \int_{-x_1/2}^{x_1/2} \int_{-y_1/2}^{y_1/2} [f_0(x, y) - f_d(x, y)]^2 \, dx \, dy$$

(7.25)

where $f_d(x, y)$ is the desired image, $f_0(x, y)$ is the restored image, and the evaluation of the above equation is subject to the physical constraint

$$|H| \leq 1 \qquad (7.26)$$

where H is the transmittance of the correcting filter.

It is emphasized that the restoration of a smeared image has application where the image was recorded without intention of image motion, that is, where a minimum exposure choice could not be made at the time the image was recorded.

7.3 RESOLVING POWER AND INFORMATION

The possibility of resolving power beyond the classical limit of an idealized optical system has been recognized by Coleman and Coleman [10], Toraldo di Francia [11], Ranchi [12,13], and Harris [14]. In this section, we discuss some of the physical limitations besides the inevitable noise of optical systems [15].

There are two long-established theorems from analytic function theory that have been found to be useful in studies of resolution beyond the classical limit. The first of these two theorems deals with the finite integral of the Fourier transform of a spatially bounded object. This theorem [16] states that the Fourier transform of a spatially bounded function is analytic throughout the entire domain of the spatial frequency plane.

The second theorem [17] states that if a function of a complex variable is analytic in the region R, it is possible, from knowledge of the function in an arbitrarily small region within R, to determine the whole function within R by means of analytic continuation. As a corollary to this theorem, it can be stated that if any two analytic functions have functional values that coincide in an arbitrarily small region in the region of analyticity, then the values of these functions must be equal everywhere throughout their common region of analyticity. This is known as the *identity theorem* or the *uniqueness theorem*.

From these two theorems, it is clear that if the spatial transfer characteristics (i.e., the spatial frequency and phase response) of an optical system are known, then for a bounded object it is possible to resolve the object with infinite precision by means of *analytic continuation*. Furthermore, from the corollary to the second theorem, the ambiguity of two close objects does not exist, and the resolved object is therefore unique.

Infinite precision of object resolution is, however, a mathematical ideal. In Section 7.4, we see that an infinitely precise solution is physically unrealizable. In fact, the amount of information gained from extension of the complex spectral density of the object, which is limited by the spatial cutoff frequency of the optical system, must come from proceedings with the analytic continuation. We also show that the degree of precision depends on the accuracy of the functional spectrum expansion. The smaller the region of spectral density known, the greater the effort required for better precision of object resolution.

Let the function $f(z)$ represent the complex spatial frequency spectrum of a bounded object (where $z = x + iy$, and x and y are the spatial frequency coordinates). If $f(z)$ is analytic throughout a given region R, and the functional value is assumed to be given over an arbitrarily small region (or an arc) within R, then by analytic continuation $f(z)$ can be uniquely determined throughout the region R.

Let A denote an arc within R on which the function $f(z)$ is given (Fig. 7.2). Since $f(z)$ is analytic in R, the value of the derivative $f'(z)$ is independent of the manner in which Δz tends to zero. Suppose z_0 denotes a point of A. Then because of the analyticity of $f(z)$, this function can be expanded by the Taylor series about z_0, at all points interior to a circle having its center at z_0 and lying within R. Thus,

$$f(z) = f(z_0) + \sum_{n=1}^{\infty} \frac{f^n(z_0)}{n!}(z - z_0)^n \quad |z - z_0| \leq r_0 \quad (7.27)$$

where $f''(z_0)$ denotes the nth-order derivative of $f(z)$ at z_0, and r_0 is the corresponding radius of the convergent circle within R.

Thus, in order to determine the functional value of $f(z)$ within $|z - z_0| \leq r_0$, we must know all the functional derivatives $f''(z_0)$. For the most general $f(z)$, of course it is not possible to write down the infinite number of derivatives required. If we assume that Eq. (7.27) converges to $f(z)$ with little error when the series is truncated at some finite $n = N$, and also that we can find approximately the associated derivatives $f''(z_0)$ (which is difficult to achieve in practice; see [18]), then the functional value of $f(z)$ within the convergent circle C_0 can be approximated by

$$f(z) \simeq f(z_0) + \sum_{n=1}^{N} \frac{f^n(z_0)}{n!}(z - z_0)^n \quad (7.28)$$

Now consider any other point P in R. Let A_1 denote a curve connecting z_0 and P and lying in the interior of R (Fig. 7.2), and let the radius r_0 be equal to the shortest distance between curve A_1 and boundary R. Let z_1 be a point on curve A_1 within circle C_0, and let C_1 be a circle with its center at z_1 and radius r_0. Then the function $f(z)$ within C_1 is approximated by

$$f(z) \simeq f(z_1) + \sum_{n=1}^{N} \frac{f^n(z_1)}{n!}(z - z_1)^n \quad (7.29)$$

By continuing this process, curve A_1 can be covered by a finite sequence of circles, C_0, C_1, C_2, ..., C_m, of radius r_0; thus, the functional value of $f(z)$ over the whole of R can be determined. However, from the well-known theorem stating that a spatially bounded object cannot be spatially frequency bounded [19], the complete extension of $f(z)$ requires an infinite sequence of circles to cover the entire spatial frequency plane. Therefore, this analytic extension of $f(z)$ also requires an infinite amount of effort (energy), and it is not physically realizable.

Equation (7.29) is the analytic continuation of Eq. (7.28). The error of this approximate extension of $f(z)$ increases as the analytic continuation proceeds. For example, a one-dimensional hypothetical spatial frequency spectrum is shown in Fig. 7.3, where p_c is the spatial cutoff frequency of the optical system. If we allow the complex spatial frequency spectrum to be bounded for all practical purposes, then by an analytic continuation similar to the foregoing, we can see that the degree of precision of object resolution increases with an increase in spatial frequency bandwidth. It is emphasized that, strictly speaking, a spatially bounded object cannot be spatially frequency bounded [19]. However, this assumption of boundedness is often made in practice. Furthermore, the amount of information gained by analytic continuation of the spatial frequency spectrum clearly comes from this effort, that is, the expenditure of energy, to obtain the derivatives $f^n(z_0)$. We note that determination of $f^n(z_0)$ is generally difficult to achieve in practice.

FIG. 7.2   Analytic continuation.

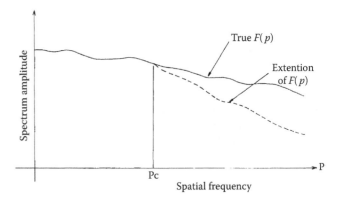

FIG. 7.3   Hypothetical spatial frequency spectrum, showing the extension of the spatial frequency spectrum beyond the diffraction limit of the system.

We have shown in this section that it is not possible to obtain infinite precision of object resolution in practice. However, considering the uniqueness theorem cited earlier, one might wonder why we need to expand the spatial frequency spectrum, since an arbitrarily small region of the spectrum is sufficient to represent an object. Unfortunately, to utilize this fact we need to have *a priori* knowledge of objects and their corresponding spatial frequency spectra. If the objects to be detected are finite in number, this may be possible. However, in practice, objects are generally not finite in number; in fact, there may be uncountably many objects. In this case, an infinitely large volume of information storage (i.e., the dictionary of their corresponding spatial frequency spectra) must be available. Such an infinitive capacity is of course physically unrealizable. Incidentally, this result is similar to that obtained by Toraldo di Francia [11]: "A two-point resolution is impossible unless one has *a priori* an *infinite* amount of information about the object."

7.4 COHERENT AND DIGITAL IMAGE ENHANCEMENT

Image enhancement by means of coherent and digital computer techniques has been shown in various applications [20,21]. We discuss in this section some of their basic distinctions. In Section 7.2, we discussed smeared image restoration by means of an inverse spatial filter. Although such an inverse filter is not physically realizable, there is an approximating technique that may approach it.

A complex spatial filter can be realized by combining an amplitude and a phase filter [5]. Such a phase filter is difficult to synthesize in practice, but these difficulties may be overcome by means of a computer-generated filter [22].

To illustrate the enhancement of a linear smeared photographic image, let us recall Eq. (7.5), where the transfer characteristic of an inverse filter is

$$H(p) = K \frac{\Delta x}{F(p)} \qquad (7.30)$$

where

$$F(p) = \Delta x \frac{\sin p(\Delta x/2)}{p(\Delta x/2)}$$

We have seen that $H(p)$ is a bipolar function with a countable number of poles. Such an inverse filter, as pointed out in Section 7.1, is not physically realizable. However, if we are willing to accept a certain restoration error, then the inverse complex filter may be written as

$$H_1(p) = A(p) \exp[i\phi(p)] \qquad (7.31)$$

where $A(p)$ and $\phi(p)$ are the amplitude and phase filter functions, respectively. The physical constraint of such a filter, in

coherent optical processing, is of course $0 \le A(p) \le 1$. The corresponding restored Fourier spectrum is

$$F_1(p) = F(p)H_1(p) \qquad (7.32)$$

Let us define the *relative degree of image enhancement* [23]:

$$D(A_m) = \frac{\int_{\Delta p} [F_1(p)/\Delta x] dp}{A_m \Delta p} \times 100\% \qquad (7.33)$$

where A_m denotes the minimum value of the amplitude filter function, and Δp is the spatial bandwidth of interest. Δp is limited by the diffraction limit of the processing system. From Eq. (7.33), we see that a perfect degree of enhancement within Δp is possible, except when A_m approaches zero. But we can also see that the Fourier spectrum also approaches zero as A_m approaches zero. Therefore, a perfect degree of enhancement, even within the diffraction limit, is not possible. It is also emphasized that in coherent optical processing, image enhancement is also limited by the diffraction limit of the imaging system (e.g., the camera). Thus, the ultimate limit is restricted by the imaging or the processing system, whichever comes first. Therefore, a perfect degree of image enhancement does not apply beyond the diffraction limit.

We see that resolution beyond the diffraction limit can be achieved by means of analytic continuation [14], or by the Karhunen–Loéve expansion [24]. Both of these techniques are very restricted, as both possess a common practical limitation: expenditure of an infinite amount of energy.

If we accept certain errors, then restoration beyond the diffraction limit can be obtained by a finite extension of the Fourier spectrum, but the amount of information gain, as noted, comes from the compensation of entropy increase in expanding the spectrum. For analytic continuation, it is the computational effort in obtaining the functional derivatives, and for the Karhunen–Loéve expansion, it is the effort in obtaining the eigenfunctions.

Now we come to the distinction between coherent and computer image restoration. In the coherent technique, restoration is basically an analog technique. The image can be processed simultaneously in both spatial and spatial frequency domains. The restoration is, however, limited by the diffraction limits of the imaging and processing systems, whichever comes first.

In computer image restoration, the spatial plane is divided into discrete variables. Image processing usually takes place by sequentially sampling the image plane. Thus, it is limited by the finite sample points, but not by the diffraction limit of the system. The significant distinction between these techniques is that the coherent system is a *passive* device and the digital system is an *active* one. In coherent spatial filtering, the physical constraint of the restoration filter is

$$|H(p,q)| \le 1 \qquad (7.34)$$

where p and q are the spatial frequency coordinates. However, in the digital filtering technique, it is possible to process the image beyond this constraint.

In concluding this section, we point out that it is too early to compare these two image restoration techniques in their present stages. However, if image restoration is meant to be within the diffraction limit, because of the high spatial resolution and relative simplicity in optical handling, then a coherent technique may possess certain advantages. However, if image restoration is to be beyond the diffraction limit, then a digital technique may be more suitable. Our conclusion is that a hybrid system (coherent–digital) may offer a better image restoration technique.

7.5 INFORMATION LEAKAGE THROUGH A PASSIVE CHANNEL

In this section, we consider spatial information transmission through a passive optical channel. It is known to optical information specialists that a coherent light field is generally complex. The information-bearing elements can be described by amplitude and phase [5] and are strictly independent variables. Answering the question as to which of these elements is more relevant in certain applications is not the intention of this discussion. However, we treat both amplitude and phase as equally relevant. We determine the information leakage [25] from the basic entropy definitions developed in Chapter 1. The result as applied to image enhancement is also discussed.

The technique of transmitting spatial information through an optical channel is shown in Fig. 7.4. Let us denote the input ensemble of the amplitude and phase signals by $\{A, \phi\}$, and the corresponding output ensemble of the amplitude and phase variables by $\{B, \theta\}$. It is noted that

$$[A, \phi] \triangleq A(x, y) \exp\left[i\phi(x, y)\right] \quad (7.35)$$

and

$$(B, \theta) \triangleq B(\alpha, \beta) \exp\left[i\theta(\alpha, \beta)\right] \quad (7.36)$$

where (x, y) and (α, β) are the corresponding input and output spatial coordinate systems. The amount of information the input signal $\{A, \phi\}$ provides, as defined by Shannon [26,27], is

$$H(A, \phi) = -\int_0^{2\pi}\int_0^\infty p(A, \phi)\log_2 p(A, \phi)\,dA\,d\phi \quad (7.37)$$

FIG. 7.4 Input–output optical information channel.

with

$$\int_0^{2\pi}\int_0^\infty p(A, \phi)\,dA\,d\phi = 1 \quad (7.38)$$

where $p(A, \phi)$ is the joint probability density of A and ϕ, with $0 < A < \infty$ and $0 \leq \phi < 2\pi$. The corresponding conditional entropy or *spatial channel equivocation* is

$$H(A, \phi/B, \theta) = -\int_0^{2\pi}\int_0^\infty\int_0^{2\pi}\int_0^\infty p(A, \phi, B, \theta)$$
$$\log_2 p(A, \phi/B, \theta)\,dA\,d\phi\,dB\,d\theta \quad (7.39)$$

with

$$\int_0^{2\pi}\int_0^\infty\int_0^{2\pi}\int_0^\infty p(A, \phi, B, \theta)\,dA\,d\phi\,dB\,d\theta = 1 \quad (7.40)$$

where $p(A, \phi/B, \theta)$ is the corresponding input–output joint probability density, and $p(A, \phi/B, \theta)$ is the *a posteriori* probability density with $0 < B < \infty$ and $0 \leq \theta \leq 2\pi$. Now the mutual information (as defined in Chapter 1) of the optical spatial channel can be determined:

$$I(A, \phi; B, \theta) = H(A, \phi) - H(A, \phi/B, \theta) \quad (7.41)$$

Similarly, one can write

$$I(A, \phi; B, \theta) = H(B, \theta) - H(B, \theta/A, \phi) \quad (7.42)$$

where

$$H(B, \theta) = -\int_0^{2\pi}\int_0^\infty p(B, \theta)\log_2 p(B, \theta)\,dB\,d\theta \quad (7.43)$$

and

$$H(B, \theta/A, \phi) = -\int_0^{2\pi}\int_0^\infty\int_0^{2\pi}\int_0^\infty p(A, \phi, B, \theta)$$
$$\log_2 p(B, \theta/A, \phi)\,dA\,d\phi\,dB\,d\theta \quad (7.44)$$

By substitution of Eqs. (7.37) and (7.39) into Eq. (7.42), we have

$$I(A, \phi; B, \theta) = \int_0^{2\pi}\int_0^\infty\int_0^{2\pi}\int_0^\infty p(A, \phi, B, \theta)$$
$$\log_2 \frac{p(A, \phi, B, \theta)}{p(A, \phi)p(B, \theta)}\,dA\,d\phi\,dB\,d\theta \quad (7.45)$$

From Eq. (7.45), it is a simple matter to show that

$$I(A,\phi; B,\theta) \geq 0 \qquad (7.46)$$

The equality of Eq. (7.46) holds if and only if

$$p(A,\phi, B,\theta) = p(A,\phi)p(B,\theta) \qquad (7.47)$$

that is, the input–output ensembles are statistically independent.

From Eq. (7.46), we learn that, on the statistical average, information transmitted through an optical spatial channel is never negative. The extreme case is a complete loss of information transmitted through the channel [$I(A\ \phi; B, \theta) = 0$]. In essence, $I(A, \phi; B, \theta) = 0$ implies that $H(A, \phi/B, \theta) = H(A, \phi)$, the channel equivocation, is equal to the information provided at the input.

With these basic properties of spatial information, we are in a position to investigate information leakage through the channel (i.e., losses during information transmission). In order to do so, we assume that the optical channel can be decomposed into finite cascaded channels [28,29], as shown in Fig. 7.5. By $\{A_n, \phi_n\}$, we denote the input signal ensemble applied to the nth cascaded channel, and by $\{A_{n=1}, \phi_{n=1}\}$ the corresponding output signal ensemble, where $n = 1, 2, ..., N$.

It can be seen that there exists a certain relationship between the input and output ensembles at each of the cascaded channels. The overall output signal ensemble (i.e., the set of $\{A_{N+1}, \phi_{N+1}\} = \{B, \theta\}$) therefore depends on the input of $\{A_1, \phi_1\} = \{A, \phi\}$. This dependent property may be described by the conditional probability densities

$$p(A_{N+1},\phi_{N+1}/A_N,\phi_N,...,A_1,\phi_1) = P(A_{N+1},\phi_{N+1}/A_N,\phi_N) \quad (7.48)$$

Similarly, by Bayes' theorem [30], a reversal relationship for these cascaded channels can be written as

$$p(A_1,\phi_1/A_2,\phi_2,...,A_{N+1},\phi_{N+1}) = P(A_1,\phi_1/A_2,\phi_2) \quad (7.49)$$

It is noted that the relationship of Eqs. (7.48) and (7.49) is correct only for *linear* cascaded channels.

As the optical information propagates through the cascaded channels, the corresponding entropy equivocations increase:

$$H(A_1,\phi_1/A_{N+1},\phi_{N+1}) \geq H(A_1,\phi_1/A_N,\phi_N) \geq \cdots \geq H(A_1,\phi_1/A_2,\phi_2)$$

$$(7.50)$$

Now let us investigate the term-wise relationship of these entropy equivocations. We have, for example,

$$H(A_1,\phi_1/A_3,\phi_3) - H(A_1,\phi_1/A_2,\phi_2)$$

$$= -\int_0^{2\pi}\int_0^{\infty}\int_0^{2\pi}\int_0^{\infty} p(A_1,\phi_1,A_3,\phi_3)$$

$$\log_2 p(A_1,\phi_1/A_3,\phi_3)dA_1\,d\phi_1\,dA_3\,d\phi_3$$

$$+ \int_0^{2\pi}\int_0^{\infty}\int_0^{2\pi}\int_0^{\infty} p(A_1,\phi_1,A_2,\phi_2)$$

$$\log_2 p(A_1,\phi_1/A_2,\phi_2)dA_1\,d\phi_1\,dA_2\,d\phi_2$$

$$(7.51)$$

Equation (7.51) can be written as

$$H(A_1,\phi_1/A_3,\phi_3) - H(A_1,\phi_1/A_2,\phi_2)$$

$$= \underset{\substack{\{A_2,\phi_2\}\\\{A_3,\phi_3\}}}{\iiiint} p(A_2,\phi_2,A_3,\phi_3) \underset{\{A_1,\phi_1\}}{\iint} p(A_1,\phi_1/A_2,\phi_2,A_3,\phi_3)$$

$$\cdot \log_2 \frac{p(A_1,\phi_1/A_2,\phi_2,A_3,\phi_3)}{p(A_1,\phi_1/A_3,\phi_3)}dA_1\,d\phi_1\,dA_2\,d\phi_2\,dA_3\,d\phi_3$$

$$(7.52)$$

Then by the well-known fact [31]

$$-\int_{-\infty}^{\infty} p(x)\log\frac{p(x)}{g(x)}\,dx \geq 0 \qquad (7.53)$$

where

$$\int_{-\infty}^{\infty} p(x)\,dx = \int_{-\infty}^{\infty} g(x)\,dx = 1$$

where $p(x)$ and $g(x)$ are probability density functions in Eq. (7.52), it can be shown that

$$H(A_1,\phi_1/A_3,\phi_3) - H(A_1,\phi_1/A_2,\phi_2) \geq 0 \qquad (7.54)$$

FIG. 7.5 Decomposition of the optical channel into finite cascaded channels.

or, equivalently,

$$H(A_1, \phi_1/A_3, \phi_3) \geq H(A_1, \phi_2/A_2, \phi_2) \geq 0 \qquad (7.55)$$

The result of Eq. (7.54) indicates that the entropy equivocation increases as the information flows through the linear cascaded channels. Then by a simple induction principle, one can show that

$$H(A_1, \phi_1/A_{N+1}, \phi_{N+1}) \geq H(A_1, \phi_1/A_n, \phi_n) \text{ for } n = 2, 3, \ldots, N$$
$$\qquad (7.56)$$

From Eq. (7.56), it can be shown that the corresponding mutual information of the cascaded channels is

$$H(A_1, \phi_1; A_2, \phi_2) \geq I(A_1, \phi_1; A_3, \phi_3) \geq \cdots \geq I(A_1, \phi_1; A_{N+1}, \phi_{N+1})$$
$$\qquad (7.57)$$

This indicates that a passive optical channel has a tendency to leak information.

The useful result of Eq. (7.57) as applied to electrical communication was first recognized by Woodward [32], and although it may seem trivial, it should not be overlooked in the application of optical information processing. It is also interesting to note that the equalities in Eq. (7.57) hold if and only if

$$p(A_1, \phi_1/A_2, \phi_2, \ldots, A_{N+1}, \phi_{N+1}) = p(A_1, \phi_1; A_{N+1}, \phi_{N+1}) \quad (7.58)$$

Equation (7.58) implies that

$$p(A_1, \phi_1/A_2, \phi_2) = p(A_1, \phi_1/A_3, \phi_3) = \cdots = p(A_1, \phi_1/A_{N+1}, \phi_{N+1})$$
$$\qquad (7.59)$$

The conditions of Eq. (7.59) imply that the cascaded channel is noiseless and undistorted. However, under certain circumstances, despite a somewhat noisy cascaded channel, the equality in mutual information of Eq. (7.57) may hold. But the basic significance of our results is that the information transmitted through a passive optical channel can never be increased. We have seen that the output information, at best, is equal to the input. However, for an active channel, the output information can be made greater than the input, but only through the expenditure of external energy (entropy).

Now we come to the problem of image enhancement. We noted in Section 7.4 that a coherent optical channel is passive. The basic constraints are primarily the diffraction limit and spatial filter synthesis:

$$|H(p, q)| \leq 1 \qquad (7.60)$$

Thus, we emphasize that a passive optical channel is not able to enhance the image beyond the information limit of the input. But it is also noted that an optical channel is by no means inferior. For example, in many optical information processings, the information obtained at the output end usually depends on certain observations, in most cases, by photodetectors or the human eye. Thus, certain information may exist, but beyond the resolution limit of the detectors. However, if it goes through certain coherent processings, this information may be made perceivable. But again, the output information does not increase beyond the input. In other words, if the information does not exist, then it will be impossible to extract it by means of a passive optical channel. However, if an active channel is employed, then in principle it is possible to enhance the information beyond the diffraction limit. The most interesting techniques for information enhancement beyond the diffraction limit, as noted, are analytic continuation and the Karhunen–Loéve expansion. However, with both these techniques the required expenditure of external energy in some cases can be enormous.

7.6 RESTORATION OF BLURRED IMAGES

The restoration of a smeared photographic image by means of coherent optical spatial filtering has been briefly mentioned elsewhere. Some of the physical constraints discovered from the standpoint of information theory have also been discussed in preceding sections. The synthesis of a complex spatial filter to reduce the effect of blurring is now discussed.

As noted, a complex spatial filter may be realized by combining an amplitude filter and a thin-film phase filter. Such a synthesis may be accomplished by means of a holographic technique. The preparation of such a phase filter has been studied by Stroke and Zech [33] for the restoration of blurred images, and by Lohmann and Paris [22] for optical data processing.

In this section, we consider the synthesis of a filter that, when combined with an amplitude filter, can be used for restoration of an image that has been blurred. The complex filtering process discussed may be able to correct some blurred images, but the result is by no means an optimum filtering.

We recall that the Fourier transform form of a linearly distorted (blurred) image is (Eq. (7.1))

$$G(p) = S(p)D(p) \qquad (7.61)$$

where $G(p)$ is the distorted image function, $S(p)$ is the undistorted image function, $D(p)$ is the distorting function of the imaging system, and p is the spatial frequency. Then the corresponding inverse filter transfer function for the restoration is

$$H(p) = \frac{1}{D(p)} \qquad (7.62)$$

As noted previously, the inverse filter function is generally not physically realizable, particularly for blurred images, due to linear motion or defocusing. If we are willing, however, to accept a certain degree of error, then an approximate inverse filter may be obtained. For example, let the transmission function of a linear smeared point image (Eq. (7.3)) be

$$f(\xi) = \begin{cases} 1 & \text{for } -\dfrac{\Delta\xi}{2} \le \xi \le \dfrac{\Delta\xi}{2} \\ 0 & \text{otherwise} \end{cases} \tag{7.63}$$

where $\Delta\xi$ is the smear length. If a transparency satisfying Eq. (7.63) is inserted in the input plane P_1 of a coherent optical data processor, as shown in Fig. 7.6, the resultant complex light field on the spatial frequency plane will be

$$F(p) = \Delta\xi \frac{\sin(p\Delta\xi/2)}{p\Delta\xi/2} \tag{7.64}$$

which is essentially the Fourier transform of the smeared point image. A plot of the Fourier spectrum given by Eq. (7.64) is shown in Fig. 7.7. It can be seen that the Fourier spectrum is bipolar.

In principle, this smeared image can be corrected by means of inverse filtering. Thus, a suitable inverse filter function is

$$H(p) = \frac{p\Delta\xi/2}{\sin\left(\dfrac{p\Delta\xi}{2}\right)} \tag{7.65}$$

The transfer characteristic associated with Eq. (7.65) is given in Fig. 7.8. The inverse filter function itself is not only a bipolar but also an infinite-poles function. Thus, the filter is not physically realizable. However, if we are willing to sacrifice some of the resolution, then an approximate filter function may be realized. In order to do so, we combine the amplitude filter of Fig. 7.8 with the independent phase filter of Fig. 7.9. The transfer function of this combination is

$$H(p) = A(p)e^{i\phi(p)} \tag{7.66}$$

If this approximated inverse filter is inserted in the spatial frequency plane of the data processor of Fig. 7.6, the restored Fourier transfer function will be

$$F_1(p) = F(p)H(p) \tag{7.67}$$

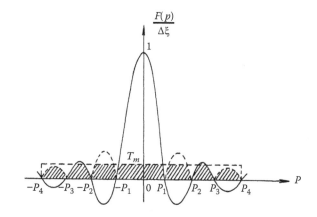

FIG. 7.7 The solid curve represents the Fourier spectrum of a linear smeared point image. The shaded area represents the corresponding restored Fourier spectrum.

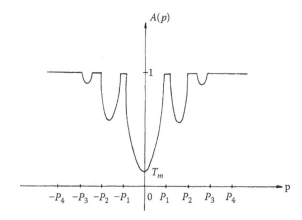

FIG. 7.8 Amplitude filter.

If we let T_m be the minimum transmittance of the amplitude filter, then the restored Fourier spectrum of the point image is the shaded spectrum shown in Fig. 7.7. We define the relative degree of image restoration (Eq. (7.33)) as

$$D(T_m) = \frac{\displaystyle\int_{\Delta p} [F(p)H(p)/\Delta\xi]\,/dp}{T_m\,\Delta p} \times 100\% \tag{7.68}$$

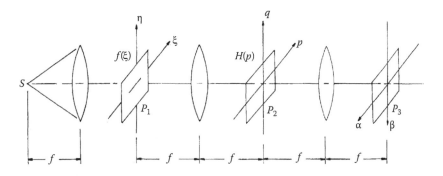

FIG. 7.6 Coherent optical data processor.

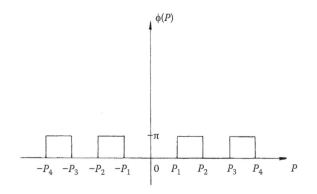

FIG. 7.9 Phase filter.

where Δp is the spatial bandwidth of interest. In Fig. 7.7, for example, $\Delta p = 2p_4$. From Eq. (7.68), we can plot the degree of image restoration as a function of T_m (Fig. 7.10). We see that a perfect restoration is approached as T_m approaches zero. However, at the same time the restored Fourier spectrum is also vanishing, and no image can be reconstructed. Thus, perfect restoration cannot be achieved in practice. These considerations aside, it seems that noise (caused by film granularity and speckling) is the major limiting factor in image restoration. In order to achieve a high degree of restoration, it appears that a lower transmittance T_m is required, and the restored Fourier spectrum will therefore be weaker. In turn, a lower signal-to-noise ratio of the restored image will result. Therefore, when considering the noise problem, it is clear that an optimum value of T_m must be obtained, at least in practice, for optimum image restoration.

As mentioned earlier in this section, a phase filter can be synthesized by means of a holographic technique (perhaps

by a computer-generated hologram). We now see how such a phase filter works in image restoration. Let us assume that the transmittance of a holographic phase filter is

$$T(p) = \frac{1}{2}\{1 + \cos[\phi(p) + \alpha_0 p]\} \qquad (7.69)$$

where α_0 is an arbitrarily chosen constant, and

$$\phi(p) = \begin{cases} \pi & \text{for } p_n \leq p \leq p_{n+1}, n = \pm 1, \pm 3, \pm 5, \ldots \\ 0 & \text{otherwise} \end{cases} \qquad (7.70)$$

With the amplitude filter added, the complex filter function can be written as

$$H_1(p) = A(p)T(p) = \frac{1}{2}A(p) + \frac{1}{4}\left[H(p)e^{i\alpha_0 p} + H^*(p)e^{-i\alpha_0 p}\right] \qquad (7.71)$$

where $H(p)$ is the approximate inverse filter function. Note also that $H(p) = H^*(p)$, because of Eq. (7.70).

Now if this complex filter $H_1(p)$ is inserted in the spatial frequency plane of Fig. 7.6, then the complex light field behind P_2 will be

$$F_2(p) = \frac{1}{2}F(p)A(p) + \frac{1}{4}\left[F(p)H(p)e^{i\alpha_0 p} + F(p)H^*(p)e^{-i\alpha_0 p}\right] \qquad (7.72)$$

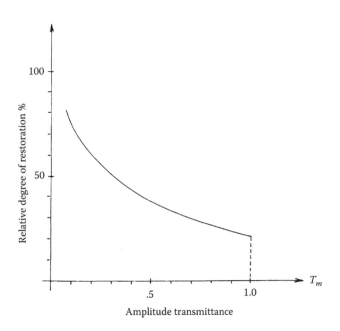

FIG. 7.10 Relative degree of restoration as a function of T_m.

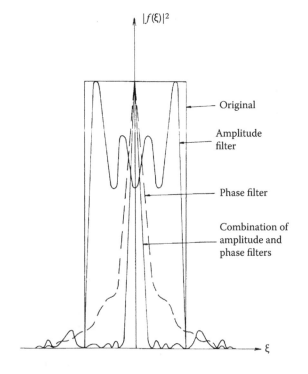

FIG. 7.11 Calculated irradiance of the restored point image.

(a)

(b)

FIG. 7.12 Image restoration with optical processing: (a) blurred image and (b) restored image.

We see that the first term of Eq. (7.72) is the restored Fourier spectrum due to the amplitude filter alone, which is diffracted onto the optical axis at the output plane P3 of Fig. 7.6. The second and third terms are the restored Fourier spectra of the smeared image; the restored images due to these terms are diffracted away from the optical axis at the output plane and centered at $\alpha = \alpha_0$ and $\alpha = -\alpha_0$, respectively. As an illustration, Fig. 7.11 shows the calculated irradiance of a restored point image blurred by linear motion, obtained with an amplitude filter, a phase filter, and a combination of both. In view of these results, we see that the phase restoration filter carries substantial weight for the restoration. This phenomenon is rather consistent with the holographic and matched filtering processes [34–36], for which the phase distribution is the major information-bearing element responsible for the image reconstruction and detection [37]. To confirm our claim, an experimental result using such a restoration filter is presented in Fig. 7.12. We see that blurred images can indeed be restored with an optical processor [38].

It is emphasized that the relative degree of restoration is with regard to the spatial bandwidth of interest. It is clear that the ultimate limit of Δp is restricted by the diffraction limits of the optical imaging and processing systems, whichever comes first. Therefore, it does not follow that a high degree of restoration implies restoration beyond the diffraction limit.

REFERENCES

1. J. Tsujiuchi, Correction of Optical Images by Compensation of Aberrations and Spatial Frequency Filtering, in E. Wolf, ed., *Progress in Optics*, vol. II, North-Holland, Amsterdam, 1963.
2. S. L. Zhuang, T. H. Chao, and F. T. S. Yu, Smeared Photographic Image Deblurring Utilizing White-Light Processing Technique, *Opt. Lett.*, vol. 6, 102 (1981).
3. G. W. Stroke, F. Furrer, and D. R. Lamberty, Deblurring of Motion-Blurred Photographs Using Extended-Range Holographic Fourier-Transform Division, *Opt. Commun.*, vol. 1, 141 (1969).
4. F. T. S. Yu, Image Restoration, Uncertainty, and Information, *J. Opt. Soc. Am.*, vol. 58, 742 (1968); *Appl. Opt.*, vol. 8, 53 (1969).
5. F. T. S. Yu, *Introduction to Diffraction, Information Processing, and Holography*, MIT Press, Cambridge, MA, 1973.
6. R. V. Churchill, *Fourier Series and Boundary Value Problems*, McGraw-Hill, New York, 1941.
7. M. Born and E. Wolf, *Principle of Optics*, 2nd ed., Pergamon, New York, 1964.
8. J. L. Synge, *Relativity*, North-Holland, Amsterdam, 1955.
9. J. L. Powell and B. Crasemann, *Quantum Mechanics*, Addison-Wesley, Reading, MA, 1961.
10. H. S. Coleman and M. F. Coleman, Theoretical Resolution Angles for Point and Line Test Objects in the Presence of a Luminous Background, *J. Opt. Soc. Am.*, vol. 37, 572 (1947).
11. G. Toraldo di Francia, Resolving Power and Information, *J. Opt. Soc. Am.*, vol. 45, 497 (1955).
12. V. Ranchi, *Optics, The Science of Vision*, New York University Press, New York, 1957.
13. V. Ranchi, Resolving Power of Calculated and Detected Images, *J. Opt. Soc. Am.*, vol. 51, 458 (1961).
14. J. L. Harris, Diffraction and Resolving Power, *J. Opt. Soc. Am.*, vol. 54, 931 (1964).
15. F. T. S. Yu, Optical Resolving Power and Physical Realizability, *J. Opt. Soc. Am.*, vol. 59, 497 (1969); *Opt. Commun.*, vol. 1, 319 (1970).
16. E. T. Whittaker and G. N. Watson, *A Course of Modern Analysis*, 4th ed., Cambridge University Press, Cambridge, MA, 1940.
17. E. A. Guilleman, *The Mathematics of Circuit Analysis*, John Wiley, New York, 1951.
18. J. F. Steffesen, *Interpolation*, Chelsea Publishing, New York, 1950.
19. R. E. A. Paley and N. Wiener, Fourier Transform in the Complex Domain, *Am. Math. Soc. Colloq.*, vol. 19, 16 (1934).
20. G. W. Stroke, Optical Computing, *IEEE Spectrum*, vol. 9, 64 (1972).
21. H. C. Andrews, *Computer Techniques in Image Processing*, Academic, New York, 1970.
22. A. W. Lohmann and D. P. Paris, Computer Generated Spatial Filters for Coherent Optical Data Processing, *Appl. Opt.*, vol. 7, 651 (1968).
23. J. Tsujiuchi, T. Honda, and T. Fukaya, Restoration of Blurred Photographic Images by Holography, *Opt. Commun.*, vol. 1, 379 (1970).
24. B. R. Frieden, Evaluation, Design and Extrapolation Methods for Optical Signals, Based on Use of the Prolate Functions, in E. Wolf, ed., *Progress in Optics*, vol. IX, North-Holland, Amsterdam, 1971.
25. F. T. S. Yu and A. Tai, Information Leakage through a Passive Optical Channel, *J. Opt. Soc. Am.*, vol. 64, 560 (1974); *Opt. Commun.*, vol. 14, 51 (1975).
26. C. E. Shannon, A Mathematical Theory of Communication, *Bell Syst. Tech. J.*, vol. 27, 379–423, 623–656 (1948).
27. C. E. Shannon and W. Weaver, *The Mathematical Theory of Communication*, University of Illinois Press, Urbana, 1949.
28. R. A. Silverman, On Binary Channels and Their Cascades, *IRE Trans. Inf. Theory*, vol. IT-1, 19–27 (1955).
29. N. Abramson, *Information Theory and Coding*, McGraw-Hill, New York, 1963, p. 113.
30. E. Parzen, *Modern Probability Theory and Its Applications*, John Wiley, New York, 1960.
31. R. M. Fano, *Transmission of Information*, MIT Press, Cambridge, MA, 1961.
32. P. M. Woodward, *Probability and Information Theory, with Applications to Radar*, Pergamon, New York, 1953.

33. G. W. Stroke and R. G. Zech, A Posteriori Image-Correcting Deconvolution by Holographic Fourier-Transform Division, *Phys. Lett.*, vol. 25A, 89 (1967).

34. F. T. S. Yu, Optimal Linearization in Holography, *Appl. Opt.*, vol. 8, 2483 (1969).

35. F. T. S. Yu, Linear Optimization in the Synthesis of Nonlinear Spatial Filter, *IEEE Trans. Inform. Theory*, vol. IT-17, 524 (1971).

36. F. T. S. Yu and S. Jutamulia, *Optical Signal Processing, Computing, and Neural Networks*, Wiley-Interscience, New York, 1992, sec. 6.

37. F. T. S. Yu, Exploitation of Incoherent Source for Partially Coherent Processing, *Opt. Eng.*, vol. 36, 1458 (1997).

38. F. T. S. Yu, Coherent and Digital Image Enhancement, Their Basic Differences and Constraints, *Opt. Commun.*, vol. 3, 440 (1971).

8 Quantum Effect on Information Transmission

Definitions of the entropy theory of information and of channel capacity were presented in previous chapters. It is noted that for discrete sources and channels, these definitions have a mathematical consistency, which is quite satisfactory from an intuitive point of view. However, a formalistic extension of the theory to a continuous channel in the optical frequency range leads to erroneous results. For instance, the capacity of a continuous additive Gaussian channel is [1] (Eq. (1.139))

$$C = \Delta v \log_2 \left(1 + \frac{S}{N} \right) \quad \text{bits/sec} \qquad (8.1)$$

where S is the average signal power, N is the average power of a white Gaussian noise, and Δv is the bandwidth of the channel.

We see that if the average noise power approaches zero, then the channel capacity approaches infinity. This is obviously contradictory to the basic physical constraints. Therefore, as the information transmission moves to high-frequency space, where the quantum effect takes place, the communication channel naturally leads to a discrete model. This is the way the quantum theory of radiation, which replaces the classical wave theory, explains the physical phenomena.

We have in previous chapters on several occasions called attention to the quantum phenomenon of information. However, a precise discussion of the quantum effect on a communication channel has not been treated. Although it was in 1929 that Szilard [2] described a basic concept of information from the statistical thermodynamics point of view, the concept was not actually pursued until 1948, by Shannon [3]. As noted, the work by Shannon was originally based on a pure mathematical standpoint and seemed to have nothing to do with the physical sciences. However, the relationship between information theory and physical science was quickly recognized by Gabor [4] in 1950, only a few years later. He advocated that information theory be treated as a branch of physics. But it was the work of Brillouin [5–9] in the earlier 1950s that successfully established a profound relationship between physical entropy and information. Since then, work on the entropy principle of information has been presented in a broad range of applications. However, it was Stern's [10,11] work in 1960 that actually explained the quantum effect on a communication channel. Stern's work essentially used Gabor's time–frequency cells (logons) in a technique for approaching the maximum entropy of a photon source. In the absence of noise, he obtained several interesting results. Following this trend, in 1962 Gordon [12] successfully derived an expression of information capacity for a narrowband channel by applying frequency representation of a one-dimensional electromagnetic signal and noise. He also discussed the information capacity of coherent amplification and heterodyne and homodyne receivers. In a paper by Lebedev and Levitin [13], information transmission by radiation through wideband and narrowband photon channels was again discussed; the effect of quantum statistics on the communication channel and the paradox of infinite channel capacity in the absence of noise were once again explained. Since Lebedev and Levitin's work closely followed the entropy theory of information, we adopt their approach in this chapter.

8.1 PROBLEM FORMULATION AND ENTROPY CONSIDERATION

We consider a communication system consisting of an information source of electromagnetic radiation. It is assumed that the information source radiates an electromagnetic signal with an average power S, and the signal propagated through the electromagnetic channel is assumed to be perturbed by additive thermal noise at temperature T. For simplicity, we assume that the electromagnetic signal is one-dimensional (i.e., the photon fluctuation is restricted to only one polarized state), in which the propagation occurs in the same direction of the wave vectors. It is emphasized that the corresponding occupation number of the quantum levels can be fully described. Obviously, these occupation quantum levels correspond to the *microsignal* structure of the information source. It is therefore assumed that these occupation numbers can be uniquely determined for those representing an input signal ensemble. We further assume that an ideal receiver (an ideal photon counter) is used at the output end of the channel. That is, the receiver is able to detect the specified frequencies of the electromagnetic signal. It is also emphasized that in practice, the interaction of the electromagnetic signal and the detector is strictly statistical; that is, a certain amount of information loss is expected at the output end of the receiver. The idealized model of the receiver we have proposed mainly simplifies the method by which the quantum effect on the communication channel can be easily calculated.

In the course of the investigation, we use the entropy principle to calculate the amount of information provided at the input end of the channel. We consider an input signal ensemble $\{a_i\}$, $i = 1, 2, ..., n$. Let $P(a_i)$ be the corresponding signal probability. If the input signal is applied to a physical system, then each $\{a_i\}$ is able to bring the system to a state of b_j, where the set $\{b_j\}$, $j = 1, 2, ..., n$, represents the *macroscopic* states of the system and each macroscopic state b_j is an ensemble of various *microscopic* states within the system. Let us denote by $P(b_j/a_i)$ the corresponding transitional probability. Then for each applied signal a_i the corresponding conditional entropy $H(B/a_i)$ is

$$H(B/a_i) = -\sum_{j=1}^{n} P(b_j/a_i) \log_2 P(b_j/a_i) \qquad (8.2)$$

Thus, the entropy equivocation (the average conditional entropy) is

$$H(B/A) = \sum_{i=1}^{n} P(a_i)H(B/a_i) \qquad (8.3)$$

Since

$$P(b_j) = \sum_{j=1}^{n} P(a_i)P(b_j/a_i) \qquad (8.4)$$

the output entropy is

$$H(B) = -\sum_{j=1}^{n} P(b_j) \log P(b_j) \qquad (8.5)$$

The corresponding mutual information provided by the physical system (treated as a communication channel) is

$$I(A; B) = H(B) - H(B/A) \qquad (8.6)$$

We see that the corresponding thermodynamic entropy is

$$\Delta S = I(A; B)k \ln 2 \qquad (8.7)$$

The entropy concept of Eq. (8.7) can be interpreted as indicating that the microstate of the system acts as a communication channel in the transmission of information. However, it is noted that the result of Eq. (8.6) is derived from a strictly stationary *ergodic* property. In practice, we see that the output signal should be time limited, and the system would have some memory. Thus, we conclude that based on the second law of thermodynamics, the physical entropy of the system (the channel) should be greater than that of the information transfer:

$$\Delta S > I(A; B)k \ln 2 \qquad (8.8)$$

Thus, as noted in a previous chapter, the inequality of Eq. (8.8) provides us with the basic connection between physical entropy and information entropy. This is the fundamental relationship between statistical physics and information theory. We use this entropy consideration in the following sections to discuss the quantum effect on a communication channel.

8.2 CAPACITY OF A PHOTON CHANNEL

We can now calculate the information capacity of a photon channel. For simplicity, we consider only the case of additive noise, that is, the case in which the noise statistic within the communication channel does not alter in the presence of a signal. At a specific frequency v, let us denote the mean quantum number of an electromagnetic signal by $\bar{m}(v)$, and the mean quantum number of a noise by $\bar{n}(v)$. Thus, the signal plus noise is

$$\bar{f}(v) = \bar{m}(v) + \bar{n}(v) \qquad (8.9)$$

Since the photon density (i.e., the mean number of photons per unit time per frequency) is the mean quantum number, the corresponding signal energy density per unit time is

$$E_s(v) = \bar{m}(v)hv \qquad (8.10)$$

where h is Planck's constant. Similarly, the noise energy density per unit time is

$$E_N(v) = \bar{n}(v)hv \qquad (8.11)$$

Since the mean quantum number of the noise (the blackbody radiation at temperature T) follows Planck's distribution (Eq. (6.9)),

$$E_N(v) = \frac{hv}{\exp(hv/kT) - 1} \qquad (8.12)$$

the corresponding noise energy per unit time (noise power) can be calculated as

$$N = \int_{\varepsilon}^{\infty} E_N(v)\,dv = \int_{\varepsilon}^{\infty} \frac{hv}{\exp(hv/kT) - 1}\,dv = \frac{(\pi kT)^2}{6h} \qquad (8.13)$$

where k is Boltzmann's constant, and the lower integral limit ε is an arbitrarily small positive constant. (*Note*: Equation (8.12) is also known as the *Bose–Einstein distribution*.)

We see that the minimum amount of entropy transfer required by the signal radiation is

$$\Delta S = \int_{0}^{N} \frac{dE_s(T')}{T'} = \int_{0}^{T} \frac{dE_s(T')}{dT'} \frac{dT'}{T'} \qquad (8.14)$$

where $E_s(T)$ is the corresponding signal energy density per unit time as a function of temperature T', and T is the temperature of the blackbody radiation. Thus, in the presence of a signal the output radiation energy per unit time (the power) can be written as

$$P = S + N \qquad (8.15)$$

where S and N are the signal and the noise power, respectively. Since the signal is assumed to be deterministic (i.e., the microstates of the signal are fully determined), the signal entropy can be considered zero. It is emphasized that the validity of this assumption is mainly based on the independent statistical nature of the signal and the noise, for which the photon statistics follow the Bose–Einstein distribution. However, it is also noted that the Bose–Einstein distribution cannot be used for the case of *fermions*, because, owing to the *Pauli exclusion principle*, the microstates of the noise are restricted by the occupational states of the signal, or vice versa. In other words, in the case of fermions, the signal and the noise can *never* be assumed to be statistically independent. For the case of Bose–Einstein statistics, we see that the amount of entropy transfer by radiation remains unchanged:

$$H(B/A) = \frac{\Delta S}{k \ln 2} \qquad (8.16)$$

Since the mutual information (information transfer) is $I(A; B) = H(B) - H(B/A)$, we see that $I(A; B)$ reaches its maximum when $H(B)$ is a maximum. Thus, for maximum information transfer, the signal should be chosen randomly. But the maximum value of entropy $H(B)$ occurs when the ensemble of the microstates of the total radiation (ensemble B) corresponds to the Gibbs distribution, which corresponds to the thermal equilibrium. Thus, the corresponding mean occupational quantum number of the total radiation also follows the Bose–Einstein distribution at a given temperature $T_e \geq T$:

$$\bar{f}(v) = \frac{hv}{\exp(hv/kT_e) - 1} \qquad (8.17)$$

where T_e can be defined as the effective temperature.
Thus, the total radiation entropy (the output entropy) can be determined by the equation

$$H(B) = \frac{1}{k \ln 2} \int_0^{T_e} \frac{dE_N(T')}{dT'} \frac{dT'}{T'} \qquad (8.18)$$

Then in accordance with the definition of channel capacity, as given in Chapter 1, we have

$$C = H(B) - H(B/A) = \frac{1}{k \ln 2} \int_T^{T_e} \frac{dE_N(T')}{dT'} \frac{dT'}{T'} \qquad (8.19)$$

where

$$E_N(T') = \frac{(\pi kT')^2}{6h} \qquad (8.20)$$

The total output radiation power is $P = S + N$, which can be expressed as

$$\frac{(\pi kTe)^2}{6h} = S + \frac{(\pi kT)^2}{6h} \qquad (8.21)$$

Therefore, one can evaluate T_e as a function of signal power S and noise temperature T:

$$T_e = \left[\frac{6hS}{(\pi k)^2} + T^2 \right]^{1/2} \qquad (8.22)$$

By substituting Eq. (8.22) into Eq. (8.19), the capacity of a photon channel can be shown:

$$C = \frac{\pi^2 kT}{3h \ln 2} \left[\left(1 + \frac{6hS}{(\pi kT)^2} \right)^{1/2} - 1 \right] \qquad (8.23)$$

Since the amount of energy required to transmit 1 bit of information is S/R, where R is the rate of information, we have

$$\frac{S}{R} = \frac{kTC \ln 2}{R} + \frac{3hC^2 (\ln 2)^2}{2\pi^2 R} \qquad (8.24)$$

It is also noted that the rate of information can never be greater than the channel capacity ($R \leq C$); thus, the minimum signal energy required to transmit the information at a rate equal to the channel capacity ($R = C$) is

$$S_{\min} = kT \ln 2 + \frac{3}{2\pi^2} hR(\ln 2)^2 \qquad (8.25)$$

Although the signal spectrum is excluded to infinity, it decreases rapidly at $hv \gg kTe$ (according to Eq. (8.17)). Thus, the bandwidth required to obtain the channel capacity of Eq. (8.23) can be estimated:

$$\Delta v \simeq \frac{kT_e}{h} = \frac{kT}{h} \left[\frac{6hS}{(\pi kT)^2} + 1 \right]^{1/2} \qquad (8.26)$$

It is noted that if the signal-to-noise ratio is high ($hv \gg kT$),

$$\frac{6hS}{(\pi kT)^2} \gg 1 \qquad (8.27)$$

then the channel capacity of Eq. (8.23) is limited by the quantum statistic:

$$C_{\text{quant}} = \frac{\pi}{\ln 2} \left(\frac{2S}{3h} \right)^{1/2} \qquad (8.28)$$

We see that this quantum channel capacity agrees with Stern's [10] result, except by a factor of $\sqrt{2}$.

However, if the signal-to-noise ratio is low ($h\nu \ll kT$),

$$\frac{6hS}{(\pi kT)^2} \ll 1 \qquad (8.29)$$

then the channel capacity of Eq. (8.23) is reduced to the classical limit:

$$C_{class} = \frac{S}{kT \ln 2} \qquad (8.30)$$

Equation (8.30) is essentially Shannon's [1,3] result for a wideband channel with additive Gaussian noise.

It is also noted that if the rate of information is low, then the minimum signal energy required (Eq. (8.25)) can be approximated by the equation

$$S_{min} \simeq kT \ln 2 \qquad (8.31)$$

This is the result we obtained in a previous chapter (Eq. (6.16)) for low-frequency observation.

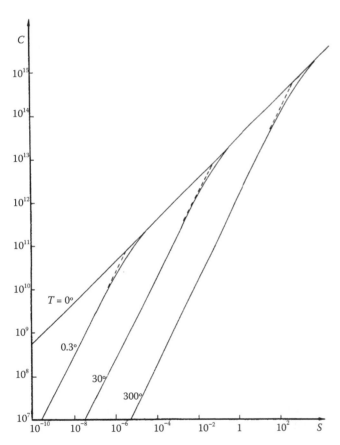

FIG. 8.1 Capacity of channel for radiation as a function of signal power, for various values of thermal noise temperature T. Dashed lines represent the classical asymptotes of Eq. (8.30).

It is also interesting to plot the capacity of the photon channel of Eq. (8.23) as a function of S for various values of thermal noise temperature T, as shown in Fig. 8.1. We see that the capacity of the channel depends on the signal-to-noise ratio. The classical limit approaches the quantum limit asymptotically, at a point of intersection, where

$$\frac{6hS}{(\pi kT)^2} = 4 \qquad (8.32)$$

The photon channel for a low signal-to-noise ratio agrees with the classical limit. However, for a high signal-to-noise ratio, the quantum channel of Eq. (8.28) appears much higher in comparison with the typical information rate in practice. This is essentially due to the idealized transmitter and receiver we have assumed in the calculation. In the following sections, a more detailed discussion of the photon channel is given, and in particular is concentrated on narrowband channels.

8.3 INFORMATIONAL THEORISTIC APPROACH

In Section 8.2, we calculated the capacity of a photon channel from the thermodynamic entropy standpoint. An identical result can also be derived from a strictly informational theoristic point of view. However, we will see that the latter approach is more mathematically inclined.

Let us now suppose that at a finite time interval of Δt, a signal embedded in additive noise is received. It is assumed that the received signal is located within a small localized spherical volume with a radius proportional to Δt. The photons of the received signal are assumed to have a discrete spectral energy:

$$E_i = h\nu_i \quad i = 1, 2, \dots \qquad (8.33)$$

where $\nu_i = i/\Delta t$.

Let the input signal mean occupational quantum number be m_i, at frequency ν_i, and let the mean occupational quantum number of the noise be n_i. Thus, the output signal (signal plus noise) for every ν_i can be described as

$$f_i = m_i + n_i \qquad (8.34)$$

Let us denote the corresponding probability distributions for m_i, n_i, and f_i by $p(m_i)$, $p(n_i)$, and $p(f_i)$, respectively. Since m_i and n_i are assumed to be statistically independent, the corresponding conditional probability can be written as

$$p(f_i/m_i) = p(f_i - m_i) \qquad (8.35)$$

The average signal energy density per unit time is therefore

$$E_S = \frac{1}{\Delta t} \sum_{i=0}^{\infty} \sum_{m_i=0}^{\infty} m_i P(m_i) h\nu_i \qquad (8.36)$$

Thus, the amount of mutual information between the input and the output radiation (photon channel) for every v_i is

$$I(f_i; m_i) = H(f_i) - H(f_i/m_i) \qquad (8.37)$$

Since the signal and the noise are assumed to be statistically independent, it is clear that

$$H(f_i/m_i) = f(n_i) \qquad (8.38)$$

Thus, the corresponding mutual information of the photon channel is

$$I(f_i; m_i) = H(f_i) - H(n_i) \qquad (8.39)$$

By substitution of the definitions of $H(f_i)$ and $H(n_i)$ in Eq. (8.39), Eq. (8.39) can be written as

$$I(f_i; m_i) = \sum_{m_i=0}^{\infty} \sum_{f_i=m_i}^{\infty} p(m_i) p(f_i/m_i) \log_2 \frac{p(f_i/m_i)}{p(f_i)} \qquad (8.40)$$

Thus, the total amount of mutual information per unit time is the ensemble average of $I(f_i; m_i)$ over i:

$$I(f; m) = \frac{1}{\Delta t} \sum_{i=0}^{\infty} I(f_i; m_i) \qquad (8.41)$$

and it can be shown that

$$I(f; m) = \frac{1}{\Delta t}[H(f) - H(n)] \qquad (8.42)$$

Let us recall that the photon fluctuation (due to thermal radiation) follows the Gibbs distribution:

$$p(n_i) = \frac{1 - \exp(-hv_i/kT)}{\exp(n_i hv_i/kT)} \qquad (8.43)$$

Then the corresponding conditional entropy equivocation can be written as

$$
\begin{aligned}
H(n_i) &= \sum_{n_i=0}^{\infty} p(n_i) \log_2 p(n_i) \\
&= \frac{hv_i}{kT[\exp(hv_i/kT) - 1]\ln 2} - \log_2[1 - \exp(-hv_i/kT)]
\end{aligned}
$$
$$(8.44)$$

Since $I(f; m)$ is maximum whenever $H(f)$ is maximum, we seek the distribution of $p(f_i)$ for which $H(f)$ is maximum.

It is noted that the average received signal power can be written as

$$\frac{1}{\Delta t} \sum_{i=1}^{\infty} \sum_{f_i=1}^{\infty} f_i hv_i p(f_i) = S + \frac{1}{\Delta t} \sum_{i=1}^{\infty} \frac{hv_i}{\exp(hv_i/kT) - 1} \qquad (8.45)$$

where S is the corresponding signal power, and $p(f_i)$ is the probability distribution of the received signal:

$$\sum_{f_i=1}^{\infty} p(f_i) = 1 \qquad (8.46)$$

Thus, the maximum entropy can be obtained by the Gibbs distribution:

$$p(f_i) = \frac{1 - \exp(-hv_i/\alpha)}{\exp(f_i hv_i/\alpha)} \qquad (8.47)$$

where α is a Lagrange multiplier having a thermal energy of $\alpha = kT_e$. We now let the time interval approach infinity, $\Delta t \to \infty$, and then $1/\Delta t$ can be written as an incremental frequency element, $1/\Delta t = dv$. Thus, the noise entropy of Eq. (8.44) can be obtained by the integral

$$H(n) \int_0^{\infty} \left\{ \frac{hv}{[\exp(hv/kT) - 1]kT \ln_2} - \log_2[1 - \exp(-hv/kT)] \right\} dv$$
$$(8.48)$$

which can be reduced to

$$H(n) = \frac{\pi^2 kT}{3h \ln_2} \qquad (8.49)$$

Similarly, the output signal entropy can be determined:

$$H(f) = -\sum_{f_i} \int_0^{\infty} p(f_i) \log_2 p(f_i) \, dv \qquad (8.50)$$

By substituting Eq. (8.47) into Eq. (8.50), one can show that

$$H(f) = \frac{\pi^2 \alpha}{3h \ln 2} \qquad (8.51)$$

It is noted that α (the Lagrange multiplier) equals kT_e, as in the previous section. By imposing the condition of Eq. (8.45), one could have obtained the photon channel capacity of Eq. (8.23):

$$C = H(f) - H(n) = \frac{\pi^2 kT}{3h \ln 2} \left[\left(1 + \frac{6hS}{(\pi RT)^2} \right)^{1/2} - 1 \right] \qquad (8.52)$$

which is identical to Eq. (8.23).

It is interesting to obtain the explicit form of the probabilistic signal distribution. Since it is assumed that the signal and the noise are statistically independent, we have the following equation for every v_i:

$$p(f_i) = \sum_{m_i=0}^{f_i} p(m_i)p(f_i - m) \qquad (8.53)$$

where $n = f - m$, and $p(m)$ and $p(n)$ denote the signal and noise probability distributions, respectively. (*Note*: To simplify the notation, the subscript of i is omitted in the following equations.)

If we let the Lagrange multiplier $\alpha = kT_e$, and substitute Eqs. (8.43) and (8.47) into Eq. (8.53), we then obtain the equations

$$p(m)\,|_{m=0} = \frac{1 - \exp(-hv/kT_e)}{1 - \exp(-hv/kT)} \qquad (8.54)$$

and

$$p(m) = \frac{[1 - \exp(-hv/kT_e)]\{1 - \exp[(hv/k)(1/T_e - 1/T)]\}}{[1 - \exp(-hv/kT)]\exp(mhv/kT_e)} \qquad (8.55)$$

We see that the signal distribution resembles, but is quite different from, the Gibbs distribution. Thus, it is incorrect to assume that the signal is statistically similar to that of thermal noise $P(n)$.

The mean occupational quantum number (the signal) can therefore be obtained by the definition

$$\bar{m} = \sum_{m=1}^{\infty} mp(m) = \frac{1}{\exp(hv/kT_e) - 1} - \frac{1}{\exp(hv/kT) - 1} \qquad (8.56)$$

Thus, we see that the signal as a function of v does not follow the Bose–Einstein distribution, in contrast to the normal assumption [12].

8.4 NARROWBAND PHOTON CHANNEL

Let us now propose a narrowband photon channel in which the effective bandwidth Δv is assumed to be very narrow compared with its center frequency v, $(\Delta v \ll v)$. In other words, the noise power is mostly concentrated within a narrow bandwidth of Δv. Thus, the corresponding power spectral density can be considered approximately uniform throughout the narrow bandwidth Δv and can be approximated:

$$N(v, T) \simeq \frac{hv}{\exp(hv/kT) - 1} \qquad (8.57)$$

The narrowband noise power as a function of temperature T can be written as

$$N(T) = N(v, T)\Delta v \qquad (8.58)$$

It is noted that, as in the previous sections, the maximum information rate can be achieved if the output signal (signal plus noise) is also a Gibbs distribution of Eq. (8.47). Thus, Eq. (8.58) provides an optimum transfer of information for a narrowband channel. Since it is assumed that photon noise is additive, the output power spectral density is

$$P(v, T) = \frac{S}{\Delta v} + N(v, T) \qquad (8.59)$$

where $S/\Delta v$ is the signal power spectral density.

Again, based on the entropy theory of information, as shown in Eq. (8.19), the capacity of a narrowband photon channel can be determined by the integral

$$C = \frac{\Delta v}{k \ln 2} \int_N^P \frac{d[N(v, T)]}{T(N, v)} \qquad (8.60)$$

By evaluating the above equation, one can show that

$$C = \Delta v \left\{ \log_2\left[1 + \frac{S}{hv\,\Delta v}(1 - e^{-hv/kT})\right] + \left(\frac{S}{hv\,\Delta v} + \frac{1}{e^{hv/kT} - 1}\right) \right.$$
$$\left. \cdot \log_2\left[1 + \frac{hv\,\Delta v(e^{hv/kT} - 1)}{S(e^{hv/kT} - 1) + hv\,\Delta v}\right] - \frac{hv/kT}{\ln 2(e^{hv/kT} - 1)}\right\} \qquad (8.61)$$

Equation (8.61) is essentially the result obtained by Gordon [12, eq. (5)].

From Eq. (8.61), we see that the channel capacity decreases monotonically with the frequency v. The monotonic behavior $C/\Delta v$, as a function of hv/kT, is plotted in Fig. 8.2 for various values of $S/kT\,\Delta v$.

Furthermore, if one lets the mean occupational quantum number m remain fixed, then the minimum number of quanta required for 1 bit of information to be transfered can be written as

$$M = \frac{\bar{m}\,\Delta v}{C} \qquad (8.62)$$

From Eq. (8.62), we see that the number of quanta remains uniform at $hv \gg kT$ (the quantum effect) but increases rapidly as the frequency decreases, $hv \gg kT$. The effect of M, the minimum required quanta, as a function of hv/kT is plotted in Fig. 8.3, for various values of $\bar{m}$. It is also interesting to note that if $hv \gg kT$, then the narrowband photon channel capacity can be reduced to the approximation

$$C \simeq \Delta v \log_2\left(1 + \frac{S}{N}\right) \quad \text{for } hv \ll kT \qquad (8.63)$$

where $N - kT\,\Delta v$ is the thermal noise power of the channel. Equation (8.63) essentially represents the well-known

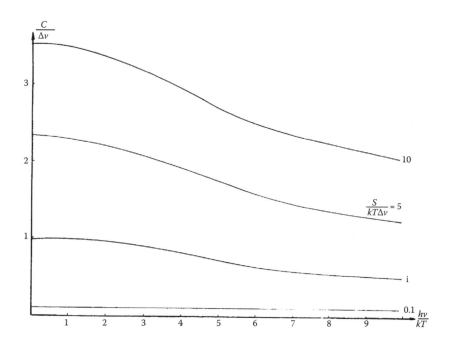

FIG. 8.2 $C/\Delta v$ as a function of hv/kT values of $S/kT\,\Delta v$, where v is the center frequency and Δv is the bandwidth of the narrowband channel.

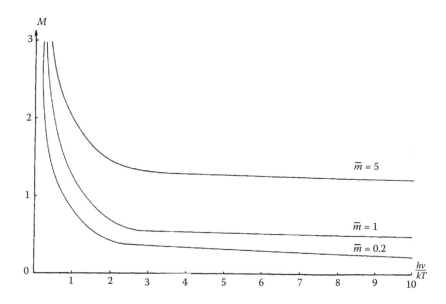

FIG. 8.3 Minimum member of quanta per bit information M as a function of hv/kT, for various values of m.

Gaussian channel described by Shannon and Weaver [1]. Let us recall that when Shannon derived this expression for a Gaussian channel, he completely disregarded the physical nature of the problem. Nevertheless, his Gaussian channel is the asymptotic result of our quantum channel of Eq. (8.61) at the low-frequency extreme ($hv \ll kT$). In other words, at low frequency ($hv \ll kT$), the quantum energy of hv can be approximated by a continuous energy spectrum for which the amplitude of the radiating field can be approximated by a Gaussian statistic. It is well known that Gaussian amplitude distribution corresponds to continuous exponential energy distribution, which follows Gibbs statistics. Moreover, we see

that as the limit of $hv/kT \to 0$, the Gibbs distribution for the noise (Eq. (8.43)) becomes

$$p(E_n) = \frac{1}{kT}\exp\left(-\frac{E_n}{kT}\right) \qquad (8.64)$$

where $E_n = nhv$. Similarly, from Eq. (8.47), as $hv/kT \to 0$, the output signal energy distribution (signal plus noise) can be shown:

$$p(E_f) = \frac{1}{kT_e}\exp\left(-\frac{E_f}{kT_e}\right) \qquad (8.65)$$

where $E_f = fhv$, $kT_e = S + N$ (the total output signal power), and T_e is the *effective output signal temperature*.

As noted in Chapter 2, a time-variable signal (in our case, a one-dimensional polarized electromagnetic signal or photon beam) with bandwidth Δv and duration Δt, where $\Delta v\, \Delta t \gg 1$, can be completely described by $2\,\Delta v\, \Delta t$ degrees of freedom. However, from the quantum theory point of view, the electromagnetic field is primarily localized in a *phase volume* of $h\, \Delta v\, \Delta t$, and the signal can be described *only* by $\Delta v\, \Delta t$ degrees of freedom, namely, the energy values assigned to various quantum levels. It is clear that the minimum sampling frequency, from the quantum standpoint, requires only $1/\Delta t$, instead of $1/2\, \Delta t$. Let us now assume that the noise amplitudes in the x and y (vertical and horizontal) directions are statistically independent and can be described by zero-mean Gaussian statistics, with the respective variances $\sigma_x^2 = \sigma_y^2 = N/2$:

$$p(x) = \frac{1}{\sqrt{\pi N}} \exp\left(-\frac{x^2}{N}\right) \tag{8.66}$$

and

$$p(y) = \frac{1}{\sqrt{\pi N}} \exp\left(-\frac{y^2}{N}\right) \tag{8.67}$$

where N is the average noise power. It can be seen that the corresponding noise energy distribution, $E_n = x^2 + y^2$, follows the Gibbs distribution:

$$p(E_n) = \frac{1}{kT} \exp\left(-\frac{E_n}{kT}\right) \tag{8.68}$$

which results from Eq. (8.64).

Similarly, we let the output signal (signal plus noise) amplitudes in the x and y directions have Gaussian distributions, with their variances equal to $(S + N)/2$. Thus, we have

$$p_f(x) = \frac{1}{\sqrt{\pi(S + N)}} \exp\left(-\frac{x^2}{S + N}\right) \tag{8.69}$$

and

$$p_f(y) = \frac{1}{\sqrt{\pi(S + N)}} \exp\left(-\frac{y^2}{S + N}\right) \tag{8.70}$$

where S is the input signal power. We see again that the output signal energy, $E_f = x^2 + y^2$, follows Gibbs statistics:

$$p(E_f) = \frac{1}{kT_e} \exp\left(-\frac{E_f}{kT_e}\right) \tag{8.71}$$

which is identical to Eq. (8.65).

Now we calculate the amount of mutual information (information transfer) per degree of freedom, first from the classical theory point of view:

$$I_{\text{class}} = H(f) - H(n)$$
$$= -\int_{-\infty}^{\infty} p_f(x) \log_2 p_f(x)\, dx + \int_{-\infty}^{\infty} p_n(x) \log_2 p_n(x)\, dx \tag{8.72}$$

By substituting Eqs. (8.69) and (8.68) into Eq. (8.72), we obtain

$$I_{\text{class}} = \frac{1}{2} \log_2\left(1 + \frac{S}{N}\right) \tag{8.73}$$

which was the formula obtained by Shannon and Weaver [1].

Similarly, from the quantum theory standpoint, the amount of information transfer per degree of freedom can be determined:

$$I_{\text{quant}} = -\int_{-\infty}^{\infty} p(E_f) \log_2 p(E_f)\, dE_f + \int_0^{\infty} p(E_n) \log_2 p(E_n)\, dE_n$$
$$= \log_2\left(1 + \frac{S}{N}\right) \tag{8.74}$$

Therefore, we see that

$$I_{\text{quant}} = 2I_{\text{class}} \tag{8.75}$$

The amount of information at the quantum limit is twice as large as that at the classical limit. But the degrees of freedom in the classical theory is $2\,\Delta v\, \Delta t$, which is twice that for the quantum limit. Therefore, we conclude that the capacity of the quantum-mechanical channel is equal to that of the classical channel:

$$C = 2\Delta v\, I_{\text{class}} = \Delta v\, I_{\text{quant}} = \Delta v \log_2\left(1 + \frac{S}{N}\right) \tag{8.76}$$

This interesting result may explain certain ambiguities and the paradoxes of the classical channel of Eq. (8.1) that Shannon calculated. We have shown that the quantum-mechanical channel coincides with the classical channel asymptotically at the lower-frequency limit, as $hv/kT \to 0$. It is noted that the quantum effect takes place in the communication channel when

$$\frac{hv}{kT} \gg 1 \tag{8.77}$$

and

$$\frac{S}{hv\,\Delta v}\left[\exp\left(\frac{hv}{kT}\right)-1\right]\gg 1 \qquad (8.78)$$

that is, when the mean occupation quantum number of the thermal noise is not only much smaller than unity, but also very small compared with the mean occupation quantum number of the input signal. Thus, we see that under the conditions of Eqs. (8.77) and (8.78), Eq. (8.61) can be approximated:

$$C \simeq \Delta v\left[\log_2\left(1+\frac{S}{hv}\right)+\frac{S}{hv}\log_2\left(1+\frac{hv}{S}\right)\right] \qquad (8.79)$$

The capacity of the channel increases as the mean occupation number $\bar{m} = S/hv$ increases. But it should be cautioned that Eq. (8.79) is valid only under the condition $\Delta v/v \gg 1$, that is, for a narrowband channel. Therefore, it is incorrect to assume that the capacity of the channel becomes infinitely large when $v \to 0$; strictly speaking, the capacity will never exceed the quantum limit, as given in Eq. (8.28).

Furthermore, for high signal power S, which implies that $S/hv\,\Delta v \gg 1$, Eq. (8.61) can be approximated by Shannon's formula:

$$C \simeq \Delta v \log_2\left(1+\frac{S}{kT_{eq}}\right) \qquad (8.80)$$

where $kT_{eq} = hv/e$, $e = 2.718\ldots$, and T_{eq} is the *equivalent noise temperature*.

It must be pointed out that according to Einstein's formula for energy fluctuations in blackbody radiation, the equivalent temperature of the quantum noise is defined as $kT_{eq} = hv$. But according to the generalized Nyquist formula, the definition of T_{eq} is $kT_{eq} = hv/2$. For the problem quantum effect in information transmission, according to Lebedev and Levitin [13], T_{eq} is defined as $kT_{eq} = hv/e$.

In concluding this section, we point out that in the case of two possible independent polarized states of radiation, the photon channel we have discussed can be considered two independent channels. If the total input signal power is assumed constant, then the overall capacity (i.e., the combination of these two channels) will be maximum if and only if the input signal power is equally divided between the channels. To obtain the overall channel capacity, one can simply replace the signal power S by $S/2$ in all the capacity equations in the previous sections and multiply the whole equation by a factor of 2. We see that in the quantum limit case of Eq. (8.28), the use of the two possible polarizations gives rise to a twofold gain in capacity. However, in the classical limit case of Eq. (8.30), we see that the channel capacity does not change after the modification.

8.5 OPTIMUM SIGNAL POWER DISTRIBUTION: A SPECIAL CASE

We illustrate the problem of distributing the signal power over the bandwidth of the channel to obtain the optimum capacity of a photon channel. For example, take the case of noise temperature as a function of frequency, $T = T(v)$.

To approach this distribution problem, we adopt the expression for the narrowband channel given in Eq. (8.61). We see that this overall capacity can be obtained by the expression

$$C_0 = \int_0^\infty \frac{C(v)}{\Delta v}\,dv \qquad (8.81)$$

where $C(v)$ is the capacity of the narrowband channel, which is a function of the center frequency v. It is also noted that the signal power density and the noise temperature are also functions of the frequency v ($S(v)$ and $T(v)$).

Since we assume that the photon noise is additive, the output power spectral density of the channel is the sum of the signal and noise spectral densities:

$$P(v) = S(v) + N(v) \qquad (8.82)$$

where $S(v)$ and $N(v)$ are the signal and noise spectral densities, respectively. The noise power spectral density can be described by the equation

$$N(v) = \frac{hv}{\exp[hv/kT(v)]-1} \qquad (8.83)$$

In order to obtain an optimum power distribution of the output signal (signal plus noise), one can apply the variational technique to the integral of Eq. (8.81) under the following conditions:

$$\int_0^\infty P(v)\,d(v) = S + \int_0^\infty \frac{hv\,dv}{\exp[hv/kT(v)]-1} \qquad (8.84)$$

where S is the input signal power.

Since the output power spectral density $P(v)$ is at least equal to or higher than that of the input,

$$P(v) \geq S(v) \qquad (8.85)$$

at the locations where the noise spectral density is higher than that of $S(v)$; $P(v)$ follows the variation in the noise spectral density $N(v)$. Thus, the distribution of the optimum output power spectral density (signal plus noise) is

$$P(v) = \begin{cases} \dfrac{hv}{\exp(hv/kT_e)-1} & T_e \geq T(v) \\[3mm] N(v) & T_e < T(v) \end{cases} \qquad (8.86)$$

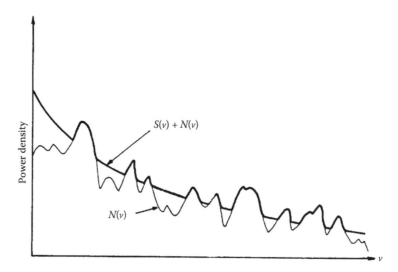

FIG. 8.4 Optimum distribution of signal power spectral density, a special case of frequency-dependent noise temperature $T(v)$. The lower curve represents the noise power spectral density, and the upper curve represents the output signal (i.e., signal plus noise) power spectral density.

where T_e is the equivalent temperature. It is clear that Eq. (8.86) is essentially the result we obtained previously. The effective output signal temperature T_e can be determined by Eq. (8.84):

$$S = \int_l [P(v) - N(v)]\, dv \qquad (8.87)$$

where l indicates integration over the interval for which $T_e \geq T(v)$.

Thus, we see that the optimum spectral distribution of the input signal power is such that all the degrees of freedom of the received signal are *heated* (or raised) to the same noise temperature T, as shown in Fig. 8.4.

Moreover, in classical theory ($hv \ll kT$) the output power spectral density is simply proportional to the temperature T; that is, the output power remains constant:

$$S + N = \text{constant} \qquad (8.88)$$

We see that Eq. (8.88) is essentially the result Shannon obtained.

In this chapter, we have treated in detail the basic quantum effect on information transmission. For such a quantum channel, for maximum entropy transfer, the output signal ensemble has been shown to follow the Gibbs distribution, for which it corresponds to the thermal equilibrium. Thus, by means of the entropy theory of information, the capacity of a photon channel can be calculated. As a result, we have shown that although the information transfer at the quantum limit is twice as large as that at the classical limit, the corresponding degree of freedom at the quantum limit is half that at the classical limit. Therefore, the capacity of the quantum channel should remain equal to that of the classical channel. We have also shown that as the frequency is lowered

($hv \ll kT$), this information capacity of the photon channel approaches asymptotically the continuous additive Gaussian channel of Shannon. As the frequency approaches infinity, the photon channel capacity approaches a finite quantum limit. Furthermore, for a frequency-dependent noise temperature, it is possible, at least in theory, to obtain an optimum channel by distributing the signal power spectral density over the frequency domain.

REFERENCES

1. C. E. Shannon and W. Weaver, *The Mathematical Theory of Communication*, University of Illinois Press, Urbana, 1949.
2. L. Szilard, Über die Entropieverminderung in Einem Thermodynamischen System bei Eingriffen Intelligenter Wessen, *Z. Phys.*, vol. 53, 840 (1929).
3. C. E. Shannon, A Mathematical Theory of Communication, *Bell. Syst. Tech. J.*, vol. 27, 379–423, 623–656 (1948).
4. D. Gabor, Communication Theory and Physics, *Phil. Mag*, vol. 41, no. 7, 1161 (1950).
5. L. Brillouin, Thermodynamics and Information Theory, *Am Sci.*, vol. 38, 594 (1950).
6. L. Brillouin, Maxwell's Demon Cannot Operate. Information Theory and Entropy I, *J. Appl. Phys.*, vol. 22, 334 (1951).
7. L. Brillouin, Physical Entropy and Information II, *J. Appl. Phys.* vol. 22, 338 (1951).
8. L. Brillouin, The Negentropy Principle of Information, *J. Appl. Phys.*, vol. 24, 1152 (1953).
9. L. Brillouin, *Science and Information Theory*, Academic, New York, 1956.
10. T. E. Stern, Some Quantum Effects in Information Channels, *IRE Trans.*, IT-6, no. 4, 435 (1960).
11. T. E. Stern, Information Rates in Photon Channels and Photon Amplifiers, *IRE Int. Con. Rec.*, pt. 4, 182 (1960).
12. J. P. Gordon, Quantum Effects in Communication Systems, *Proc. IRE*, vol. 50, 1898 (1962).
13. D. S. Lebedev and L. B. Levitin, Information Transmission by Electromagnetic Field, *Inf. Control*, vol. 9, 1 (1966).

9 Coherence Theory of Optics

9.1 ASPECTS OF COHERENCE

In the widespread application of coherent light, it is appropriate to discuss the principles of coherence in radiation [1–3]. If the radiations from two point sources maintain a fixed phase relation between them, they are said to be *mutually coherent*. An extended source is coherent if all points of the source have fixed phase differences between them.

In classical electromagnetic radiation, it is usually assumed that the electric and magnetic fields are always measurable at any time and position. In these cases, no accounting of coherence theory is needed. However, there are situations where this assumption cannot be made, and it is often helpful to apply coherence theory to the problem. For example, if we want to determine the diffraction pattern caused by the radiation from several sources, we cannot obtain an exact result unless the degree of coherence among the sources is known. In such a situation, however, it is desirable to obtain an ensemble average to represent the *statistically* most likely result from any such combination of sources. It may thus be more useful to provide a statistical description than to follow the dynamical behavior of a wave field in detail.

Let us assume an electromagnetic wave field propagating in space, as depicted in Fig. 9.1, where $u_1(t)$ and $u_2(t)$ denote the instantaneous wave disturbances at positions 1 and 2, respectively. We choose the second-order moment as the quantity to be averaged, as given by

$$\Gamma_{12}(\tau) \triangleq \langle u_1(t+\tau)u_2^*(t) \rangle \qquad (9.1)$$

which is known as the *mutual coherence function*, where the asterisk denotes the complex conjugate and τ denotes a time delay variable. The angle brackets represent the time ensemble as written by

$$\Gamma_{12}(\tau) = \lim_{T\to\infty} \frac{1}{2T} \int_{-T}^{T} u_1(t+\tau)u^*(t)\,dt \qquad (9.2)$$

It is apparent that the mutual coherence function is essentially the temporal *cross-correlation function* between $u_i(t)$ and $u_2(t)$.

The complex degree of coherence can be defined as

$$\gamma_{12}(\tau) \triangleq \frac{\Gamma_{12}(\tau)}{[\Gamma_{11}(0)\Gamma_{22}(0)]^{1/2}} \qquad (9.3)$$

in which $\Gamma_{11}(\tau)$ and $\Gamma_{22}(\tau)$ are the *self-coherence functions* of $u_1(t)$ and $u_2(t)$, respectively.

The normalized mutual coherence function $\gamma_{12}(\tau)$ can be measured by applying Young's experiment, as shown in Fig. 9.2, in which Σ represents an extended monochromatic source. As light falls on a diffraction screen at a distance r_{10} from the source, with two small pinholes Q_1 and Q_2 separated by a distance d at the screen, an interference pattern is observed at the observation screen. We shall adopt a measurable quantity, defined by *Michelson's visibility* of the fringes, as given by

$$v \triangleq \frac{I_{\max} - I_{\min}}{I_{\max} + I_{\min}} \qquad (9.4)$$

where $I_{\max}$ and $I_{\min}$ are the maximum and minimum intensities of the fringes.

As the source size Σ shrinks, we see that the visibility of the fringes increases. By varying the distance d between the pinholes, the visibility varies as d increases, as shown in Fig. 9.3. The variation of visibility can in fact be used as a measure of *spatial coherence* on the diffraction screen.

On the other hand, as the point of observation P moves away from the center of the observing screen, visibility decreases as the path difference $\Delta r = r_1 - r_2$ increases, until it becomes zero. This effect also depends on the monochromaticity of the source. The visibility of the fringes has been found to be appreciable only within a path difference of

$$\Delta r \approx \frac{2\pi c}{\Delta\omega} \qquad (9.5)$$

which is known as the *coherent length* of the source, where c denotes the light velocity and $\Delta\omega$ is the spectral width of the source.

It is, however, not necessary to have completely coherent light to produce an interference pattern, but under the right conditions, such a pattern may be obtained from an incoherent source. This effect is called *partial coherence*.

Since $u_1(t)$ and $u_2(t)$ can be subjected to the scalar wave equation, as given by

$$\nabla^2 u = \frac{1}{c^2}\frac{\partial^2 u}{\partial u^2} \qquad (9.6)$$

the intensity at P can be shown as

$$I_P = \langle u_P(t)u_P^*(t) \rangle = I_1 + I_2 + 2\operatorname{Re}\left\langle c_1 u_1\left(t - \frac{r_1}{c}\right)c_2^* u_2^*\left(t - \frac{r_2}{c}\right)\right\rangle$$

$$(9.7)$$

where I_1 and I_2 are the intensities from $u_1(t)$ and $u_2(t)$, respectively.

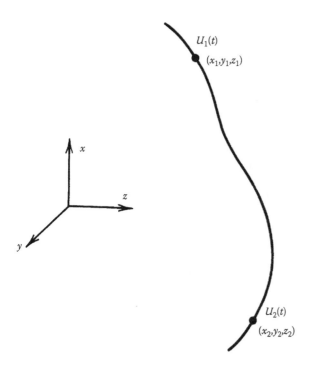

FIG. 9.1 Complex wave front in space.

FIG. 9.2 Young's experiment. The Σ is an extended but nearly monochromatic source.

By letting $t_1 = r_1/c$ and $t_2 = r_2/c$, and $\tau = t_1 - t_2$, the preceding equation becomes

$$I_p = I_1 + I_2 + 2c_1 c_2^* \mathrm{Re}[\Gamma_{12}(\tau)] \qquad (9.8)$$

By allowing $|c_1|^2 \Gamma_{22}(0) = I_2$, and $|c_1|^2 \Gamma_{11}(0) = I_1$, called the *best condition*, the intensity can be written as

$$I_p = 2I[1 + |\gamma_{12}(\tau)| \cos \phi_{12}(\tau)] \qquad (9.9)$$

in which we see that

$$v = |\gamma_{12}(\tau)| \qquad (9.10)$$

In other words, under the best conditions the visibility of the fringes is a measure of the degree of coherence.

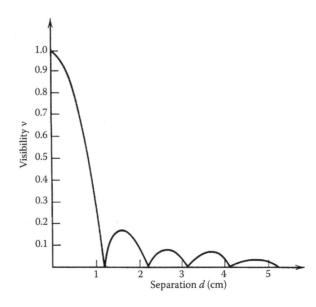

FIG. 9.3 Visibility as a function of pinhole separation.

9.2 SPATIAL AND TEMPORAL COHERENCE

In the preceding section, we saw that the term *spatial coherence* is applied to the effects due to the source size. For example, if we look at two points that are at equal light path distances from a point source, the radiations reaching these points will be exactly the same. Then the mutual coherence will be equal to the self-coherence at either point, that is,

$$\Gamma_{12}(Q_1, Q_2, \tau) = \langle u(Q_1, t + \tau) u^*(Q_2, t) \rangle = \Gamma_{11}(\tau) \qquad (9.11)$$

As the source is enlarged, we can no longer claim an equality of mutual coherence and self-coherence. This lack of complete coherence is a *spatial* effect. On the other hand, the *temporal coherence* is an effect that is due to the finite spectral width of the source. The coherence is complete for strictly monochromatic radiation but becomes only partial as other wavelengths are added. It is never possible to separate completely spatial and temporal coherence, but it is well to name them and point out their significances.

9.2.1 SPATIAL COHERENCE

By utilizing Young's experiment, the angular size of a spatially incoherent source (e.g., an extended source) as it relates to spatial coherence can be determined. Let us now consider the paths of two light rays from a point on a linearly extended source ΔS, as they pass through the two narrow slits Q_1 and Q_2 depicted in Fig. 9.4. If we let $r_{10} \gg d$, the intensity distribution at observation screen P can be written as

$$I_p(\alpha) = I_1 + I_2 + 2\sqrt{I_1 I_2} \cos\left[k\left(\frac{d}{r_{10}} \xi + \frac{d}{r_{20}} \alpha \right) \right] \qquad (9.12)$$

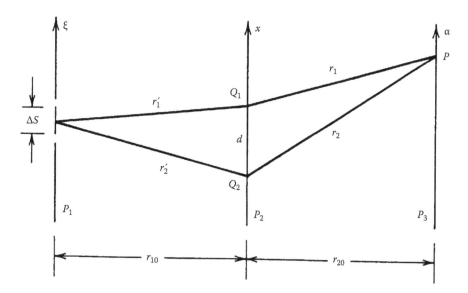

FIG. 9.4   Measurement of spatial coherence.

which describes a set of parallel fringes along the α axis. For a one-dimensional extended source of size ΔS, the resulting fringe pattern can be shown as

$$
\begin{aligned}
I'_p(\alpha) &= I_1 \Delta S + I_2 \Delta S + 2\sqrt{I_1 I_2} \int_{\Delta S} \cos\left[k\left(\frac{d}{r_{10}}\xi + \frac{d}{r_{20}}\alpha \right) \right] d\xi \\
&= \Delta S \left\{ I_1 + I_2 + 2\sqrt{I_1 I_2}\, \mathrm{Sinc}\left[\left(\frac{\pi d}{\lambda r_{10}} \right) \Delta S \right] \cos\left(k\frac{d}{r_{20}}\alpha \right) \right\}
\end{aligned}
$$

(9.13)

as the fringes vanish at

$$ d = \frac{\lambda r_{10}}{\Delta S} \qquad (9.14) $$

Thus, we see that the region of spatial coherence is determined by the visibility of the interference fringes. We see that the coherence region increases as the source size decreases or as the distance of the source increases. In other words, the degree of spatial coherence increases as the distance of the propagated wave increases.

9.2.2   Temporal Coherence

It is possible to split a light wave into two paths, to delay one of them, and then to recombine them to form an interference fringe pattern. In this way, we can measure the temporal coherence of the light wave. In other words, the degree of temporal coherence is the measure of the cross-correlation of a wave field at one time with respect to another wave field at a later time. Thus, the definition of temporal coherence can also refer to *longitudinal coherence* as opposed to

transversal (i.e., spatial) coherence. The maximum difference in the optical path lengths of two waves derived from a source is known as the *coherent length* of that source. Since spatial coherence is determined by the wave field in the transversal direction and temporal coherence is measured along the longitudinal direction, spatial coherence and temporal coherence describe the degree of coherence of a wave field within a spatial volume.

The *Michelson interferometer* is one of the techniques commonly used to measure the temporal coherence of a light source, as shown in Fig. 9.5. The beam splitter BS divides the light beam into two paths, one of which goes to mirror M_1 and the other goes to mirror M_2. By varying the path length (i.e., τ) to one of the mirrors, the variation of the interference fringes at screen P can be observed. Thus, the measurement of the visibility of the interference fringes corresponding to the measurement of the degree of temporal coherence $\gamma_{11}(\tau)$, at a time delay γ, can be obtained.

Since the physical nature of a light source affects the coherency of wave propagation, the spectral width of the source affects the temporal coherence. The time interval Δt, during which the wave propagated is coherent, can be approximated by

$$ \Delta t \approx \frac{2\pi}{\Delta \omega} \qquad (9.15) $$

for which the coherence length of the light source can be written as

$$ \Delta r = \Delta t c \approx \frac{2\pi c}{\Delta \omega} \qquad (9.16) $$

where c is the velocity of light, and $\Delta \omega$ is the spectral bandwidth of the light source.

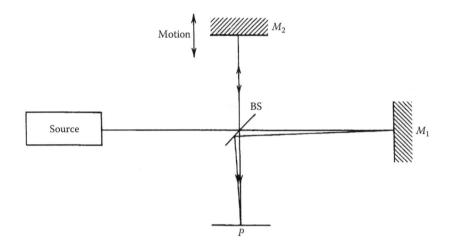

FIG. 9.5   Measurement of temporal coherence. BS, beam splitter; M, mirror; P, observation screen.

9.3   COHERENT AND INCOHERENT PROCESSING

Consider a hypothetical optical system [4] as depicted in Fig. 9.6. The light emitted by the source Σ is assumed monochromatic. To determine the light field at the output plane, we let this complex light distribution at the input plane, due to $d\Sigma$, be $u(x, y)$. If the complex amplitude transmittance of the input plane is $f(x, y)$, the complex light field immediately behind the signal plane would be $u(x, y) f(x, y)$. By assuming the optical system is linear and spatially invariant, the complex light field at the output plane of the system, due to $d\Sigma$, can then be written as

$$g(\alpha, \beta) = [u(x, y) f(x, y)] * h(x, y) \qquad (9.17)$$

where the asterisk denotes the convolution operation, and the corresponding intensity distribution, due to Σ, is given by

$$I(\alpha, \beta) = \iint |g(\alpha, \beta)|^2 \, d\Sigma \qquad (9.18)$$

which can be written as

$$I(\alpha, \beta) = \iiiint \Gamma(x, y; x', y') h(\alpha - x, \beta - y) h^*(\alpha - x', \beta - y')$$
$$\cdot f(x, y) f^*(x', y') dx \, dy \, dx' \, dy'$$

$$(9.19)$$

where

$$\Gamma(x, y; x', y') = \iint_{\Sigma} u(x, y) u^*(x', y') d\Sigma \qquad (9.20)$$

is the *spatial coherence function*, which is also known as the *mutual intensity function.*

Let us now choose two arbitrary points Q_1 and Q_2 at the input plane. If r_1 and r_2 are the distances from Q_1 and Q_2 to $d\Sigma$, the respective complex light disturbances at Q_1 and Q_2, due to $d\Sigma$, can be written as

$$u_1(x, y) = \frac{[I(\xi, \eta)]^{1/2}}{r_1} e^{ikr_1} \quad u_2(x', y') = \frac{[I(\xi, \eta)]^{1/2}}{r_2} e^{ikr_2} \quad (9.21)$$

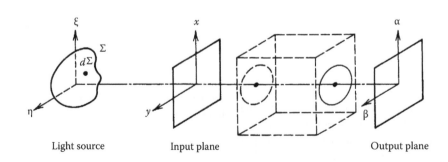

FIG. 9.6   Hypothetical optical signal processing system.

where $I(\xi, \eta)$ is the intensity distribution of the light source. By substituting Eq. (9.21) into Eq. (9.20), we have

$$\Gamma(x, y; x', y') = \iint_{\Sigma} \frac{I(\xi, \eta)}{r_1 r_2} \exp[ik(r_1 - r_2)] d\Sigma \quad (9.22)$$

When the light rays are paraxial, $r_1 - r_2$ can be approximated by

$$r_1 - r_2 \approx \frac{1}{r}[\xi(x - x') + \eta(y - y')] \quad (9.23)$$

where r is the distance between the source and the signal plane. Then Eq. (9.23) becomes

$$\Gamma(x, y; x', y') = \frac{1}{r^2} \iint I(\xi, \eta)$$
$$\times \exp\left\{i\frac{k}{r}[\xi(x - x') + \eta(y - y')]\right\} d\xi \, d\eta \quad (9.24)$$

which is essentially the inverse Fourier transform of the source intensity distribution. Equation (9.24) is also known as the *Van Cittert–Zernike theorem.*

Let us now consider two extreme cases. In one, we let the light source become infinitely large, that is, $I(\xi, \eta) \cong K$; Eq. (9.24) then becomes

$$\Gamma(x, y; x', y') = K_1 \delta(x - x', y - y') \quad (9.25)$$

which describes a completely *incoherent* illumination, where K_1 is a proportionality constant.

On the other hand, if we let the light source be vanishingly small, that is, $I(\xi, \eta) \cong K\delta(\xi, \eta)$, Eq. (9.24) becomes

$$\Gamma(x, y; x', y') = K_2 \quad (9.26)$$

which describes a completely *coherent* illumination, where K_2 is an arbitrary constant. In other words, a monochromatic point source describes a strictly coherent regime, whereas an extended source describes a strictly incoherent radiation.

For the completely incoherent illumination, the output intensity distribution can be shown as

$$I(\alpha, \beta) = \iint |h(\alpha - x, \beta - y)|^2 |f(x, y)|^2 \, dx \, dy \quad (9.27)$$

from which we see that the incoherent optical system is *linear in intensity.*

On the other hand, under the strictly coherent illumination as described in Eq. (9.26), Eq. (9.19) becomes

$$I(\alpha, \beta) = g(\alpha, \beta)g^*(\alpha, \beta) = \iint h(\alpha - x, \beta - y)f(x, y) \, dx \, dy$$
$$\cdot \iint h^*(\alpha - x', \beta - y')f^*(x', y') \, dx' \, dy' \quad (9.28)$$

from which we can see that the coherent optical system is *linear in complex amplitude.*

9.4 EXPLOITATION OF COHERENCE

In the preceding section, we saw that a coherent processor is capable of carrying out the complex amplitude processing, which offers myriad applications [5,6]. However, coherent optical processing also suffers from coherent artifact noise, which limits its processing capabilities. To alleviate these limitations, we discuss methods to exploit the coherence contents from an incoherent source for complex amplitude (or coherent) processing. Since all physical sources are neither strictly coherent nor strictly incoherent, it is possible to extract the inherent coherence contents from an incoherent source for coherent processing.

9.4.1 EXPLOITATION OF SPATIAL COHERENCE

We begin our discussion on the exploitation of spatial coherence from an extended incoherent source. A conventional optical signal processor is shown in Fig. 9.7, in which an

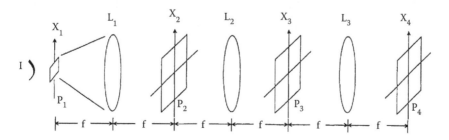

FIG. 9.7 Incoherent source optical processor. I, incoherent source; L_1, collimating lens; L_2 and L_3, achromatic transformation lenses; P_1, source-encoding mask; P_2, input plane; P_3, Fourier plane; P_4, output plane.

extended light source is used. With reference to the spatial coherence function of Eq. (9.24), we have

$$\Gamma\left(\mathbf{x}_2 - \mathbf{x}_2'\right) = \iint \gamma(\mathbf{x}_1) \exp\left[i2\pi \frac{\mathbf{x}_1}{\lambda f}(\mathbf{x}_2 - \mathbf{x}_2')\right] d\mathbf{x}_1 \quad (9.29)$$

From this result, we see that the spatial coherence function and the source-encoding intensity form a Fourier transform pair, as given by

$$\begin{aligned} \gamma(\mathbf{x}_1) &= \mathbf{F}\left[\Gamma\left(\mathbf{x}_2 - \mathbf{x}_2'\right)\right] \\ \Gamma\left(\mathbf{x}_2 - \mathbf{x}_2'\right) &= \mathbf{F}^{-1}[\gamma(\mathbf{x}_1)] \end{aligned} \quad (9.30)$$

where $\mathbf{F}$ denotes the Fourier transform operation. In other words, if a specific spatial coherence function is needed, the corresponding source-encoding function can be obtained by the Fourier transformation. Note that the source-encoding function $\gamma(x_1)$ can consist of apertures of different shapes or slits, but it should be a positive real function that satisfies the following physically realizable constraint $\mathbf{x}$:

$$0 \le \gamma(\mathbf{x}_1) \le 1 \quad (9.31)$$

9.4.2 Exploitation of Temporal Coherence

Since the scale of the Fourier transform is linearly proportional to the wavelength of the light source, it is apparently not capable of (or inefficient at) using a broadband source. To do so, a narrow-spectral-band (i.e., temporally coherent) source is needed. In other words, the spectral spread of the input object should be confined within a small fringe spacing of the spatial filter, which is given by

$$\frac{p_m f \Delta\lambda}{2\pi} \ll d \quad (9.32)$$

where d is the fringe spacing of the spatial filter, p_m is the upper angular spatial frequency content of the input object, f is the focal length of the transform lens, and $\Delta\lambda$ is the spectral bandwidth of the light source. Then the spectral width (or the temporal coherence requirement) of the light source should satisfy the following constraint:

$$\frac{\Delta\lambda}{\lambda} \ll \frac{\pi}{h_0 p_m} \quad (9.33)$$

where λ is the center wavelength of the light source, $2h_0$ is the size of the input object transparency, and $2h_0 = (\lambda f)/d$.

There are, however, techniques to extract temporal coherence content from a broadband source. One of the simplest and most efficient methods is by dispersing the Fourier spectra in the Fourier plane, which can be obtained by placing a spatial sampling grating at the input plane P_2. For example, if the input object is sampled by a phase grating as given by

$$f(x_2)T(x_2) = f(x_2)\exp(ip_0 x_2) \quad (9.34)$$

then the corresponding Fourier transform would be

$$F(p,q) = F\left(x_3 - \frac{\lambda f}{2\pi}p_0\right) \quad (9.35)$$

in which we see that $F(p, q)$ is smeared into rainbow colors along the x_3 in the Fourier plane. Thus, a high degree of temporal coherence of the Fourier spectrum within a narrow-spectral-band filter can be obtained. In other words, by simply increasing the sampling frequency p_0 at the input domain P_2, a high degree of temporal coherent Fourier spectra can be obtained, as given by

$$\frac{\Delta\lambda}{\lambda} \cong \frac{4p_m}{p_0} \ll 1 \quad (9.36)$$

9.4.3 Complex Amplitude Processing

Let us describe a broad-spectral-band processor that exploits the coherent contents for complex amplitude processing. By referring to the incoherent source processor of Fig. 9.7, an extended white light source is used. First, let us look at the spatial coherence requirement, for which a source-encoding mask is used. We illustrate the complex signal detection, in which we assume that the target is much smaller than the entire input plan. Thus, the spatial coherence requirement should be imposed by the size of the target rather than the entire input spatial domain P_2. To ensure a physically realizable source-encoding mask, the required spatial coherence that covers the input target can be written as

$$\Gamma\left(\left|r_2 - r_2'\right|\right) = \frac{J_1\left[(\pi/h_0)\left|r_2 - r_2'\right|\right]}{(\pi/h_0)\left|r_2 - r_1'\right|} \quad (9.37)$$

where J_1 is a first-order Bessel function of the first kind and h_0 is the size of the target. The corresponding spatial coherence as a function of $\left|r_2 - r_2'\right|$ is sketched in Fig. 9.8a, such that the source-encoding function is obtained as

$$\gamma(|r_1|) = \text{cir}\left\{\frac{|r_1|}{w}\right\} \quad (9.38)$$

where $w = (f\lambda)/h_0$ represents the diameter of a circular aperture, f is the focal length of the collimating lens, and λ is the wavelength.

We now consider the next illustration for linear image deblurring. Since smeared image deblurring is a one-dimensional

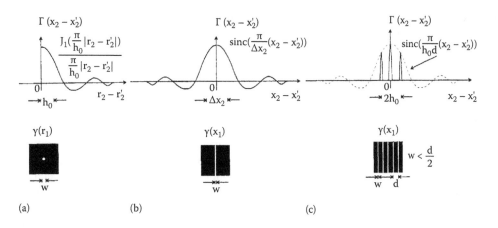

FIG. 9.8 Source encoding and spatial coherence. Γ, spatial coherence function; γ, source-encoding mask. (a) Correlation detection. (b) Image deblurring. (c) Image subtraction.

processing operation, inverse filtering takes place with respect to the smeared length of the blurred object. Thus, the required spatial coherence depends on the smeared length instead of the entire input plane. If we assume that a spatial coherence function is given by

$$\Gamma\left(x_2, x_2'\right) = \text{Sinc}\left\{\frac{\pi}{\Delta x_2}\left(x_2, x_2'\right)\right\} \qquad (9.39)$$

then the source-encoding function can be shown as

$$\gamma(x_1) = \text{rect}\left(\frac{x_1}{w}\right) \qquad (9.40)$$

where Δx_2 is the smeared length, $w = (f\lambda)/(\Delta x_2)$ is the slit width of the encoding aperture as shown in Fig. 9.8b, and

$$\text{rect}\left(\frac{x_1}{w}\right) = \begin{cases} 1 & -\dfrac{w}{3} \le x_1 \le \dfrac{w}{2} \\ 0 & \text{otherwise} \end{cases} \qquad (9.41)$$

We now consider the third example for image subtraction. Since the spatial coherence depends on the corresponding point pair of the images to be subtracted, a strictly broad spatial coherence function is not required. In other words, a point-pair spatial coherence function is needed. To ensure the physically realizability of the source-encoding function, we assume the point-pair spatial coherence function is given by

$$\Gamma\left(x_2 - x_2'\right) = \frac{\sin\left[\left(N\pi/h_0\right)\left(x_2 - x_2'\right)\right]}{N \sin\left[\pi/h_0\right)\left(x_2 - x_2'\right)\right]} \text{Sinc}\left[\frac{\pi w}{h_0 d}\left(x_2 - x_2'\right)\right] \qquad (9.42)$$

where $2\,h_0$ is the main separation of the two input image transparencies. As $N \gg 1$ and $w \ll d$, the preceding equation represents a sequence of narrow pulses that occur at

$\left(x_2 - x_2'\right) = n h_0$, where n is a positive integer. Thus, we see that a high degree of coherence of the point-pair function can be obtained. By Fourier transforming the preceding equation, the source-encoding function can be shown as

$$\gamma(x_1) = \sum_{n=1}^{N} \text{rect}\left(\frac{x_1 - nd}{w}\right) \qquad (9.43)$$

where w is the slit width, and $d = (\lambda f)/h_0$ is the separation between the slits. By plotting the preceding equation, the source-encoding mask is in fact represented by N equally spaced narrow slits, as shown in Fig. 9.8c.

We now investigate the temporal coherence requirement, for which a sampling grating is used to disperse the Fourier spectrum in the Fourier plane. Let us consider the temporal coherence requirement for two-dimensional complex signal detection. By referring to Eq. (9.36), a high degree of temporal coherence can be achieved by using a higher sampling frequency. We assume that the Fourier spectrum dispersion due to the sampling grating is along the x direction and the narrow-spectral-band matched filter is limited by the wavelength spread of Eq. (9.33), that is,

$$\Delta\lambda = \frac{\pi\lambda}{10 h_0 p_m} \qquad (9.44)$$

Thus, we see that the spatial width (or size) of the narrow-spectral-band filter is given by

$$\Delta w = \frac{f\lambda}{\pi} p_m \qquad (9.45)$$

In view of Eq. (9.36), the sampling frequency can be determined as

$$p_0 = \frac{4\lambda p_m}{\Delta\lambda} \qquad (9.46)$$

In other words, the higher the sampling frequency used, the more efficient the processing performance would be. We see that a larger set of temporal matched filters can be used, which results in higher-output image irradiance. However, the apparent price paid is using large achromatic transform lenses, which tend to be more expensive.

9.4.4 INCOHERENT SUPERPOSITION

In the preceding section, we saw that spatial and temporal coherence can be exploited by spatial encoding and spectral dispersion of an incoherent source. We see that complex amplitude processing can be carried out with a set of narrow-spectral-band filters, as given by

$$H_n(p_n, q_n; \lambda_n) \quad \text{for } n = 1, 2, \dots, N, \qquad (9.47)$$

where (p_n, q_n) represents the angular frequency coordinates and λ_n is the center wavelength of the narrow-width filter. Then it can be shown that the output light intensity would be the incoherent superposition of the filtered signals, as given by

$$I(x, y) \cong \sum_{n=1}^{N} \Delta\lambda_n \mid f(x, y; \lambda_n) * h(x, y; \lambda_n) \mid^2 \qquad (9.48)$$

where the asterisk denotes the convolution operation, $f(x, y; \lambda_n)$ represents the input signal illuminated by λ_n, $\Delta\lambda_n$ is the narrow spectral width of the narrow-spectral-band filter, and $h(x, y; \lambda_n)$ is the spatial impulse response of $H(p_n, q_n; \lambda_n)$, that is,

$$h(p_n, q_n; \lambda_n) = \mathbf{F}^{-1}[H(p_n, q_n; \lambda_n)] \qquad (9.49)$$

In view of the preceding result, the incoherent source processor can indeed process the information in complex amplitude as a coherent processor. Since the output intensity distribution is the sum of mutually incoherent image irradiances, the annoying coherent artifact noise can be suppressed.

However, if the signal processing is a one-dimensional operation, then the processing can be carried out with a fan-shaped broadband filter to compensate the scale variation of the Fourier spectra. The output intensity distribution can be shown as

$$I(x, y) = \int_{\Delta\lambda} \mid f(x, y; \lambda) * h(x, y; \lambda) \mid^2 d\lambda \qquad (9.50)$$

where the integral is over the entire spectral band of the light source. Again, we see that the output irradiance is obtained by incoherent superposition of the entire spectral band light source, by which we can in principle get rid of the coherent artifact noise.

9.5 REMARKS

The merit of using coherent light for optical data processing, in contrast to using incoherent light, must be the complex amplitude processing. Thus, the coherent processing system has a much larger information processing capacity than the incoherent processing system.

One of the major advantages of using the broadband incoherent source for data processing must be the elimination of coherent artifact noise; Dennis Gabor regarded it as the number one enemy in optical signal processing. Although the system under incoherent illumination in principle cannot process the information in complex amplitude as does the coherent system, temporal and spatial coherence contents from an incoherent source can be exploited for complex amplitude processing. In other words, to exploit the spatial coherence content at the input domain, one can spatially encode the light source. And to exploit a high temporal coherence content at the Fourier domain, one can disperse the Fourier spectral in rainbow color. The advantage of exploiting the coherence content from an incoherent source is that the information can be processed in complex amplitude as it is by a coherent processor, and at the same time, the coherent artifact noise can be suppressed as it is processed with an incoherent processor.

REFERENCES

1. M. Born and E. Wolf, *Principle of Optics*, 2nd rev. ed., Pergamon Press, New York, 1964.
2. F. T. S. Yu and S. Jutamulia, *Optical Signal Processing, Computing and Neural Networks*, Wiley-Interscience, New York, 1992.
3. D. Gabor, Laser Speckle and Its Elimination, *IBM J. Res. Dev.*, vol. 14, 509–574 (1970).
4. F. T. S. Yu, *White Light Signal Processing*, Wiley-Interscience, New York, 1985.
5. S. T. Wu and F. T. S. Yu, Image Subtraction with Encoded Extended Incoherent Source, *Appl. Opt.*, vol. 20, 4082–4088 (1988).
6. F. T. S. Yu, Exploitation of Incoherent Source for Partially Coherent Processing, *Opt. Eng.*, vol. 36, 1458–1464 (1997).

10 Wavelet Transforms with Optics

Two major questions concerning this chapter may be asked: What is a wavelet? Why is it interesting for solving signal processing problems? These are crucial questions that a signal analyst would want to know the answers to, and the answer may be summarized as follows: wavelet representation is a versatile technique having many physical and mathematical insights with great potential applications to signal and image processing. In other words, wavelets can be viewed as a new basis for representing signals and images that can be used for signal analysis and image synthesis.

10.1 ASPECTS OF WAVELET TRANSFORM

For nonstationary signal processing, the natural way to obtain joint time–frequency resolution of a signal is to take the Fourier transform of the signal within a time window function. This is called a *short-time Fourier transform* (STFT), where the size of the window is assumed to be invariant with respect to the frequency content of the signal [1–4]. On the other hand, if the size of the window changes as the analyzing frequency changes, then the transform is known as a *wavelet transform* (WT) [4–7].

The expression of the STFT can be written as [1–3]

$$\mathrm{STFT}(\tau,\omega) = \int x(t)h*(t-\tau)\exp(-i\omega t)\,dt \quad (10.1)$$

where $h(t)$ is an analyzing window function, ω is a selected analyzing frequency, and τ is an arbitrary time shift. Notice that if $h(t)$ is a Gaussian function, then the transform is also known as the *Gabor transform*. The STFT has been widely used in signal processing, such as time-varying signal analysis and filtering, spectral estimation, and signal compression [1–4]. Usually, the STFT offers very good performances for signals having uniform energy distribution within an analyzing window. Thus, the selection of the analyzing window size is critically important for achieving an optimum joint time–frequency resolution. However, an apparent drawback that prevents the STFT from wider application is that the invariant size of the analyzing window leads to a lower time resolution for nonuniformly distributed signals, such as transient, discontinuous signals. Therefore, to overcome this shortcoming, the WT can be used as given by [4–7]

$$\mathrm{WT}(\tau,a) = \frac{1}{\sqrt{a}}\int x(t)\psi*\left(\frac{1-\tau}{a}\right)dt \quad (10.2)$$

where a is a scaling factor, $\psi(t)$ is called the *mother wavelet*, and τ is an arbitrary time shift. Notice that the shape of the *mother wavelet* $\psi(t)$ shrinks as the scaling factor a decreases,

while it dilates as a increases. The shrunk and dilated wavelets are also known as the *daughter wavelets* [5,6]. Thus, to have a better time resolution, a narrower WT window should be used for higher-frequency content. In principle, the WT suffers from the same time–frequency resolution limitation as the STFT; that is, the time resolution and the frequency resolution cannot be resolved simultaneously as imposed by the inequity

$$\Delta t\,\Delta\omega \ge 2\pi \quad (10.3)$$

where Δt and $\Delta\omega$ can be written as

$$\Delta t = \frac{\int |h(t)|\,dt}{|h(0)|} \quad \text{and} \quad \Delta\omega = \frac{\int |H(\omega)|\,d\omega}{|H(0)|}$$

in which $H(\omega)$ is the Fourier transform of $h(t)$. The product of Δt and $\Delta\omega$ is known as the *logon*. In practice, window functions having smaller cells have better joint time–frequency localization. The *Gaussian window function* is the best one in the sense of meeting the lower bound of the inequality. However, the Gaussian function lacks either biorthogonality or orthogonality, which is the constraint of window functions for perfect signal reconstruction. Perfect reconstruction is one of the objectives of the STFT and the WT, such as nonlinear filtering, signal compression, and synthesis.

10.2 FOURIER DOMAIN PROCESSING (FDP)

We start from the basic definitions of the semicontinuous STFT and the semicontinuous WT for which general window functions are given [8]. There are two types of STFT and WT, namely, complex-valued and real-valued types. To discuss the WT, we begin with the semicontinous STFT, defined as

$$\mathrm{STFT}(\tau,n) = \frac{1}{\sqrt{2\pi}}\int x(t)h*(t-\tau)\exp(-in\omega_0 t)\,dt \quad (10.4)$$

where n is an integer, $n\omega_0$ is the selected analyzing frequency, and τ is an arbitrary time shift. In Fourier domain representation, Eq. (10.4) can be written as

$$\mathrm{STFT}(\tau,n) = \frac{1}{\sqrt{2\pi}}\exp(-in\omega_0\tau)$$
$$\times \int X(\omega)H*(\omega-n\omega_0)\exp(i\omega\tau)\,d\omega \quad (10.5)$$

where $X(\omega)$ and $H(\omega)$ are the Fourier transforms of $x(t)$ and $h(t)$, respectively, and the super arterial asterisk denotes the complex conjugate. Since $h(t)(-in\omega_0 t)$ can be regarded as a band-pass filter centered at $n\omega_0$, $\text{STFT}(\tau, n)$ is actually a multifilter system in which the bandwidth of the filters are invariant in the entire frequency domain. Although any smooth function with finite energy can be used as an analyzing window, windows having lower uncertainty are preferred. On the other hand, in nonlinear filtering, the perfect reconstruction of a signal is the main issue, such that the window function $h(t)$ is subjected to some physical constraints.

The inverse operation of the STFT can be written as

$$x(t) = \frac{1}{\sqrt{2\pi}} \sum_{n=-\infty}^{+\infty} \int \text{STFT}(\tau,n) s(t-\tau) \exp(in\omega_0 t)\, d\tau \quad (10.6)$$

which can also be expressed in terms of frequency domain, such as

$$x(t) = \frac{1}{\sqrt{2\pi}} \sum_{n} \int F\{\text{STFT}(\tau,n)\}_\tau S(\omega - n\omega_0)\exp(i\omega t)\, d\omega$$

$$(10.7)$$

where $s(t)$ is a given synthesis window function, its Fourier transform is given as $S(\omega)$, and $F\{\}_\tau$ represents the Fourier transform with respect to τ. For a perfect reconstruction, the relationship between $h(t)$ and $s(t)$ can be formed by the following biorthogonality constraint [3,4]:

$$\sum_{n=-\infty}^{+\infty} H^*(\omega - n\omega_0) S(\omega - n\omega_0) = C \quad (10.8)$$

where C is a constant that need not be unity.

It is well known that $s(t)$ cannot be uniquely determined and there are infinite choices for it that satisfy the constraint of Eq. (10.8). A logical choice for a symmetrical forward and inverse operation as described in Eqs. (10.4) and (10.6) is that $s(t) = h(t)$, for which $h(t)$ is called an orthogonal function. For the other choices of $s(t) \neq h(t)$, $h(t)$ and $s(t)$ are called *biorthogonal functions*. There is, however, an attractive biorthogonal pair of $h(t)$ and $\delta(t)$ (i.e., $s(t) = \delta(t)$) that is very suitable for the optical implementation of the STFT since it results in a simple inverse operation, where $\delta(t)$ is a Dirac delta function.

We notice that by choosing $s(t) = \delta(t)$ or $s(t) = h(t)$, Eq. (10.8) can be written in terms of the biorthogonality and orthogonality constraints as follows:

$$\sum_{n=-\infty}^{+\infty} H^*(\omega - n\omega_0) = C \quad \text{where } s(t) = \delta(t) \quad (10.9a)$$

$$\sum_{n=-\infty}^{+\infty} |H(\omega - n\omega_0)|^2 = C \quad \text{where } s(t) = h(t) \quad (10.9b)$$

Although the Gaussian function has been shown to be one of the best window functions maximally concentrated in time and frequency, it cannot perfectly reconstruct a signal as referred to the constraints of Eq. (10.9); that is, the Gaussian function lacks the biorthogonality. A simple window $h(t)$ that satisfies the constraints of Eq. (10.9) is a sinc function for which its Fourier transform is given by

$$H(\omega) = \text{rect}\left(\frac{\omega}{\omega_0}\right) = \begin{cases} 1 & |\omega| \leq \omega_0 \\ 0 & |\omega| > \omega_0 \end{cases}$$

Nevertheless, the high-order sidelobes of a sinc window function vanish rather slowly as a function of t. By referring to the constraints of Eq. (10.9a), a general form of $H(\omega)$ can be constructed as given by

$$H(\omega) = \begin{cases} \sin^2\left[\frac{\pi}{2}v\left(\frac{\omega_1 + \omega}{\omega_1 - \omega_2}\right)\right] & -\omega_2 \leq \omega \leq -\omega_1 \\ 1 & -\omega_1 \leq \omega \leq -\omega_1 \\ \cos^2\left[\frac{\pi}{2}v\left(\frac{\omega - \omega_1}{\omega_2 - \omega_1}\right)\right] & \omega_1 \leq \omega \leq \omega_1 \\ 0 & |\omega| > \omega_2 \end{cases}$$

$$(10.10)$$

which forms a biorthogonal window function in the time domain, where v is a function that satisfies $v(x) + v(1 - x) = 1$, $0 \leq x \leq 1$. An example of $v(x)$, as shown in Fig. 10.1, is given as

$$v_n(x) = \sin^2\left[\frac{\pi}{2}v_{n-1}(x)\right] \quad (10.11)$$

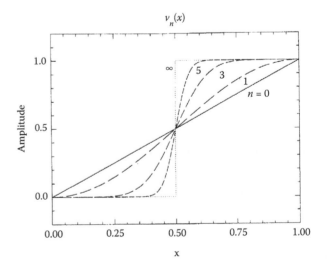

FIG. 10.1 v function as n increases.

where $v_0(x) = x$, and $n > 1$ is an integer. We see that as $n \to \infty$, $v_n(x)$ approaches a step function that will produce a rectangular $H(\omega)$. We should mention that a square root version of Eq. (10.10) satisfies the constraint of Eq. (10.9b), which in turn produces an orthogonal window function.

Referring to the variations of $v(x)$ in Fig. 10.1, the Fourier domain representation of a set of typical window functions $h(t)$ is shown in Fig. 10.2a, as given by

$$H_1(\omega) = \text{rect}(2\omega)$$

$$H_2(\omega) = \begin{cases} \cos^2\left(\dfrac{\pi}{2}\omega\right) & |\omega| \le 1 \\ \\ 0 & \text{others} \end{cases}$$

$$H_3(\omega) = \begin{cases} \cos^2\left(\dfrac{\pi}{2}\omega\right) & |\omega| \le 1 \\ \\ 0 & \text{others} \end{cases}$$

$$H_4(\omega) = \exp(-\pi\omega^2)$$

The corresponding time domain windows are

$$h_1(t) = \frac{\sin(t/2)}{t/2}$$

$$h_2(t) = \frac{\pi \cos(t)}{(\pi/2)^2 - t^2}$$

$$h_3(t) = \frac{\pi^2 \sin(t)}{t(\pi^2 - t^2)}$$

$$h_4(t) = \exp\left(-\frac{t^2}{4\pi}\right)$$

In view of the frequency–time transformation relationship shown in Fig. 10.2, we see that the window $h_3(t)$ exhibits a close resemblance to the Gaussian window $h_4(t)$, in addition to the Fourier domain (i.e., $H_3(\omega) \sim H_4(\omega)$). One of the merits of using $h_3(t)$ is that it is a biorthogonal function, which can simplify the inverse operation since the signal synthesis is just the summation of all the STFT signals from each analyzing filter band, as can be seen from Eq. (10.9a). Notice that the sinc window $h_1(t)$ shows the disadvantage of slow vanishing sidelobes, but it is a practical filter of rectangular form in the Fourier domain. One of the interesting properties of the $h_1(t)$ is having both orthogonality and biorthogonality in terms of Eqs. (10.9a) and (10.9b). The window $h_2(t)$ is orthogonal in terms of the constraint of Eq. (10.9b) and has comparable joint time–frequency resolution with the window $h_3(t)$.

A quantitative comparison among these window functions in terms of joint time–frequency resolution cell is tabulated in Table 10.1. We can see that the cell (i.e., $\Delta t\,\Delta\omega$) of the biorthogonal window $h_3(t)$ almost meets the lower bound of the inequity equation (10.3). The resolution cell of the orthogonal window $h_2(t)$ is also close to the lower bound of the uncertainty

(a)

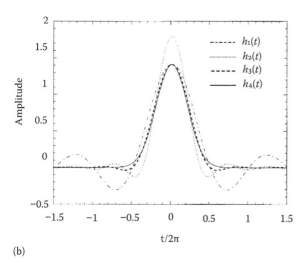

(b)

FIG. 10.2 Window functions. (a) Fourier domain representations: $H_1(\omega)$, $H_2(\omega)$, $H_3(\omega)$, and $H_4(\omega)$. (b) Windows: $h_1(t)$, $h_2(t)$, $h_3(t)$, and $h_4(t)$.

TABLE 10.1

Joint Time–Frequency Resolution Cell $\Delta t\,\Delta\omega$ of the Window Functions

Resolution Cell	Windows			
	$h_1(t)$	$h_2(t)$	$h_3(t)$	$h_4(t)$
$\Delta\omega$	1.0	1.27	1.0	1.0
Δt	20.28	7.40	6.54	2π
$\Delta t\,\Delta\omega$	20.28	9.43	6.54	6.28

principle, but almost one and half times larger than that of $h_3(t)$. On the other hand, because of its ideal cutoff frequency, the window $h_1(t)$ has the largest cell, which could be regarded as the upper bound of the joint time–frequency resolution cell for nontrivial smooth window functions.

One of the objectives is to design biorthogonal windows that would provide a smaller cell of joint time–frequency resolution. In view of Fig. 10.2, the biorthogonal window $h_3(t)$ happens to be an excellent approximation to the Gaussian window and almost meets the lower bound of the uncertainty inequity. Furthermore, an accompanying advantage of the window $h_3(t)$ is the simplification of the inverse STFT.

10.3 WAVELET TRANSFORM

A semicontinuous WT can be written as

$$WT(\tau,n) = a_0^n \int x(t)\psi * \left[a_0^n(t-\tau)\right]dt \qquad (10.12)$$

and its Fourier domain representation is

$$WT(\tau,n) = \frac{1}{2\pi}\int X(\omega)\psi * [a_0^{-n}\omega]\exp(i\omega\tau)d\omega \qquad (10.13)$$

where $a_0 > 1$ is a scaling factor (i.e., $a = a_0^{-n}$), n is an integer, $\psi(t)$ is the mother wavelet, and $\Psi(\omega)$ is its Fourier transform. Equation (10.12) is somewhat different from Eq. (10.2), where the normalization factor $1/\sqrt{a}$ (i.e., $a_0^{n/2}$) is used instead of a_0^n (i.e., $1/a$). We notice that the modification of Eq. (10.12) would simplify the optical implementation of the WT in the FDP. As has been pointed by Telfer and Szu [9], the frequency bias problem can in fact be alleviated by using the normalizing scaling factor $1/a$. Similar to the STFT, the WT can also be regarded as a multifilter system in the Fourier domain by which a signal can be decomposed into different spectral bands. Although the WT uses narrower and wider band filters to analyze the lower- and higher-frequency components, respectively, the operation is essentially similar to that of the STFT.

To meet the admissibility condition of the WT, $\psi(t)$ has to be a band-pass filter; however, for signals having rich low-frequency components, a scaling function $\phi(t)$ to preserve the low-frequency contents is needed, for which the scaling transform of the signal is given by

$$ST(\tau) = \int x(t)\phi * (t-\tau)dt \qquad (10.14)$$

Thus, the inverse operation of the WT can be written as

$$x(t) = \int ST(\tau)s_\phi(t-\tau)d\tau$$

$$+ \sum_{n=-\infty}^{\infty}\int WT(\tau,n)s_\psi\left[a_0^n(t-\tau)\right]a_0^n\,d\tau \qquad (10.15)$$

and its Fourier domain representation is given by

$$x(t) = \frac{1}{2\pi}\int F\{ST(\tau)\}S_\phi(\omega)\exp(i\omega t)d\omega$$

$$+ \frac{1}{2\pi}\sum_{n=-\infty}^{\infty}\int F\{WT(\tau,n)\}_\tau S_\psi(a_0^{-n}\omega)\exp(i\omega t)d\omega \qquad (10.16)$$

where $s_\phi(t)$ and $s_\psi(t)$ are the synthesis scaling function and the wavelet function, respectively, and $S_\phi(\omega)$ and $S_\psi(t)$ are the corresponding Fourier transforms. If the WT is used for signal synthesis, then for a perfect reconstruction, $\phi(t)$ and $\psi(t)$ must satisfy the following biorthgonality and orthogonality constraints:

$$\Phi*(\omega) + \sum_{n=-\infty}^{+\infty}\Psi*\left(a_0^{-n}\omega\right) = C \text{ for } s_\phi(t) = \delta(t) \text{ and } s_\psi(t) = \delta(t) \qquad (10.17a)$$

$$\left|\Phi(\omega)\right|^2 + \sum_{n=-\infty}^{+\infty}\left|\Psi(a_0^{-n}\omega)\right|^2 = C \text{ for } s_\phi(t) = \phi(t) \text{ and } s_\psi(t) = \psi(t) \qquad (10.17b)$$

Similar to Eq. (10.10) for the STFT, a general form of $\Psi(\omega)$ that satisfies the biorthogonality constraints of Eq. (10.17a) is given by

$$\Psi(\omega) = \begin{cases} 0 & \omega < \omega_0 \\ \sin^2\left[\dfrac{\pi}{2}v\left(\dfrac{\omega-\omega_0}{\omega_0(a_0-1)}\right)\right] & \omega_0 \leq \omega \leq a_0\omega_0 \\ \cos^2\left[\dfrac{\pi}{2}v\left(\dfrac{\omega-a_0\omega_0}{\omega_0 a_0(a_0-1)}\right)\right] & a_0\omega_0 \leq \omega \leq a_0^2\omega_0 \\ 0 & \omega > a_0^2\omega_0 \end{cases} \qquad (10.18)$$

for which the scaling function $\Phi(\omega)$ can be shown as

$$\Phi(\omega) = \cos^2\left[\dfrac{\phi}{2}v\left(\dfrac{|\omega|}{(a_0+1)\omega_0/2}\right)\right] \quad |\omega| \leq (a_0+1)\omega_0/2 \qquad (10.19)$$

where the function $v(x)$ has the same definition for STFT as shown in Fig. 10.1. Thus, we see that $\Psi(\omega)$ forms a biorthogonal wavelet, and its square root version $\sqrt{\Psi(\omega)}$ is known as *Mayer's wavelet* [10], which is orthogonal in terms of the constant of Eq. (10.17b).

Figure 10.3a shows the Fourier domain representations of a set of wavelets similar to the windows shown in Fig. 10.2, as given by

$$\Psi_1(\omega) = \begin{cases} 1 & 1.5 \leq \omega \leq 3 \\ 0 & \text{others} \end{cases}$$

$$\Psi_2(\omega) = \begin{cases} \sin[(\omega-1)\pi/2] & 1 \leq \omega \leq 2 \\ \cos[(\omega-2)\pi/4] & 2 \leq \omega \leq 4 \\ 0 & \text{others} \end{cases}$$

$$\Psi_3(\omega) = \begin{cases} \sin^2[(\omega-1)\pi/2] & 1 \leq \omega \leq 2 \\ \cos^2[(\omega-2)\pi/4] & 2 \leq \omega \leq 4 \\ 0 & \text{others} \end{cases}$$

$$\psi_4(\omega) = \begin{cases} \exp\left[-\pi(\omega-2)^2\right] & \omega \leq 2 \\ \exp\left[-\pi(\omega-2)^2/4\right] & \omega \geq 2 \end{cases}$$

where the scaling factor a_0 is assumed to be 2. The real parts of the complex-valued $\psi_1(t)$, $\psi_2(t)$, $\psi_3(t)$, and $\psi_4(t)$ are plotted in Fig. 10.3b–d, respectively, in which we see that the biorthogonal window $\psi_3(t)$ is an excellent approximation to the Gaussian function $\psi_4(t)$ in both the Fourier and time domains. Therefore, the wavelet $\psi_3(t)$ has the advantage of having a

(a)

(c)

(b)

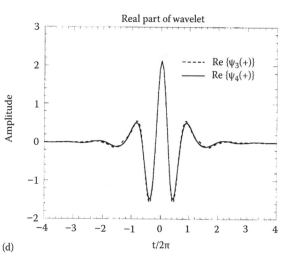

(d)

FIG. 10.3 (a) Fourier domain representations, $\Psi_1(\omega)$, $\Psi_2(\omega)$, $\Psi_3(\omega)$, and $\Psi_4(\omega)$, of the complex-valued wavelets. (b) $\psi_1(t)$. (c) $\psi_2(t)$, (d) $\psi_3(t)$, and $\psi_4(t)$.

smaller joint time–frequency resolution cell and biorthogonality, which can simplify the inverse operation of the WT. Also, $\psi_2(t)$ can be used as an orthogonal wavelet that has good joint time–frequency resolution. However, wavelet $\psi_1(t)$ is often used in practice since its Fourier transform is a rectangular form, which is rather convenient for FDP in making the masks of spatial filter banks.

Although the preceding discussions were limited to wavelets that have forms similar to those of the window functions, as discussed for the STFT, in practice, the WT has more solutions than the STFT. Namely, one can select particular wavelets for specific applications.

10.4 OPTICAL IMPLEMENTATIONS

Figure 10.4a shows an optical setup for the implementation of the STFT and WT for one-dimensional signal processing [11], in which we assume that the synthesis function $s(t) = \delta(t)$. Thus, one should use biorthogonal windows and wavelets for the STFT and WT, respectively, for perfect reconstruction. First, an input signal is displayed on a spatial light modulator (SLM), which is placed at the front focal plane of a one-dimensional Fourier transform lens CL1; then a set of filter banks are placed at the Fourier plane P2. Thus, the STFT (or WT) signals can be obtained in the back focal plane of the CL3 by capturing the spectrogram, that is, intensity of the STFT with CCDl. The reconstructed signal can be obtained at the output plane P4 by summing all the STFT (or WT)

signals diffracted from the filter banks. We notice that real-time signal filtering can be realized by simply inserting a filter at plane P3. Let us assume that the desired filter is $F(\tau, n)$; then the reconstructed signal can be shown as

$$x'(t) = \begin{cases} \sum_n \text{STFT}(\tau,n)F(\tau,n) & \text{for STFT} \\ \text{ST}(\tau) + \sum_n \text{WT}(\tau,n)F(\tau,n) & \text{for WT} \end{cases} \quad (10.20)$$

Figure 10.4b shows an optical setup in which we assume that the orthogonal windows (i.e., $s(t) = h(t)$) (or orthogonal wavelets, i.e., $s(t) = \psi(t)$) are used. Notice that the optical configuration of Fig. 10.4b is very similar to that of Fig. 10.4a, except for the inverse operation. By virtue of the reciprocity principle, the inverse operation can be accomplished by placing a phase conjugate mirror (PCM) behind plane P3. The major advantage of the PCM is the self-alignment; therefore, the filter alignment can be avoided. As the return phase conjugate signal STFT* (τ, n) (or WT* (τ, n)) is arrived at in plane P2, it will be subjected to Fourier transformation. By inserting a desired filter at plane P3, real-time signal filtering can be obtained. We note that the filter at plane P3 is proportional to $[F(\tau, n)]^{1/2}$, since the signal has to go through the filter twice. Thus, the reconstructed signal at plane P4 will be

$$x_t^*(t) = \begin{cases} \sum_n \int F\{\text{STFT}*(\tau,n)F(\tau,n)\}_\tau \, H*(\omega - n\omega_0)\exp(i\omega t)\,d\omega, \text{for STFT} \\ \int F\{\text{ST}*(\tau) + F(\tau)\}\Phi*(\omega)\exp(i\omega t)\,d\omega \\ \quad + \sum_n \int F\{\text{WT}*(\tau,n)F(\tau,n)\}_\tau \Psi*(a_0^n\omega)\exp(i\omega t)\,d\omega, \text{for WT} \end{cases}$$

which can be observed with a charge-coupled device (CCD) camera.

Although our preceding discussions are limited to one-dimensional signal processing, the basic concepts can be extended to two-dimensional image processing. This goal can be accomplished by either changing the alignment of filter banks or using photorefractive crystal filters and photorefractive crystal PCMs.

10.5 SIMULATIONS

Simulations of Fourier domain filter banks for complex-valued STFT and WT are shown in Fig. 10.5a and b, respectively. The filters centered at different frequencies are arranged from top to bottom. We see that the bandwidths of the filters are the same for the STFT, but vary in terms of the center frequencies for the WT (known as constant Q-filters). We have omitted the filters for negative frequency components since the spectrogram and scalegram from the negative frequency components are the same as those from the positive components. A test

(a)

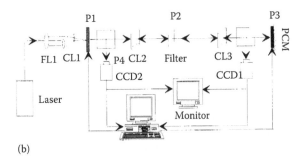

(b)

FIG. 10.4 Optical architectures for implementations of the STFT and WT: (a) for biorthogonal windows or wavelets (i.e., $s(t) = \delta(t)$) and (b) for orthogonal windows or wavelets (i.e., $s(t) = h(t)$ or $\psi(t)$). FL = Fourier Transform Lens; CL, Cylindrical Lens.

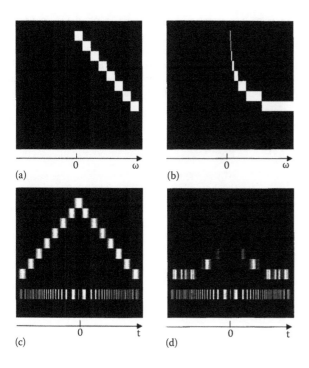

FIG. 10.5 Computer simulations of the complex-valued STFT and the complex-valued WT of a chirp with a transient signal at the center: (a) filter banks for the complex-valued STFT, (b) filter banks for the complex-valued WT, (c) spectrogram of the test signal shown at the bottom, and (d) scalegram.

FIG. 10.6 Computer simulations of the real-valued STFT and the real-valued WT of a chirp with a transient signal at the center: (a) filter banks for the real-valued STFT, (b) filter banks for the real-valued WT, (c) spectrogram of the test signal shown at the bottom, and (d) scalegram.

signal that includes a chirp and a transient signal are shown at the bottom of Fig. 10.5c and d, as given by

$$x(t) = \sin\left[\frac{\pi}{256}(t-127)^2\right] + \cos\left[\frac{3\pi}{4}(t-127)\right]\exp\left[-\frac{|(t-127)|}{2}\right],$$

$$0 \le t < 256$$

The corresponding spectrogram and the scalegram are shown at the top portion of Fig. 10.5c and d, in which we see that the STFT gives a relatively good time–frequency resolution for the chirp signal and a weaker response for the transient signal (located at the center). On the other hand, the WT provides a higher time resolution for the transient signal, but relatively poorer performance for the uniformly distributed chirp signal.

Real-valued STFT and WT are also simulated, and their filter banks are shown in Fig. 10.6a and b, respectively. Figure 10.6c and d shows the spectrogram and scalegram using the same test signal as in the complex-valued case. In view of these results, we see that the real-valued STFT and the real-valued WT provide the same time resolution as those of the complex-valued STFT and the complex-valued WT, respectively. However, the spectrogram and the scalegram are modulated by interference fringes, which would sometimes cause the loss of some details and difficulty in interpretation.

10.6 REMARKS

We have discussed the semicontinuous STFT, the semicontinuous WT, and their optical implementation. We have shown that one of the best substitutions for the Gaussian function in the Fourier domain is a squared sinusoid function that can form a biorthogonal window function in the time domain. A couple of optical architectures for the STFT and WT are illustrated. Although our discussions are confined to one-dimensional signal processing, the basic concepts can be easily extended to multidimensional signal processing.

REFERENCES

1. D. Gabor, Theory of Communication, *J. Inst. Electr. Eng.*, vol. 93, 429–457, 1946.
2. L. Cohen, Time-Frequency Distribution—A Review, *Proc IEEE*, vol. 77, 941–981, 1989.
3. F. Hlawatsch and G. F. Boudreaux-Bartels, Linear and Quadratic Time-Frequency Signal Representations, *IEEE Singal Proc. Mag.*, vol. 9, 21–67, 1992.
4. A. N. Akansu and R. A. Haddad, *Multiresolution Signal Decomposition*, Academic Press, New York, 1992, p. 291.
5. R. A. Haddad, A. N. Akansu, and A. Benyassine, Time-Frequency Localization in Transforms, Subbands, and Wavelets: A Critical Review, *Opt. Eng.*, vol. 32, 1411–1429, 1993.
6. O. Rioul and M. Vetterli, Wavelets and Signal Processing, *IEEE Signal Proc. Mag.*, vol. 8, 14–38, 1991.

7. C. H. Chui, *Wavelets: A Tutorial in Theory and Applications*, Academic Press, New York, 1992, p. 3.

8. I. Daubechies, The Wavelet Transform, Time-Frequency Localization and Signal Analysis, *IEEE Trans. Inf. Theory*, vol. 36, 961–1005, 1990.

9. B. Telfer and H. Szu, New Wavelet Transform Normalization to Remove Frequency Bias, *Opt. Eng.*, vol. 31, 1830–1834, 1992.

10. M. B. Ruskai, G. Beylkin, R. Coifman, I. Daubchies, S. Mallat, Y. Meyer, and L. Raphael, *Wavelets and Their Applications*, Jones and Bartelett Publishers, Boston, 1992, p. 241.

11. F. T. S. Yu and G. Lu, Short-Time Fourier Transform (STFT) and Wavelet Transform (WT) Using Fourier Domain Processing, *Appl. Opt.*, vol. 33, 5262–5270, 1994.

11 Pattern Recognition with Optics

The roots of optical pattern recognition can be traced back to Abbe's work in 1873 [1], which led to the discovery of spatial filtering. However, optical pattern recognition was not appreciated until the complex spatial filtering of Vander Lugt in 1964 [2]. Since then, techniques, architectures, and algorithms have been developed to construct efficient optical correlators for pattern recognition. The objective of this chapter is to discuss the optical architecture and techniques as applied to pattern recognition. Basically, there are two approaches in optical implementation, namely, the correlation approach and the neural net approach. In the correlation approach, there are two frequently used architectures, the Vander Lugt correlator (VLC) and the joint transform correlator (JTC).

11.1 OPTICAL CORRELATORS

Optical pattern recognition can be accomplished by either Fourier domain or spatial domain filtering. Correlators that use Fourier domain filtering are known as VLCs, and spatial domain filtering is frequently used (JTC). The basic distinction between them is that VLC depends on a Fourier domain filter, whereas JTC depends on a spatial domain filter. In other words, the complex spatial detection of Vander Lugt is input scene *independent*, whereas the joint transform is input scene *dependent* [3].

It is apparent that a pure optical correlator has drawbacks that make certain tasks difficult or impossible to perform. The first problem is that optical systems are difficult to program, in the sense of a general-purpose digital computer. A second problem is that accuracy is difficult to achieve in Fourier optics. A third problem is that optical systems cannot easily be used to make decisions. Even simple decision making cannot be performed optically without the intervention of electronics. However, many optical deficiencies happen to be strong points in the electronic counterpart. For example, accuracy, controllability, and programmability are obvious traits of digital computers. Thus, combining optical systems with electronic counterparts often produces better processors, as shown in Figs. 11.1 and 11.2, in which spatial light modulators (SLMs) are used for object and spatial filter devices. One of the important aspects of these hybrid optical architectures is that decision making can be done by the computer.

To illustrate the adaptivity of the preceding JTC hybrid optical correlator, we provide an interesting application to perform autonomous target tracking. The idea is to correlate the target in the current frame with the target in the previous frame. This makes the hybrid JTC adaptive by constantly updating the reference target with the dynamic input scene. For example, two sequential scenes of a moving object are displayed on the SLM of Fig. 11.3, with the previous and current frames positioned in the upper and lower halves of the SLM, as given by

$$f_{t-1}\left(x - x_{t-1}, y - y_{t-1} - \frac{\alpha}{2}\right) \text{ and } f_t\left(x - x_{t-1} - \delta x, y - y_{t-1} - \delta_y + \frac{\alpha}{2}\right)$$

(11.1)

where $2a$ is the height of the display unit, t and $t-1$ represent the current and previous time frames, and $(\delta x, \delta y)$ is the relative translation of the object. Then the complex light field at the Fourier domain can be shown to be

$$T(u,v) = F_{t-1}(u,v)\exp\left\{-i2\pi\left[ux_{t-1} + v\left(y_{t-1} + \frac{\alpha}{2}\right)\right]\right\}$$
$$+ F_1(u,v)\exp\left\{-i2\pi\left[u(x_{t-1} + \delta x) + v\left(y_{t-1} + \delta y - \frac{\alpha}{2}\right)\right]\right\}$$

(11.2)

If the detected $|T(u,v)|^2$ is sent back to the SLM, the output light distribution can be written as

$$C(x,y) = F^{-1}\{|T(u,v)|^2\} = R_{t,t}(x,y) + R_{t-1,t-1}(x,y)$$
$$+ R_{t,t-1}(x - \delta x, y + \delta y - \alpha) + R_{t-1,t}(x - \delta x, y - \delta y + \alpha)$$

(11.3)

where

$$R_{m,n}(x,y) = F^{-1}\left\{F_m(u,v)F_n^*(u,v)\right\} = \int\int f_m(u,v)f_n^*(n-x,v-y)\,du\,dv$$

represents the correlations of f_m and f_n, which are diffracted at $x_1 = \delta x$, $y_1 = \delta y - a$ and $x_2 = \delta x$, $y_2 = -\delta y + \alpha$, respectively. If the angular and scale tolerances of the JTC are approximately ±5° and ±10% [4], and the motion of the target is relatively slow compared with the processing cycle of the correlator, then f_{t-1} correlates strongly with f_t. Two high-intensity correlation peaks are diffracted into the output plane, to locations given by

$$x_t = x_{t-1} + x_1 \quad y_t = y_{t-1} + y_1 + \alpha \tag{11.4}$$

We note that a simple C language program can then be used to evaluate the target position, for which a block diagram of the system configuration is depicted in Fig. 11.4, for reference.

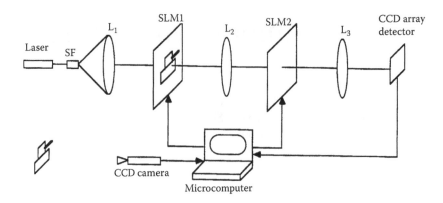

FIG. 11.1 Hybrid optical VLC. SF, spatial filter; L, lens.

FIG. 11.2 Hybrid optical JTC. BS, beam splitter; M, mirror; L, lens.

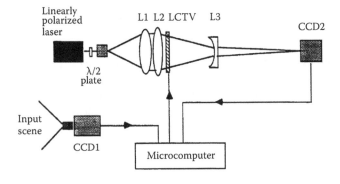

FIG. 11.3 Single SLM JTC. L, lens.

A major benefit of this technique is its adaptivity, for example, a situation where a camera mounted on a moving space vehicle is focused on a fixed target on the ground for automatic navigation. As the space vehicle approaches the target, the detected scene changes continuously: the target size appears larger, and its orientation and shape change due to the

FIG. 11.4 Optical–digital interface diagram.

motion of the vehicle. Using computer-aided design graphics, a three-dimensional treelike model was created as a simulated target on the ground. Nine image sequences simulating the changing scene viewed by a moving space vehicle are shown in part in Fig. 11.5a. The JTC tracking system has little difficulty in correlating targets from different frames, even though

FIG. 11.5 (a) Sequence of nine images recorded simulating the exact scenes as captured by a camera mounted on a moving space vehicle, with images 1, 5, and 9 shown here. (b) Tracked positions of the ground target as seen from the vehicle's coordinate frame.

the targets in the first and last frames look different. Figure 11.5b shows the tracked locations of the target as seen from the vehicle.

11.2 OPTICAL DISK–BASED CORRELATOR

Recent improvements in real-time addressable SLMs have made pattern recognition with optics much more realistic. A single-input image, however, might require correlation with a huge library of reference images. The lack of large-capacity information storage devices, which provide high-speed readout, is still a major obstacle to practical optical correlators. Optical disks (ODs), developed in recent years as mass storage media, are excellent candidates for this task. An OD-based JTC is shown in Fig. 11.6, in which a target captured by a video camera is displayed on SLM1. A beam expansion and reduction system is then used to reduce the size of the input image m times. The reference image stored on the OD is read out in parallel and magnified m times by another set of expansion lenses so that the reference image and input target are the same size. The joint transform is done by transform lens FL1, and the joint transform power spectrum (JTPS) is recorded on the write side of optically addressed SLM2. After FL2 Fourier transforms the JTPS, the resulting correlation is recorded by a high-speed charge-coupled device (CCD) camera located at the output plane. The OD advances to the next reference image, and so on, until a match is found.

A transparent, write-once OD is used in this architecture [5]. To record a library of images, a laser beam is focused to an approximately 0.5 μm spot on the photosensitive layer of the OD to record each pixel of the image sequentially. This method results in the reference images being recorded in binary form. Images with grayscale can be recorded by area modulation of the spots on the OD and read out using the differential interferometric technique.

For example, if binary images of 200 × 200 pixels are recorded, each image will occupy a 0.2 × 0.2 mm² block. If 0.01 mm spacing is allowed between adjacent blocks, then more than 27,000 reference images can be recorded on a 120 mm diameter OD. Because the block size is in the order of a fraction of a millimeter, the minute phase variation caused by a nonuniform thickness in plastic cover of the OD can be neglected. To estimate the sequential access time, we assume that the average revolution speed is 1122 rev/min, the average radius is 40 mm, and all the images are scanned sequentially on each consecutive band of the OD. Then the access time is approximately 40 μsec per block. Assuming an optically addressed ferroelectric liquid crystal, SLM is used as the square-law detector, the response time of which is between 10 and 155 μsec. Completing one correlation process should take 40–155 μsec. This is equivalent to performing more than 6400 correlations per second, a number that can hardly be matched using current electronic technology.

11.3 PHOTOREFRACTIVE-BASED CORRELATOR

Although SLMs can be used to display complex spatial filters, current state-of-the-art SLMs are low-resolution and low-capacity devices. On the other hand, photorefractive (PR) materials offer real-time recording, high resolution, and massive storage capacity. However, thick PR material has a limited shift-invariant property due to Bragg diffraction [6]. Although a thinner crystal can be used, the diffraction efficiency and storage capacity will be substantially reduced. Nevertheless, high-storage capacity, high-diffraction efficiency, and large shift invariance can be achieved by using a reflection-type wavelength-multiplexed PR matched filter, as shown in Fig. 11.7. The matched filter is recorded by combining the Fourier spectrum of an object beam $q_1(x_1)$ with a reference plane wave from the opposite direction. Correlation can be done by inserting an input object $q_2(x_2)$ at plane x, by

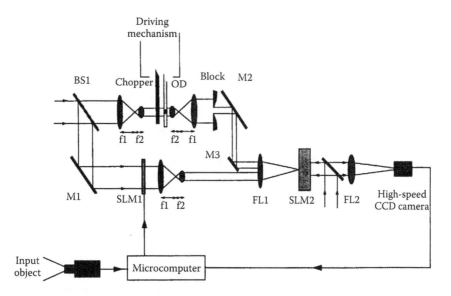

FIG. 11.6 OD-based JTC. BS, beam splitter; FL, Fourier transform; M, mirror.

FIG. 11.7 Wavelength-multiplexed reflection-type matched filter correlator. BS, beam splitter; P, polarizer.

which the output correlation can be observed at plane x_3. To separate the reading beam $q_2(x_2)$ from the writing beam, the reading beam can be made orthogonally polarized to the writing beam by a polarized beam splitter. By using the coupled mode wave theory, the correlation peak intensity as a function of shift variable S can be shown to be

$$R(S) = \left| \int q_1^*(x_1) q_1(x_1) \mathrm{Sinc}\left[-\frac{\pi D}{n\lambda} \frac{S(x_1 - S)}{f^2} \right] dx_1 \right|^2 \qquad (11.5)$$

where D is the thickness of the crystal. The shift tolerance figure of merit (FOM) is

$$(\mathrm{FOM})_{RC} = X S_{\max} = \frac{4n\lambda}{D} f^2 - S_{\max}^2 \qquad (11.6)$$

where $S_{\max}$ denotes the maximum allowable shift. Plots of FOMs, along with the VLC and JTC, are shown in Fig. 11.8, in which we see that the reflection-type PR correlator performs better. It has a higher shift tolerance, approximately one order higher than the VLC. The wavelength-multiplexed filter also offers higher and more uniform wavelength selectivity than the angular-multiplexed technique [7].

FIG. 11.8 FOMs for different types of correlators.

11.4 OPTICAL NEURAL NETWORKS

Electronic computers can solve computational problems thousands of times faster and more accurately than the human brain. However, for cognitive tasks, such as pattern recognition and understanding and speaking a language, the human brain is much more efficient. In fact, these tasks are still beyond the reach of modern electronic computers. A neural network (NN) consists of a collection of processing elements, called *neurons*. Each neuron has many input signals, but only one fanned-out signal to many pathways connected to other neurons. These pathways interconnect with other neurons to form a network. The operation of a neuron is determined by a transfer function that defines the neuron's output as a function of the input signals. Every connection entering a neuron has an adaptive coefficient called a *weight* assigned to it. The weight determines the interconnection strength between neurons, and they can be changed through a learning rule that modifies the weights in response to input signals and the transfer function. The learning rule allows the response of the neuron to change with time, depending on the nature of the input signals. This means that the network *adapts* itself to the environment and organizes the information within itself, a type of *learning*.

Generally speaking, a one-layer NN of N neurons has N^2 interconnections. The transfer function of a neuron can be described by a nonlinear relationship, such as a step function, making the output of a neuron either 0 or 1 (binary), or a *sigmoid* function, which gives rise to analog values. The operation of the ith neuron in the network can be represented by a *retrieval equation*, as given by

$$u_t = f\left(\sum_{j=1}^{N} T_{ij} u_j - \theta_i \right) \qquad (11.7)$$

where u_i is the activation potential of the ith neuron, T_{ij} is the *interconnection weight matrix* (IWM) (associative memory) between the jth neuron and the ith neuron, and f is a nonlinear processing operator.

Light beams propagating in space will not interfere with each other, and optical systems generally have large-space bandwidth products. These are the primary features that prompted the optical implementation of NNs. The first optical implementation of NNs was proposed by Psaltis and Farhat [8]. Since then, a score of optical neural net architectures have been proposed [9]. A typical hybrid optical NN using a liquid crystal television (LCTV) SLM is shown in Fig. 11.9, in which the lenslet array provides the interconnection of the IWM to the input pattern. The transmitted light field after LCTV2 is collected by an imaging lens, focusing at the lenslet array and imaging onto a CCD array detector. The array of detected signals is sent to a thresholding circuit, and the final pattern can be viewed at the monitor or sent back for the next iteration. The data flow is primarily controlled by the microcomputer, for which the LCTV-based neural net is indeed an *adaptive* NN.

One of the most frequently used neural net models is the Hopfield model [10]. The model utilizes an associative

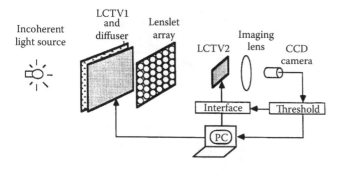

FIG. 11.9 LCTV optical NN.

memory retrieval process equivalent to an iteratively thresh-olded matrix–vector outer-product expression as given by

$$V_i \rightarrow 1 \qquad if \sum_{j=1}^{N} T_{ij}V_j \geq 0$$

$$V_i \rightarrow 0 \qquad if \sum_{j=1}^{N} T_{ij}V_j < 0 \qquad (11.8)$$

where V_i and V_j are binary output and binary input patterns, respectively, and the associative memory operation is written as

$$T_{ij} = \begin{cases} \sum_{m=1}^{n} \left(2V_i^m - 1\right)\left(2V_j^m - 1\right) & \text{for } i \neq j \\ 0 & \text{for } i = j \end{cases} \qquad (11.9)$$

where V_i^m and V_j^m are ith and jth elements of the mth binary vector, respectively.

One of the important aspects of the Hopfield model is the ability to retrieve distorted and partial patterns. For example, a set of English letters was stored in a Hopfield neural net, as shown in Fig. 11.10a. The positive and negative parts of the memory matrix are given in Fig. 11.10b and c, respectively. If a partial pattern of A is presented to the Hopfield net, a reconstructed pattern of A converges by iteration and is shown in Fig. 11.1. The construction of the Hopfield NN is through *intrapattern association*, which ignores the association among the stored exemplars. The alternative is called *interpattern association* (IPA) [11]. By using a simple logic operation, an IPA NN can be constructed. For example, suppose we have three overlapping patterns as given in the Venn diagram of Fig. 11.11, where common and special subspaces are defined. If one uses the logic operations

$$I = A \wedge \overline{(B \vee C)}$$

$$II = B \wedge \overline{(A \vee C)} \qquad V = (B \wedge C) \wedge \overline{A}$$

$$III = C \wedge \overline{(A \vee B)} \qquad VI = (C \wedge A) \wedge \overline{B} \qquad (11.10)$$

$$IV = (A \wedge B) \wedge \overline{C} \qquad VII = (A \wedge B \wedge C) \wedge \overline{\varphi}$$

(a)

(b) (c)

(d)

FIG. 11.10 Results from a Hopfield model: (a) training set; (b and c) positive and negative weight matrices, respectively; and (d) reconstructed results.

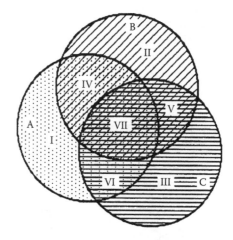

FIG. 11.11 Common and special subspaces.

then an IPA neural net can be constructed. If the interconnection weights are assigned to 1, –1, and 0, for excitory, inhibitory, and null interconnections, then a tristate IPA neural net can be constructed. For example, in Fig. 11.12a, pixel 1 is the common pixel among patterns A, B, and C; pixel 2 is the common pixel between A and B; and pixel 3 is the common pixel between A and C; whereas pixel 4 is the special pixel, which is also an exclusive pixel with respect to pixel 2. When the logic operations are applied, a tristate NN can be constructed as shown in Fig. 11.12b, with the corresponding IPA IWM shown in Fig. 11.11c.

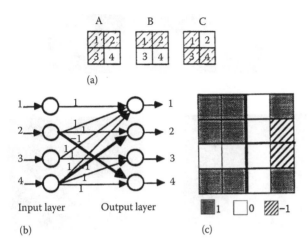

(a)

(b)

(c)

FIG. 11.12   Construction of IPA NN: (a) three reference patterns, (b) one-layer neural net, and (c) IWM.

Pattern translation can be accomplished using the *hetero-association* IPA neural net [12]. Using similar logic operations among input–output (translation) patterns, a heteroassociative IWM can be constructed. For example, an input–output (translation) training set is given in Fig. 11.13a. Using the logic operations, a heteroassociation neural net can be constructed as shown in Fig. 11.13b, where Fig. 11.13c is its IWM. To illustrate the optical implementation, an input–output training set is shown in Fig. 11.14a. The positive and negative parts of the heteroassociation IWMs are depicted in Fig. 11.14b. If a partial Arabic numeral 4 is presented to the optical neural net, a translated Chinese numeral is obtained, as shown in Fig. 11.14c. Thus, the heteroassociation neural net can indeed translate patterns.

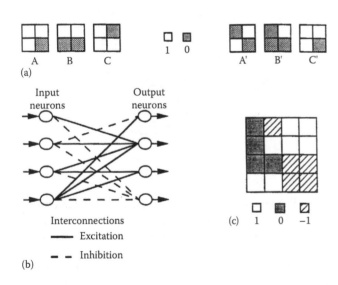

(a)

(b)

(c)

FIG. 11.13   Construction of a heteroassociation IPA NN: (a) input–output training sets, (b) heteroassociation neural net, and (c) heteroassociation IWM.

(a)

(b)

(c)

FIG. 11.14   Pattern translation: (a) Arabic-Chinese training set, (b) heteroassociation IWM (positive and negative parts), and (c) partial Arabic numeric to be translated to Chinese numeric.

11.5   COMPOSITE FILTERS

Classical spatial matched filters are sensitive to rotational and scale variances. A score of approaches to developing composite distortion-invariant filters [13] have been reported. Among them, the synthetic discriminant function (SDF) filter [14,15] has played a central role. The idea of SDF can be viewed as a linear combination of classical matched filters, as given by

$$h(x, y) = \sum_{n=1}^{N} a_n f_n(x, y) \tag{11.11}$$

where $\{f_n(x, y)\}$ represents a set of patterns with N different perspective views, and the coefficients $a_n(n = 1, 2, \ldots, N)$ can be determined by solving a set of N simultaneous equations.

The correlation peak value (CPV) between $h(x, y)$ and an arbitrary training image $f_n(x, y)$ can be shown to be

$$h(x, y) * f_n(x, y)|_{x=0, y=0} = \iint h^*(x, y) f_n(x, y) \, dx \, dy$$

$$= \iint \sum_{n=1}^{N} a_n^* f_n^*(x, y) f_m(x, y) \, dx \, dy$$

$$= \sum_{n=1}^{N} a_n^* R_{nm} = C \tag{11.12}$$

where C represents a prespecified constant, which is independent of the perspective view of $f_n(x, y)$. In other words, the CPV from an SDF filter is independent of the training images. Figure 11.15 shows a flowchart for an SDF filtering system, where the SDF filter is off-line synthesized while the correlation is on-line processed.

There are techniques proposed to improve the performance of the classical SDF filter [16]. Nevertheless, it is not the intention here to describe all those techniques. Instead, two of the simplest approaches will be mentioned: (1) higher noise tolerance can be achieved if the variance of the SDF filter is minimized, and (2) sharper correlation peaks can be obtained if the average correlation energy of the SDF filter is minimized.

Simulated annealing algorithms (SAAs) have also recently been applied to pattern recognition [17,18]. A spatial domain bipolar filter obtained by SAA [19] can be directly implemented on an input phase-modulating SLM in a JTC. To demonstrate the performance of a bipolar composite filter (BCF), the out-of-plane-oriented (T72) tanks shown in Fig. 11.16a are used as a training set. The constructed BCF using the SAA is shown in Fig. 11.16b. An input scene to the JTC is shown in Fig. 11.17c, of which the BCF in Fig. 11.16b is used, and the corresponding output correlation distribution obtained is plotted in Fig. 11.17b. We see that target T72 tanks can indeed be extracted from the noisy background.

11.6   REMARKS

The synthesis of Fourier domain matched filters using holographic techniques was first initiated by Vander Lugt in 1964 [2]. Since then, VLCs have been intensively studied, improved,

(a)

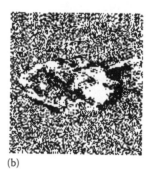

(b)

FIG. 11.16   (a) Out-of-plane rotation T72 tank images. (b) BCF to detect T72 tank.

and modified by advances in electronic-addressable SLMs and nonlinear PR devices. A couple of years ago, a different kind of correlator, called the JTC, was demonstrated by Weaver and Goodman [20] and independently by Rau [21]. However, almost no application of the JTC was reported prior to 1984, despite its inclusion in several texts on optical processing. This may have been due to a lack of adequate interface devices, such as SLMs and CCD cameras. In 1984, Yu and Lu [3,22] were the first to take advantage of newly available SLMs and demonstrated a programmable JTC. This resulted in the rebirth of the JTC.

Pattern recognition can also be approached using simulated NNs. The first optical implementation of NNs was proposed by Psaltis and Farhat in 1985 [8]. Since then, neuropattern recognition has seen to a wide variety of optical implementations.

FIG. 11.15   Black box representation of SDF correlator.

(a)

(b)

FIG. 11.17   (a) Input scene. (b) Output correlation distribution.

REFERENCES

1. E. Abbe, Beiträge Zur Theorie des Mikroskops und der Mikroskopischen Wahrnehmung, *Arch. Mikrosk. Anat.*, vol. 9, 413–463 (1873).
2. A. Vander Lugt, Signal Detection by Complex Spatial Filtering, *IEEE Trans. Inf. Theory*, vol. IT-10, 139–145 (1964).
3. F. T. S. Yu and X. J. Lu, A Real-Time Programmable Joint Transform Correlator, *Opt. Commun.*, vol. 52, 10–16 (1984).
4. A. D. Gara, Real-Time Tracking of Moving Objects by Optical Correlation, *Appl. Opt.*, vol. 18, 172–174 (1979).
5. F. T. S. Yu, E. C. Tam, T. W. Lu, E. Nishihara, and T. Nishikawa, Optical Disk-Based Joint Transform Correlator, *Appl. Opt.*, vol. 30, 915–916 (1991).
6. F. T. S. Yu, S. Yin, and Z. H. Yang, Thick Volume Photorefractive Crystal Wavelength-Multiplexed Reflection-Type Matched Filter, *Optical Memory and Neural Networks*, vol. 3, 207–214 (1994).
7. S. Yin, H. Zhou, M. Wen, J. Zhang, and F. T. S. Yu, Wavelength-Multiplexed Holographic Construction Using a Ce:Fe: Doped LiNbO$_3$ Crystal with a Tunable Visible-Light Diode Laser, *Opt. Commun.*, vol. 101, 317–321 (1993).
8. D. Psaltis and N. Farhat, Optical Information Processing Based on an Associative-Memory Model of Neural Nets with Thresholding and Feedback, *Opt. Lett.*, vol. 10, 98–100 (1985).
9. F. T. S. Yu, T. Lu, X. Yang, and D. A. Gregory, Optical Neural Network with Pocket-Sized Liquid-Crystal Televisions, *Opt. Lett.*, vol. 15, 863–865 (1989).
10. J. J. Hopfield, Neurons with Graded Responses Have Collective Computational Properties Like Those of Two-State Neurons, *Proc. Natl. Acad. Sci. U.S.A.*, vol. 81, 3088–3092 (1984).
11. T. Lu, X. Xu, S. Wu, and F. T. S. Yu, A Neural Network Model Using Inter-Pattern Association (IPA), *Appl. Opt.*, vol. 29, 284–288 (1990).
12. F. T. S. Yu, T. Lu, and X. Yang, Optical Implementation of Hetero-Association Neural Network with Inter-Pattern Association Model, *Int. J. Opt. Comput.*, vol. 1, 129–140 (1990).
13. H. J. Caulfield and R. Haimes, Generalized Matched Filtering, *Appl. Opt.*, vol. 19, 181–183 (1980).
14. D. Casasent, Unified Synthetic Discriminant Function Computational Formulation, *Appl. Opt.*, vol. 23, 1620–1627 (1984).
15. B. V. K. Vijaya Kumar, Minimum Variance Synthetic Discriminant Functions, *J. Opt. Soc. Am.*, vol. A3, 1579–1584 (1986).
16. A. Mahalanobis, B. V. K. Vijaya Kumar, and D. Casasent, Minimum Average Correlation Energy Filters, *Appl. Opt.*, vol. 26, 3633–3640 (1987).
17. S. Kirpatrick, C. D. Gelatt, and M. P. Vecchi, Optimization by Simulated Annealing, *Science*, vol. 220, 671–680 (1983).
18. M. Kim, M. R. Feldman, and C. Guest, Optimum Encoding of Binary Phase-Only Filters with a Simulated Annealing Algorithm, *Opt. Lett.*, vol. 14, 545–547 (1989).
19. S. Yin, M. Lu, C. Chen, F. T. S. Yu, T. Hudson, and D. McMillen, Design of a Bipolar Composite Filter Using Simulated Annealing Algorithm, *Opt. Lett.*, vol. 20, 1409–1411 (1995).
20. C. S. Weaver and J. W. Goodman, A Technique for Optically Convolving Two Functions, *Appl. Opt.*, vol. 5, 1248–1249 (1966).
21. J. E. Rau, Detection of Differences in Real Distributions, *Opt. J. Soc. Am.*, vol. 56, 1490–1494 (1966).
22. F. T. S. Yu and S. Jutamulia, *Optical Pattern Recognition*, Cambridge University Press, Cambridge, 1998.

12 Computing with Optics

Reaffirmation of optics parallelism and the development of picosecond and femtosecond optical switches have thrust optics into the nontraditional area of digital computing. The motivation primarily comes from the need for higher-performance general-purpose computers [1]. However, a computer with parallelism requires complicated interconnections, which is difficult to achieve by using wires or microcircuitries. Since both parallelism and interconnection are the inherent trait of optics, it is reasonable to look into the development of a general-purpose optical computer. In this chapter, we restrict ourselves to a discussion of a few topics of computing with optics.

12.1 LOGIC-BASED COMPUTING

All optical detectors are sensitive to light intensity. Binary numbers 0 and 1 represent dark and bright states. Since we cannot physically generate 1's (bright) from 0's (dark), there are some difficulties that occur when a logical 1 has to be output from 0's (e.g., NOR, XNOR, and NAND). Nevertheless, the shadow-casting method [2] can solve these problems by simply initially encoding 1 and 0 in a dual-rail form.

The shadow-casting logic essentially performs all 16 Boolean logic functions based on the combination of NOT, AND, and OR. For example, 1 and 0 are encoded with four cells, as shown in Fig. 12.1a. The spatially encoded formats A and B are placed in contact, which is equivalent to an AND operation, as shown in Fig. 12.1b. This superposed format is the input-coded image of the optical logic array processor and is set at the input plane. Four spatially distributed light-emitting diodes (LEDs) are employed to illuminate the encoded input. The shadows from each LED will be cast onto an output screen, as shown in Fig. 12.2, for which a decoding mask is needed to extract the true output. The shadow casting is essentially a selective OR operation among AB, $A\bar{B}$, $\bar{A}B$ and $\bar{A}\bar{B}$. If the on–off states of the LEDs are denoted by α, β, γ, and δ (where on is 1 and off is 0), the shadow-casting output can be expressed as follows:

$$G = \alpha(AB) + \beta(A\bar{B} + \gamma(\bar{A}B) + \delta(\overline{AB}) \qquad (12.1)$$

which is the intensity at the overlapping cells. The complete combination for generating the 16 Boolean functions is tabulated in Table 12.1. A schematic diagram of a hybrid optical logic array processor is depicted in Fig. 12.3, in which the encoded inputs A and B are displayed on LCTV1 and LCTV2, respectively. However, LCTV3 (mask) is employed to determine the values of α, β, γ, and δ. The OR operation is performed electronically rather than optically. This has the advantage that no space is needed for shadow-casting NOR.

The problems with this system are (1) the encoding operation would slow down the whole process; (2) if the OR operation is performed in parallel, a large number of electronic OR gates and wire intercommunications are required; and (3) if the logic arrays are sequentially performed, a longer processing time is required.

All the shadow-casting logic processors use encoded inputs. If the coding is done by electronics, the overall processing speed will be substantially reduced. On the other hand, the optical output is also an encoded pattern that requires a decoding mask to obtain only the true output. Although the decoding process is parallel and takes no significant processing time, the decoding mask does change the format of the output. A noncoded shadow-casting logic array that is free from these difficulties is shown in Fig. 12.4.

An electronically addressed spatial light modulator (SLM), such as a liquid crystal television (LCTV), can be used to write an image format. The negation can be done by rotating the LCTV's analyzer (by 90°) without alternating the addressing electronics or software processing. The electronic signal of input A is split into four individual SLMs. Two of them display A, and the other two display $\bar{A}$. Input B follows a similar procedure to generate two B's and two $\bar{B}$'s. The products of AB, $A\bar{B}$, $\bar{A}B$ and $\bar{A}\bar{B}$ are straightforwardly obtained by putting two corresponding SLMs up to each other. Finally, beam splitters combine AB, $A\bar{B}$, $\bar{A}B$ and $\bar{A}\bar{B}$. The logic functions are controlled by the on–off state (α, β, γ, and δ) of the illuminating light (S1, S2, S3, and S4), as illustrated in Table 12.1.

The implementation of these logic functions using LCTVs is straightforward, since no electronic signal modification or software data manipulation is required to encode inputs A and B. Although it seems that more SLMs are needed (four times as many), there is no increase in the space–bandwidth product of the system. In the original shadow-casting logic array, four pixels are used to represent a binary number 1 or 0, while a binary number can be represented by only one pixel in the noncoded system. The use of four sets of individual SLMs is to utilize fully the capability of LCTV to form a negative image format by simply rotating the analyzer by 90°. This method can eliminate the bottleneck of the shadow-casting optical parallel logic array, which is introduced by the coding process.

12.2 OPTICAL INTERCONNECTS AND SHUFFLING

In addition to logic processing, a digital computer also requires interconnections between processors. It is apparent that when the computer structure becomes more complex, wire interconnections between processors are increasingly difficult to implement. The application of free-space optical interconnection is

FIG. 12.1 (a) Encoded input patterns. (b) Product of the input patterns in the shadow-casting logic array processor.

FIG. 12.3 Shadow-casting logic processor using cascaded LCTVs.

FIG. 12.2 Original shadow-casting logic array processor. The shadows of the encoded input generated by four LEDs are superimposed.

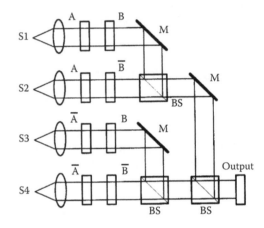

FIG. 12.4 Noncoded shadow-casting logic array processor. BS, beam splitter; M, mirror; S, light source.

TABLE 12.1
Generation of 16 Boolean Functions

	α	β	γ	δ
F0	0	0	0	0
F1	1	0	0	0
F2	0	1	0	0
F3	1	1	0	0
F4	0	0	1	0
F5	1	0	1	0
F6	0	1	1	0
F7	1	1	1	0
F8	0	0	0	1
F9	1	0	0	1
F10	0	1	0	1
F11	1	1	0	1
F12	0	0	1	1
F13	1	0	1	1
F14	0	1	1	1
F15	1	1	1	1

and among cabinets. For instance, the implementation of perfect shuffle is a typical example of optical interconnect, as will be described below.

Interconnection networks are developed to construct a concurrent processing system by interconnecting a large number of processors and memory modules. The perfect shuffle [3] is an interconnection primitive between local processors for parallel computation. The perfect shuffle is one of the most useful and universal regular interconnects. It is commonly used for sorting, interconnecting array processors, permutation networks, and special algorithms, such as the fast Fourier transform (FFT). An 8-bit perfect shuffle is shown schematically in Fig. 12.5. An optical perfect shuffle can be implemented by diffractive and refractive optics [4], which requires imaging components such as lenses, prisms, gratings, mirrors, and beam splitters, as demonstrated in Fig. 12.6.

For this particular geometry, only two and four windows are required on the first and second mask. The source (left side) and the sink (right side) arrays have the same structure with spacing a. The distance from source to sink is $7a$. It is convenient to use a geometrical analysis to determine the position and size of the windows, as well as the location of the two masks. To minimize diffraction effects and to maximize the power transferred, large windows are preferred. However,

very likely to replace complicated wire interconnections. In fact, optical interconnect is the most modest, realistic, and near-term application to electronic computers. For example, optical interconnects can be implemented by lenses, fibers, integrated optics, and holograms. The implementations can be used for clock distribution and data communication at different levels, for example, intrachips, interchips, between boards,

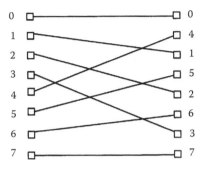

FIG. 12.5   An 8-bit perfect shuffle.

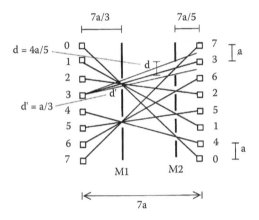

FIG. 12.6   Optical perfect shuffle based on shadow casting.

to keep the system free from cross talk, the width of the window on the second mask must be $d < 4a/5$, and that of the first mask $d' < a/3$. The method can be easily extended to three-dimensional interconnects by using a two-dimensional source and sink arrays. However, the drawback is the power loss due to the fan-out and absorption by the masks.

Another demonstration of a perfect shuffle using an imaging system is depicted in Fig. 12.7a [5]. Two sets of data are designated by $s_1(x, y)$ and $s_2(x, y)$, which are separated by a distance 2β at the input plane. With coherent illumination, the complex light distribution at the spatial frequency plane can be described as

$$E(p,q) = S_1(p,q)e^{-\beta q} + S_2(p,q)e^{i\beta q} \qquad (12.2)$$

where $S_1(p, q)$ and $S_2(p, q)$ are the Fourier spectra of the input data $s_1(x, y)$ and $s_2(x, y)$, respectively, and (p, q) are the spatial frequency coordinate system. To implement a perfect shuffle, a sinusoidal grating is inserted in the Fourier plane, as given by

$$H(q) = \frac{1}{2}\left(1 + \cos\left[(\beta + \Delta\beta)q\right]\right) \qquad (12.3)$$

where $\Delta\beta$ is a small displacement constant. Thus, the output light field distribution can be shown as

$$\begin{aligned} I(x,y) &= \frac{1}{2}s_1(x,y-\beta) + \frac{1}{2}s_2(x,y+\beta) \\ &\quad + \frac{1}{4}\left[s_1(x,y+\Delta\beta) + s_2(x,y-\Delta\beta)\right] \\ &\quad + \frac{1}{4}s_1(x,y-2\beta-\Delta\beta) + \frac{1}{4}s_2(x,y+2\beta+\Delta\beta) \end{aligned}$$

$$(12.4)$$

in which we see that $S_1(x, y)$ and $S_2(x, y)$ are shifted in opposite directions. If the incremental shift $\Delta\beta$ is equal to the separation between the processing data, S_1 and S_2, then it would be perfectly interlaced, as shown in Fig. 12.7b.

12.3   MATRIX–VECTOR MULTIPLICATION

The optical matrix–vector multiplier can be implemented as shown in Fig. 12.8. The elements of the pattern vector are entered in parallel by controlling the intensities of N LEDs. Spherical and cylindrical lenses L_1 and L_2 combine to image the LED array horizontally onto the matrix mask M, which consists of $N \times N$ elements. The combination of cylindrical

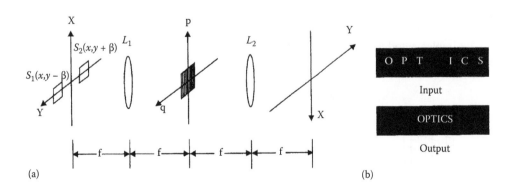

FIG. 12.7   (a) Optical perfect shuffle based on Fourier optics. (b) Input–output result. (From Q. W. Song and F. T. S. Yu, *Appl. Opt.*, 27, 1222, 1988.)

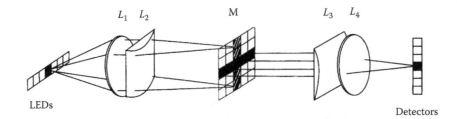

L_1 L_2 M L_3 L_4

LEDs

Detectors

FIG. 12.8   Schematic diagram of an optical matrix–vector multiplier.

and spherical lenses L_3 and L_4 collects all the light from a given row and brings it into focus on one element of the photodetectors. Since each photodetector measures the value of one element of the output vector, we see that it essentially performs the matrix–vector multiplication. This configuration can be further simplified by fabricating line-shaped LEDs and line-shaped detector arrays, as depicted in Fig. 12.9. Note that the LED array can be replaced by an SLM with uniform illumination.

Important problems can be solved with iterative multiplication of the vector by the matrix, which includes finding eigenvalues and eigenvectors, solving simultaneous linear equations, computing the discrete Fourier transform (DFT), and implementing neural networks. For example, a neural network consists of several layers of neurons, in which two successive neuron layers are connected by an interconnection net. If the neuron structure in a layer is represented by a vector, then the interconnection can be represented by a matrix. For example, the Hopfield neural network [6] uses an associative memory retrieval process, which is essentially a matrix–vector multiplier, as given by

$$V_i \rightarrow 1 \quad \text{if} \sum T_{ij}V_i \geq 0$$
$$V_i \rightarrow 0 \quad \text{if} \sum T_{ij}V_i < 0 \tag{12.5}$$

where V_i and V_j are the output and input binary vectors, respectively, and T_{ij} is the interconnection matrix.

We note that for a binary matrix, the matrix–vector multiplier is in fact a *crossbar switch*. The crossbar switch is a general switching device that can connect any N input to any N output; this is called *global connection*. Crossbar switches are usually not implemented in electronic computers because they would require N^2 individual switches, but

they are used in telephone exchanges. On the other hand, an optical crossbar interconnected signal processor would be very useful for performing FFTs, convolution, and correlation operations by taking advantage of the reconfigurability and parallel processing of crossbar interconnection [7]. Also, the optical crossbar switch can be employed to implement a programmable logic array (PLA) [8]. The electronic PLA contains a two-level, AND–OR circuit on a single chip. The number of AND and OR gates and their inputs is fixed for a given PLA. A PLA can be used as a read-only memory (ROM) for the implementation of combinational logic.

12.4   SYSTOLIC PROCESSOR

The engagement matrix–vector multiplier is in fact a variation of a systolic processor, as illustrated in Fig. 12.10 [9]. The components of vector B are shifted into the multiplier–adder modules starting at time t_0. Subsequent vector components are clocked in contiguously at t_1 for b_2, t_2 for b_3, and so on. At time t_0, b_1 is multiplied with a_{11} in module 1. The resultant b_1a_{11} is retained within the module to be added to the next product. At time t_1, b_1 is shifted to module 2 to multiply with a_{21}. At the same time, b_2 enters module 1 to multiply with a_{12}, which forms the second product of the output vector component. Consequently, module 1 now contains the sum $b_1a_{11} + b_2a_{12}$. This process continues until all the output vector components have been formed. In all, $2N - 1$ clock cycles that employ N multiplier–adder modules are required. The main advantage of a systolic processor is that optical matrix–vector multiplication can be performed in the high-accuracy digital mode [10].

FIG. 12.9 Optical matrix–vector multiplier using line-shaped LEDs and detectors.

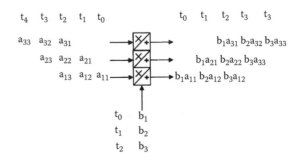

FIG. 12.10   Conceptual diagram of engagement systolic array processor. (From H. T. Kung, *Computer*, 15, 37, 1982.)

A discrete linear transformation (DLT) system can be characterized by an impulse response h_{mn}. The input–output relationship of such a system can be summarized by

$$g_m = \sum f_n h_{mn} \qquad (12.6)$$

Since the output g_m and the input f_n can be considered vectors, the preceding equation can be represented by a matrix–vector multiplication, where h_{nm} is known as a *transform matrix*. Thus, the different DLTs would have different matrices. The DFT is one of the typical examples for DLT, as given by

$$F_m = \frac{1}{N} \sum f_n \exp \frac{-i2\pi mn}{N} \qquad (12.7)$$

where $n = 0, 1, \ldots, N$, and

$$h_{mn} = \exp \frac{-i2\pi mn}{N} \qquad (12.8)$$

is also known as the *transform kernel*. To implement the DFT transformation in an optical processor, we present the complex transform matrix with real values, for which the real transform matrices can be written as

$$Re[h_{mn}] = \cos \frac{2\pi mn}{N} \quad I_m[h_{mn}] = \sin \frac{2\pi mn}{N} \qquad (12.9)$$

which are the well-known *discrete cosine transform* (DCT) and *discrete sine transform* (DST).

The relationship between the real and imaginary parts of an analytic signal can be described by the *Hilbert transform*. The discrete Hilbert transform (DHT) matrix can be written as

$$h_{mn} = 2 \frac{\sin^2 \pi(m-n)/2}{\pi(m-n)} \quad m - n \neq 0$$
$$= 0 \qquad\qquad m - n = 0 \qquad (12.10)$$

Another frequently used linear transform is the Chirp-Z transform, which can be used to compute the DFT coefficients. The discrete Chirp-Z transform (DCZT) matrix can be written as

$$h_{mn} = \exp \frac{i\pi(m-n)^2}{N} \qquad (12.11)$$

Since the DLT can be viewed as the result of a digital matrix–vector multiplication, systolic processing can be used for the implementation. By combining the systolic array processing technique and the *two-complement representation*, a DLT can be performed with a digital optical processor. Compared with the analog optical processor, the technique offers high

accuracy with a low error rate. Also, it is compatible with other digital processors.

Two's complement representation can be applied to improving the accuracy of matrix multiplication. Two's complement numbers provide a binary representation of both positive and negative values, and facilitate subtraction by the same logic hardware that is used in addition. In a *b*-bit two's complement number, the most significant bit is the sign. The remaining $b - 1$ bits represent its magnitude. An example for DCT is given in Fig. 12.11, in which the digital transform matrix has been encoded in two's complement representation with a systolic engagement format.

12.5 MATRIX–MATRIX PROCESSING

The matrix–matrix multiplier is a mathematical extension from the matrix–vector multiplier. In contrast, the implementation of the matrix–matrix multiplier requires a more complex optical setup. Matrix–matrix multipliers may be needed to change or process matrices that will eventually be used in a matrix–vector multiplier. Matrix–matrix multiplication can be computed by successive outer-product operations as follows:

$$\begin{bmatrix} a_{11} & a_{12} & a_{13} \\ a_{21} & a_{22} & a_{23} \\ a_{31} & a_{32} & a_{33} \end{bmatrix} \begin{bmatrix} b_{11} & b_{12} & b_{13} \\ b_{21} & b_{22} & b_{23} \\ b_{31} & b_{32} & b_{33} \end{bmatrix}$$

$$= \begin{bmatrix} a_{11} \\ a_{21} \\ a_{31} \end{bmatrix} \begin{bmatrix} b_{11} & b_{12} & b_{13} \end{bmatrix} + \begin{bmatrix} a_{12} \\ a_{22} \\ a_{32} \end{bmatrix} \begin{bmatrix} b_{21} & b_{22} & b_{23} \end{bmatrix}$$

$$+ \begin{bmatrix} a_{13} \\ a_{23} \\ a_{33} \end{bmatrix} \begin{bmatrix} b_{31} & b_{32} & b_{33} \end{bmatrix}$$

$$(12.12)$$

Since the outer product can be expressed as

$$\begin{bmatrix} a_{11} \\ a_{21} \\ a_{31} \end{bmatrix} \begin{bmatrix} b_{11} & b_{12} & b_{13} \end{bmatrix} = \begin{bmatrix} a_{11}b_{11} & a_{11}b_{12} & a_{11}b_{13} \\ a_{21}b_{11} & a_{21}b_{12} & a_{21}b_{13} \\ a_{31}b_{11} & a_{31}b_{12} & a_{31}b_{13} \end{bmatrix}$$

$$(12.13)$$

it can be obtained by simply butting two SLMs. Equation (12.13) can be realized optically as depicted in Fig. 12.12, in which each pair of SLMs performs the multiplication of the outer products, while the beam splitters perform the addition of the outer products. Thus, we see that the whole combinational operation can be completed in one cycle. However, if the addition is performed electronically in a sequential operation, then only a pair of SLMs is required. If the multiplication is performed sequentially, then more than one cycle operation is needed. This approach is known as systolic processing. The obvious advantage of systolic

Discrete cosine transform (DCT)

$$(h_{m,n}) = \begin{pmatrix} \cos[(2\pi/3)*0*0] & \cos[(2\pi/3)*0*1] & \cos[(2\pi/3)*0*2] \\ \cos[(2\pi/3)*1*0] & \cos[(2\pi/3)*1*1] & \cos[(2\pi/3)*1*2] \\ \cos[(2\pi/3)*2*0] & \cos[(2\pi/3)*2*1] & \cos[(2\pi/3)*2*2] \end{pmatrix}$$

$$= \begin{pmatrix} 1 & 1 & 1 \\ 1 & -1/2 & -1/2 \\ 1 & -1/2 & -1/2 \end{pmatrix} = \begin{pmatrix} 00100 & 00100 & 00100 \\ 00100 & 11110 & 00010 \\ 00100 & 11110 & 11110 \end{pmatrix}$$

FIG. 12.11   Transformation matrix for discrete cosine transform using two's complement representation.

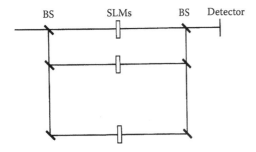

FIG. 12.12   Matrix–matrix multiplier based on outer product. BS, beam splitter.

FIG. 12.13   Matrix–matrix multiplier based on systolic array.

processing is that fewer SLMs are needed. However, the trade-off is the increase in processing time.

Figure 12.13 illustrates an example of systolic array matrix operation. Two transmission-type SLMs, A and B, are placed close together at the input plane. By successively shifting the A and B systolic array matrix formats, one can obtain the product of matrices A and B using a time-integrating charge-coupled device (CCD) detector at the output plane. Although only two SLMs are required, the computation time needed for performing an $n \times n$ matrix–matrix multiplication of b-bit

number would be $(2nb - 1) + (n - 1)$ times that needed by using an outer-product processor.

Figure 12.14 shows an example of a systolic outer-product processing. The optical processor consists of $n \times n = 3 \times 3 = 9$ pieces of SLM with $b \times b = 5 \times 5$ pixels each. Note that the systolic array representation of matrices A and B differs from the systolic formats described previously. By sequentially shifting the row and column elements of A and B into the SLMs, we can implement the $a_{ij}b_{kl}$ multiplication at each step, with an outer-product operation that has been performed in the 5×5 SLM. The result can be integrated in relation to time with a CCD detector at the output plane. Since more SLMs are employed in parallel, the matrix–matrix multiplication can be completed in fewer steps.

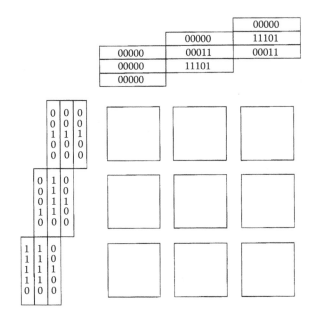

FIG. 12.14   Matrix–matrix multiplication based on the systolic outer-product method.

12.6   EXPERT SYSTEM AND ARTIFICIAL INTELLIGENCE

An expert system is one of the most common applications of artificial intelligence (AI). Expert systems have been developed to assist complex planning, scheduling tasks, diagnosing diseases, locating mineral deposits, and so forth [11–17]. A popular approach to an expert system is to use IF–THEN rules. These rules say that if a certain kind of situation arises, then a certain kind of action can be taken. There are two basic control strategies for applying the IF–THEN rules. The first strategy is to select a goal to be achieved and then to scan the rules backward to find those consequences that can achieve the goal. This is known as *goal-driven backward chaining*. Most existing expert systems use such a backward-chaining strategy. A different strategy is to scan forward through the rules until one is found whose antecedents match assertions in the database. The rule is applied, updating the database, and the scanning resumes. This process continues until either a goal state is reached or no applicable rules are found. This strategy is known as *data-driven forward chaining*. The data-driven forward-chaining strategy may approach the ideal parallel system, and would be suitable for an optical inference engine. In other words, the *sequential* reasoning electronic inference engine is replaced by a *parallel* reasoning optical inference engine. A relatively simple architecture for an optical inference engine based on parallel propositional logic is described as follows [12]. The problem to be solved by an expert system is usually represented by an AND/OR graph, which is a network combining a number of IF–THEN rules. Basically, only AND and OR gates are required to performed the inferences. A simple example of an AND/OR graph is shown in Fig. 12.15. LCTVs can be used to implement the parallel data-driven forward-chaining strategy for the AND/

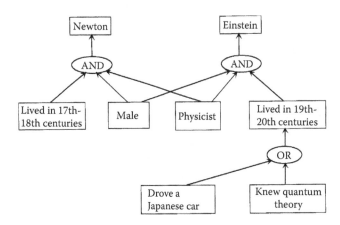

FIG. 12.15   Example of AND/OR graph for an expert system.

OR graph shown in the figure. Needless to say, a hybrid optical expert system utilizing LCTVs to perform inference processing can be employed.

Let us consider that off-pixel and on-pixel LCTVs have 1 and 2 transmittances in an arbitrary unit. The threshold level for the resultant transmittance can be set to determine the logic function as either AND or OR. For example, if the threshold is set at 2, and OR function is obtained, and the AND function can be obtained by setting the threshold equal to 3. The implementation of an AND/OR graph is illustrated in Fig. 12.15, in which three LCTVs are arranged in tandem to form three-input AND or OR gates. The resultant transmittances are listed in Table 12.2, for which we see that the threshold selections at 2 and 6 result in OR and AND functions, respectively. Notice that it is a hybrid system; a computer is used to store, update the knowledge base, write symbols (logic patterns) on the LCTVs, and control the whole operation. The search based on the AND/OR graph shown in Fig. 12.15 may be completed in two cycles. Each cycle performs one to two inference rules in parallel. The knowledge based is updated after each cycle. An LCTV display consists of about 140×120 pixels. To perform less than 140 inference rules in parallel, a vertical line can be used instead of a pixel to represent a fact. A cylindrical lens can be used to focus the vertical line onto an element of the array detector. This optical setup further introduces fault tolerance due to defects associated with the LCTV and noise. The detected signal is fed back into a computer to be thresholded

TABLE 12.2
Resultant Transmittance of Three LCTVs

LCTV1	LCTV2	LCTV3	Resultant	OR	AND
1 (off)	1 (off)	1 (off)	1	Off	Off
1 (off)	1 (off)	2 (on)	2	On	Off
1 (off)	2 (on)	1 (off)	2	On	Off
1 (off)	2 (on)	2 (on)	4	On	Off
2 (on)	1 (off)	1 (off)	2	On	Off
2 (on)	1 (off)	2 (on)	4	On	Off
2 (on)	2 (on)	1 (off)	4	On	Off
2 (on)	2 (on)	2 (on)	8	On	On

and to complete one cycle. Let us detail the operation of the preceding expert system in the following.

For example, the given facts are DROVE A JAPANESE CAR, KNEW QUANTUM THEORY, PHYSICIST, MALE, and LIVED IN 17TH–18TH CENTURIES, as the inputs to the computer to generate a symbolic representation. The intermediate results from a cycle may be used to update the knowledge base that is needed for the subsequent inference cycles. For instance, results from the first cycle are used to generate symbols for the second cycle. For a given example—DROVE A JAPANESE CAR = NO; KNEW QUANTUM THEORY = YES; MALE = YES; LIVED IN 17TH–18TH CENTURIES = NO—the generated symbols are illustrated in Fig. 12.16. If a logic gate is operated by only two inputs, then the other input is set off, and the thresholds for the OR and AND functions are 2 and 3, respectively. The first cycle will update the knowledge base with the fact LIVED IN 19TH–20TH CENTURIES = YES, which shows that the fact EINSTEIN is finally YES, while the other alternative is NO.

In the aforementioned example, when the fact MALE is YES, the chances for NEWTON and EINSTEIN are the same. In contrast, medical diagnostic expert systems are usually based on the probabilistic principle. If a symptom is positive, there are uneven probabilities for different diseases. The optical expert system [13] for diagnosis is used; the database is to establish a list of *a priori* probabilities $p(e_i/h_j)$ that an event of symptom e_i will be present given a specified hypothesis or illness h_j. The expert system is then used to compute the *a posteriori* probabilities $p(h_j/e_i)$ of a specified illness h_j, given a specified symptom e_i. After the patient has answered a number of questions, the illness or illnesses with the highest probabilities are the most likely hypotheses, as described by the following *Bayes' rule*:

$$p(h_j/e_i) = \frac{p(e_i/h_j)p(h_j)}{p(e_i/h_j)p(h_j) + p(e_i/\bar{h}_j)p(\bar{h}_j)} \quad (12.14)$$

where $\bar{h}_i$ is not $-h_j$, and $p(\bar{h}_j) - 1 - p(h_j)$. The rule can be extended to multiple events and circumstances. Two terms in the denominator (one is identical to the numerator) can be obtained using optical matrix–vector multipliers. If a response to the *i*th event has been received, a unit signal is entered at this input, while all other inputs remain zero. Notice that these inputs form the input vector, while the *a priori* probabilities $p(e_i/h_i)$ form the interconnection matrix. The elements of the output vector are multiplied by the corresponding $p(h_i)$, which are known from statistics of the region and the patient profile.

Both the deterministic and probabilistic examples given in the preceding are based on relatively simple proposition calculus, in which a true or false fact is simply a statement without any arguments. Optical expert systems based on predicate calculus have been proposed [14,15]. These systems are based on the production system paradigm, where facts and rules constitute a knowledge base. Facts define the relationships between objects, and rules define the procedures for deriving new relations from existing ones. PROLOG (which has been selected as the basis for the Japanese fifth-generation computers) is a popular A1 language that implements the production system paradigm.

A1 systems generally consist of a knowledge base and an inference engine. The inference engine accepts as input a set of facts and a set of rules from the knowledge base and one or more queries. The output of the inference engine is a set of specific conclusions that are logically inferred from the facts and rules in response to the queries. For example, a set of data could be a set of names of people.

$$D = \{\text{Ann, Beth, Carol, David, Edward, Frank}\} \quad (12.15)$$

A set of possible relationships for D among them is given as

$$R = \{\text{married-to, mother-of, father-of,}$$

$$\text{son-of, daughter-of, child-of, is-male, is-female}\} \quad (12.16)$$

(a)

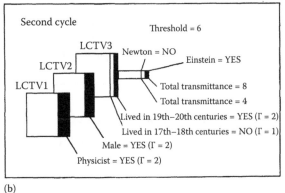

(b)

FIG. 12.16 Operation of inference. Symbols are generated on LCTVs at (a) the first cycle and (b) the second cycle. Black stripes represent on-pixels (transmittance = 2); white stripes represent off-pixels (transmittance = 1).

The data objects and relationships link among them is a collection of facts and rules that relate the elements of D and R, for instance,

David married-to Ann
Edward married-to Beth
David father-of Carol
Edward father-of Frank

The query addressed to this knowledge base is "Who is the mother of Carol?" or ?mother-of Carol. The PROLOG code for these expressions is given as follows:

married-to (David, Ann)
married-to (Edward, Beth)
father-of (David, Carol)
father-of (Edward, Frank)
and the query is
?-mother-of (X, Carol)

In an optical inference engine [14], mapping templates are used to store the relationships between the data objects, and are thus defined by the facts. Conclusions are assigned to queries by applying these mapping templates to the data objects in the order prescribed by the rules. The married-to and father-of facts can be represented by two-dimensional mapping templates, as shown in Fig. 12.17a and b, respectively. Operation of the ?mother-of Carol query is performed as follows.

An LCTV is used to display the mapping template for the father-of facts, that is, Fig. 12.17a. The query data Carol will turn on a horizontal line of C to illuminate the LCTV, which is in fact a two-dimensional optical AND operation between the horizontal line. The transmittance from the LCTV will be low, except that the element (D, C) being high indicates that David is the father of Carol. The bright spot at (D, C) is then expanded optically (using a cylindrical lens) to form a vertical line, and is projected onto another LCTV showing the mapping template of the married-to facts, that is, Fig. 12.17b. Consequently, the transmittance from the second LCTV shows only one bright spot at (D, A). The bright spot is now expanded horizontally so that it can be detected by the detector A. This indicates the conclusion is Ann.

It is interesting to notice that the mapping template method can also be performed using an algebraic matrix operation [15] for which the married-to facts can be represented by the following binary matrix:

$$[MT] = \begin{array}{c} A \\ B \\ C \\ D \\ E \\ F \end{array} \begin{array}{cccccc} A & B & C & D & E & F \\ \left[\begin{array}{cccccc} 0 & 0 & 0 & 0 & 0 & 0 \\ 0 & 0 & 0 & 0 & 0 & 0 \\ 0 & 0 & 0 & 0 & 0 & 0 \\ 1 & 0 & 0 & 0 & 0 & 0 \\ 0 & 1 & 0 & 0 & 0 & 0 \\ 0 & 0 & 0 & 0 & 0 & 0 \end{array} \right] \end{array}$$

and the father-of fact is represented by another binary matrix:

$$[FO] = \begin{array}{c} A \\ B \\ C \\ D \\ E \\ F \end{array} \begin{array}{cccccc} A & B & C & D & E & F \\ \left[\begin{array}{cccccc} 0 & 0 & 0 & 0 & 0 & 0 \\ 0 & 0 & 0 & 0 & 0 & 0 \\ 0 & 0 & 0 & 0 & 0 & 0 \\ 0 & 0 & 1 & 0 & 0 & 0 \\ 0 & 0 & 0 & 0 & 0 & 1 \\ 0 & 0 & 0 & 0 & 0 & 0 \end{array} \right] \end{array}$$

The query ?mother-of Carol can be answered by the rule

X mother-of Carol, IF Y married-to X AND Y father-of Carol

(12.17)

which involves two facts connected by a rule. One may consider that, in general, a rule is a function of facts. A new fact can be concluded from the function, through a sequence of logic operations applied to the known facts. In the matrix-encoding method, logic operations between relations are carried out by thresholding the matrix–matrix product for the AND operation and by thresholding the matrix–matrix sum for the OR operation. Equation (12.17) can be used to generate new facts of mother-of as follows:

$$[MO] = [MT][FO] \qquad (12.18)$$

which yields

$$[MO] = \begin{array}{c} A \\ B \\ C \\ D \\ E \\ F \end{array} \begin{array}{cccccc} A & B & C & D & E & F \\ \left[\begin{array}{cccccc} 0 & 0 & 1 & 0 & 0 & 0 \\ 0 & 0 & 0 & 0 & 0 & 1 \\ 0 & 0 & 0 & 0 & 0 & 0 \\ 0 & 0 & 0 & 0 & 0 & 0 \\ 0 & 0 & 0 & 0 & 0 & 0 \\ 0 & 0 & 0 & 0 & 0 & 0 \end{array} \right] \end{array}$$

FIG. 12.17 Templates showing the relationship of (a) father-of and (b) married-to.

To get the conclusion of the query ?mother-to Carol, a matrix–vector multiplication is performed, in which the input vector is [A = 0, B = 0, C = 1, D = 0, E = 0, F = 0.] The output vector is obtained as follows:

$$
\begin{bmatrix}
0 & 0 & 1 & 0 & 0 & 0 \\
0 & 0 & 0 & 0 & 0 & 1 \\
0 & 0 & 0 & 0 & 0 & 0 \\
0 & 0 & 0 & 0 & 0 & 0 \\
0 & 0 & 0 & 0 & 0 & 0 \\
0 & 0 & 0 & 0 & 0 & 0
\end{bmatrix}
\begin{bmatrix}
0 \\ 0 \\ 1 \\ 0 \\ 0 \\ 0
\end{bmatrix}
=
\begin{matrix} A \\ B \\ C \\ D \\ E \\ F \end{matrix}
\begin{bmatrix}
1 \\ 0 \\ 0 \\ 0 \\ 0 \\ 0
\end{bmatrix}
$$

$$(12.19)$$

Since the output vector is [A = 1, B = 0, C = 0, D = 0, E = 0, F = 0], the conclusion is Ann. Thus, we see that the optical matrix–vector and matrix–matrix multipliers can also be applied to AI.

12.7   REMARKS

We have shown that computing can be carried out with optics; however, the proposed optical systems are basically emulating the electronic counterpart for computing. Unfortunately, the available electro-optic devices are still way below the required specifications for this application. The success of optical computing is very much dependent on the future development of these devices. However, such devices will take years to develop or may never be developed. Unless the issues of devices can be solved, it can hardly be claimed that we actually have an optical computer to compete.

Nevertheless, there is an alternative approach that remains to be seen. Instead of emulating the electronics for computing, one can search for a new type of mathematics, or develop one, that is more suitable for optics to compute. It is my belief that the behavior of optics and electronics is quite different. For instance, light is a better carrier for information transmission, while electrons interact more efficiently. To ask optics to behave like electronics for computing may not be the most efficient approach.

REFERENCES

1. M. J. Quinn, *Designing Efficient Algorithms for Parallel Computers*, McGraw-Hill, New York, 1987.
2. J. Tanida and Y. Ichioka, Optical Logic Array Processor Using Shadowgrams, *J. Opt. Soc. Am.*, vol. 73, 800 (1983).
3. C. L. Wu and T. Y. Feng, The Universality of the Shuffle-Exchange Network, *IEEE Trans. Comput.*, vol. C-30, 324 (1981).
4. A. W. Lohmann, W. Stork, and G. Stucke, Optical Perfect Shuffle, *Appl. Opt.*, vol. 25, 1530 (1986).
5. Q. W. Song and F. T. S. Yu, Generalized Perfect Shuffle Using Optical Spatial Filtering, *Appl. Opt.*, vol. 27, 1222 (1988).
6. J. J. Hopfield, Neural Networks and Physical Systems with Emergent Collective Computational Abilities, *Proc. Natl. Acad. Sci. U.S.A.*, vol. 79, 2554 (1982).
7. A. D. McAulay, Optical Crossbar Interconnected Digital Signal Processor with Basic Algorithms, *Opt. Eng.*, vol. 25, 82 (1986).
8. R. Arrathoon, Historical Perspectives: Optical Crossbars and Optical Computing, *Proc. SPIE*, vol. 752, 2 (1987).
9. H. T. Kung, Why Systolic Architecture? Computer, vol. 15, 37 (1982).
10. H. J. Caulfield, W. T. Rhodes, M. J. Foster, and S. Horvitz, Optical Implementation of Systolic Array Processing, *Opt. Commun.*, vol. 40, 86 (1981).
11. P. S. Guilfoyle, Systolic Acousto-Optic Binary Convolver, *Opt. Eng.*, vol. 23, 20 (1984).
12. F. Hayes-Roth, D. A. Waterman, and D. B. Lenat, eds., *Building Expert Systems*, Addison-Wesley, Reading, MA, 1983.
13. S. Jutamulia, G. Storti, and X. Li, Expert Systems Based on LCTV AND/OR Logic, *Opt. Laser Technol.*, vol. 21, 392 (1989).
14. A. D. McAulay, Real-Time Optical Expert Systems, *Appl. Opt.*, vol. 25/26, 1927 (1987).
15. C. Warde and J. Kottas, Hybrid Optical Inference Machines: Architectural Considerations, *Appl. Opt.*, vol. 25, 946 (1986).
16. J. Y. Jau, F. Kiamilev, Y. Fainman, S. Esener, and S. H. Lee, Optical Expert System Based on Matrix-Algebraic Formulation, *Appl. Opt.*, vol. 27, 5170 (1988).
17. F. T. S. Yu and S. Jutamulia, *Optical Signal Processing, Computing, and Neural Networks*, Wiley-Interscience, New York, 1992.

13 Communication with Fiber Optics

Exploitation of information transmission by light is one of the oldest and most important aspects of communication for all the living species on Earth. In fact, information transfer from one place to another can be easily carried by light from very close range, such as microscopic observation, to millions of miles by telescopic lenses. Due to the discovery of high-coherence light sources of lasers and the advent of high-quality fiber-optic technologies, communication with optics can be carried out beyond the transoceanic distances with small-sized optical fiber. The carrier frequency of fiber-optic communications can be as high as hundreds of terahertz in the visible or near-infrared range of the electromagnetic spectrum. In fact, the fiber-optic communication carrier frequency is about five orders higher than that of the conventional microwave communication technology, which is in about the 1 Ghz region. The objective of this chapter is to briefly describe the basic aspects of communication with fiber optics.

13.1 ASPECTS OF FIBER-OPTIC COMMUNICATION

Basically, the capacity of communication is limited by the bit rate–distance product BL, where B is the bit rate and L is the distance of communication before the signal diminishes to an unrealistic level. In other words, L is the distance between the repeaters. In order to appreciate the significance of using fiber-optic technology for communication, BL (which represents the transmission capacity) for various emerging telecommunication technologies is plotted as a function of calendar year, as shown in Fig. 13.1. We see that light-wave technology offers a much higher transmission capacity than the microwave communication. Added with the recent development of the erbium-doped fiber amplifier, the capacity can be as high as 10^{12}–10^{15} (bits/sec)-km.

The root for using light-wave technology for long-distance transmission could have started from the beginning of the discovery of the laser in the 1960s. The breakthrough research on low-transmission-loss fiber materials in the 1970s also provided a big surge application on long-distance communication. The transmission loss of fiber optics can be reduced to about 20 dB/km near the 1 μm wavelength region. In addition, by the recent development of compact GaAs semiconductor lasers, fiber-optic communication has a worldwide commitment to develop highly efficient long-distance communication systems. Although the implementations of fiber-optic communication were postulated [1] as early as the 1970s, it was the actual experimental demonstration in the 1980s that showed that transmission can be implemented over 4000 km, by compensating the

fiber loss through stimulated Raman scattering [2]. Along with the recent development of erbium-doped fiber amplifiers, fiber-optic communication systems can in principle transmit information over 1000 km at 10 Gb/sec and over 350 km at 20 Gb/sec [3]. Mention should also be made that by using a recirculating loop configuration, solitons at 2.4 Gb/sec can be maintained over 12,000/km [4]. Although communication technology with fiber optics is barely $2\frac{1}{2}$ decades old, it has reached the stage of worldwide practical implementations.

In a practical communication, information is generally transmitted in *analog* form, for example, speeches and images. These informational signals can in turn be carried by coded *digital* signals, to improve the signal-to-noise ratio (SNR) transmission. An example of analog-to-digital signal transmission is illustrated in Fig. 13.2, in which the signal sampling is determined by the *Nyquist* rate. The next step is to encode the sampled value into binary digital form, for which each sampled value is quantized into 1 and 0 binary formats. The number of m bits sampled value is related to the quantized signal levels m:

$$M = 2^m \quad \text{or} \quad m = \log_2 M \tag{13.1}$$

The bit rate associated with the *pulse code modulation* (PCM) is

$$B = mf_s \geq (2\Delta f)\log_2 M \tag{13.2}$$

where f_s is Shannon's sampling frequency:

$$fs \geq \Delta f \tag{13.3}$$

and Δf is the bandwidth of the analog signal. Since this quantization also adds noise, known as *quantization noise*, to this coded signal the number of quantized levels should be limited by

$$M \geq \frac{A_{\max}}{A_N} \tag{13.4}$$

where $A_{\max}$ is the highest amplitude of the analog signal, and A_N is the *root mean square* (rms) noise amplitude. Notice that the ratio $A_{\max}/A_N$ is known as the *dynamic range* of the quantization and is related to the SNR given by

$$SNR = 20\log_{10}\frac{A_{\max}}{A_N} \tag{13.5}$$

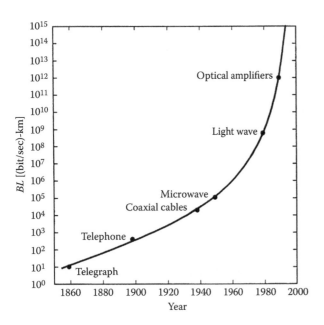

FIG. 13.1   Bit rate–distance vs. calendar year of emerging communication technologies.

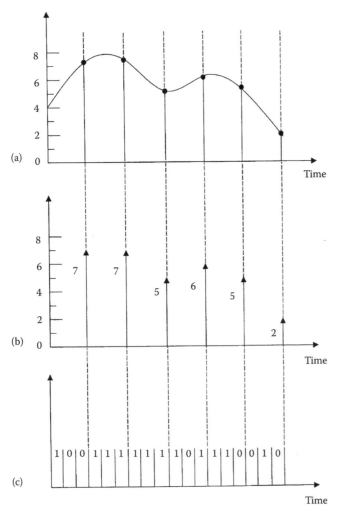

FIG. 13.2   Analog-to-digital signal conversion: (a) analog signal, (b) sampled signal, and (c) digital signal.

In view of the preceding PCM bit rate, we see that

$$B > \frac{\Delta f(SNR)}{3} \tag{13.6}$$

where SNR is in decibel units.

Let us now consider a digital voice channel, which is operated at 64 kb/sec. Since most optical fiber communication systems can handle a 150 Mb/sec bit rate, it is possible to transmit several digital coded signals simultaneously in the same fiber channel, namely, by using either *time-division multiplexing* (TDM) or *wavelength-division multiplexing* (WDM) techniques. In other words, either multisignals can be transmitted by interleaving the channel signals to form a composite bit stream for TDM transmission, or the channel signals can be spaced along the wavelength (or frequency) domain for WDM transmission, as illustrated in Fig. 13.3.

It is apparent that fiber-optic communications differ from other communication technologies. Aside from its higher carrier frequency typically in the range of 100 Thz, as in contrast with microwave communication, its carrier frequency is in the order of 10 Ghz. Thus, one would expect fiber-optic communication to be at least 10^5 times the transmission capacity than is microwave communication. By taking a conservative estimation, fiber-optic communication has the potential to carry about a 1 Tb/sec bit rate. In view of the current state-of-the-art optical communication systems, which are operating at about a 1 Gb/sec bit rate, it is trivial to understand that fiber-optic communication would have great potential for further improvement.

To summarize our discussion, a block box diagram fiber-optic communication is shown in Fig. 13.4, in which analog signals, from this information source, are first encoded in digital form. By converting the digital signals in light pulses, the information can be transmitted through the optical fiber. At the receiving end, the light pulses are then properly decoded and transformed into analog signals, prior to being sent to the user. The basic difference, compared with the usual communication channels, is that the information is transmitted by coded light pulses through fiber optics. Although fiber-optic communication can be applied in various fields of information transfer, it has been mostly developed for telecommunication usages. There are two types of telecommunications, namely, *long-haul* or *short-haul*, depending on long or short distances. Long-haul communication is mostly applied to intercity communication, which has a distance ranging from 50 to 100 km or even longer. Short-haul communication is typically used for intracity or local loop traffic, which is less than 10 km. Long-haul telecommunication systems require high-capacity trunk lines, which are benefitted mostly by the use of fiber-optic light-wave technologies. Short-haul telecommunication uses a single-channel light-wave system. For applications, the cost-effectiveness is not generally the concern.

(b)

FIG. 13.3   Samples of (a) TDM and (b) WDM.

Information
source   Encoder   Transmitter   Receiver   Decoder   User

Optical
fiber

FIG. 13.4   Block box representation of fiber-optic communication.

13.2   OPTICAL FIBER STRUCTURES

The structure of a typical optical fiber is illustrated schematically in Fig. 13.5. The center of the fiber is a cylindrical core
with refractive index η_1, surrounded by a layer of material,
called the cladding, with a lower refractive index η_2. The
light waves are restricted by the cladding within the core and
propagate along the fiber. The outer jacket layers not only pro-
tect the fiber from moisture and abrasion but also provide the
strength that is required practically.

The core and the cladding are made of either silicate material (e.g., glass) or plastic. Three major combinations of these
two types of material are used to make optical fibers: plastic
core with plastic cladding, glass core with plastic cladding,
and glass core with glass cladding. A plastic core is generally made of polystyrene or polymethyl methacrylate, while
plastic cladding is typically made of silicone or Teflon. For
glass cores and claddings, the silica must be extremely pure;
however, very small amounts of dopants, such as boron,

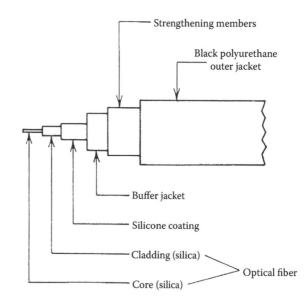

FIG. 13.5   Structure of an optical fiber.

germanium, or phosphorous, may be added to change the refractive indices. In some claddings, boron oxide is often added to silica to form borosilicate glass.

In comparison with glass fibers, plastic fibers are flexible, inexpensive, and easy to install and connect. Furthermore, they can withstand greater stress and weigh 50% less than glass fibers. However, they do not transmit light as efficiently. Due to their considerably high losses, they are used only for short-haul communication (such as networks within buildings) and some fiber sensors. Since glass core fibers are much more widely used than plastic ones, subsequent references to fibers are to glass fibers. Glass fibers, although slightly heavier than plastic fibers, are much lighter than copper wires. For example, a 40 km long glass fiber core weighs only 1 kg, whereas a 40 km long copper wire with a 0.32 mm outer diameter weighs about 30 kg. In a fiber-optic communication network, information is transmitted by means of the propagation of electromagnetic waves along the core of the optical fiber. To minimize the transmission loss, the electromagnetic wave must be restricted within the fiber core, and not be allowed to leak into the cladding. As we will see, the refractive index of the core η_2 must be greater than that of the cladding η_2 if such a requirement is to be met.

The diameters of the core and cladding determine many of the optical and physical characteristics of the fiber. For example, the diameter of a fiber should be large enough to allow splicing and the attachment of connectors. However, if the diameter is too large, the fiber will be too stiff to bend and will take up too much material and space. In practice, the diameter of fiber cores ranges from 5 to 500 μm, and outer diameters of fiber claddings vary from 100 to 700 μm. To keep the light wave within the fiber core, the cladding must have a minimum thickness of one or two wavelengths of the light transmitted. The protective jackets may add as much as 100 μm to the fiber's total diameter.

Optical fibers with the minimum core diameter have the advantage of possessing a much broader bandwidth. However, their small size may cause severe handling and connection problems. In fact, for any given fiber, tight tolerances are necessary, because even a slight variation in the dimensions can cause significant changes in optical characteristics. The tolerance for the core diameter is typically about ±2 μm.

Although ordinary glass is brittle and easily broken or cracked, glass fibers usually have high tensile strength and are able to withstand hard pulling or stretching. The toughest fibers are as strong as stainless steel wires of the same diameter, and have the same tensile strength as copper wires with twice the diameter. For example, 1 km lengths of these fibers have withstood pulling forces of more than 500,000 lb/in.[2] before breaking. A 10 m long fiber can be stretched by 50 cm and still spring back to its original shape, and a fiber with a 400 μm diameter can be bent into a circle with a radius as small as 2 cm.

In order to produce fibers with such tenacity, manufacturers have to keep the glass core and cladding free from microscopic cracks on the surface or flaws in the interior. When a fiber is under stress, it can break at any one of these flaws. Flaws can develop during or after manufacturing. Even a tiny particle of dust or a soft piece of Teflon can lead to a fatal scratch on the surface of the fiber core or cladding. To prevent such abrasion, manufacturers coat the fiber with a protective plastic (or organic polymer) jacket immediately after the fiber has been drawn. This jacket also protects the fiber surface from moisture and cushions the fiber when it is pressed against irregular surfaces. The cushioning reduces the effect of small random bends (i.e., so-called microbends), which would otherwise cause transmission losses. Moreover, the jacket compensates for some of the contractions and expansions caused by temperature variations.

Although a single fiber is used in some applications, a number of fibers can be tightly packed to form a fiber cable. The cable is optimally designed such that it has little or no stress among fibers themselves.

13.3 FIBER-OPTIC TRANSMISSION

Optical fibers can be considered cylindrical waveguides. There are two boundaries to be considered in an optical fiber: one at the core–cladding interface, and the other at the cladding–air (or cladding–jacket) interface, as shown in Fig. 13.6, in which we assume that the core has a higher refractive index than the cladding. This figure shows how light rays are coupled into and kept within the optical fiber. If the incident

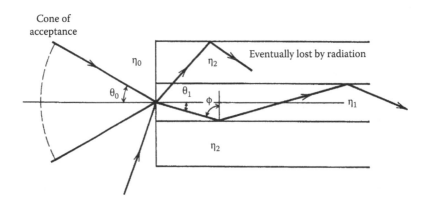

FIG. 13.6 Acceptance angle for an optical fiber. φ is the incident angle at the core–cladding boundary.

angle of a light ray at the end of the fiber core is less than the angle of acceptance,

$$\theta_a = \sin^{-1} \sqrt{\frac{\eta_1^2 - \eta_2^2}{\eta_0}} \qquad (13.7)$$

then the light ray will be totally reflected at the core–cladding boundary, and it will be guided along the fiber case. Light rays at greater incident angles may be guided by the cladding. Since the loss in the cladding material is usually much higher than that of the core, the light rays guided along the cladding layer will eventually diminish due to absorption. Therefore, information can be transmitted only along the core of an optical fiber.

In order to efficiently couple light rays into an optical fiber, it is necessary to generate a cone of light that matches the acceptance angle of the fiber. This can be achieved by using a condenser lens with a suitable numerical aperture, as shown in Fig. 13.7. When small-sized semiconductor lasers or light-emitting diodes (LEDs) are used, they can also be put in close proximity to the fiber, as depicted in Fig. 13.8. In this direct coupling scheme, the light source and the fiber are generally linked by index-matching liquid to reduce the losses. The coupling efficiency depends on the degree of matching between the numerical apertures of the light source and the fiber.

After the light has become properly coupled into the fiber, two physical mechanisms take place during the transmission: *attenuation* and *dispersion*. Attenuation is the reduction in light intensity as light propagates through the fiber, which is mainly due to the chemical impurities in the fiber core. If the absorption resonance of these impurities is within the range of optical frequencies being transmitted, there will be considerable attenuation through absorption. A tremendous research and development effort has been made to reduce the amounts of these impurities. At present, optical fibers with an attenuation of less than 0.5 dB/km have been made, and an attenuation of 3 dB/km has become standard for fiber communication networks.

There are, however, two other sources of attenuation, both of which are related to the effect of bending, as illustrated in Figs. 13.9 and 13.10. These diagrams depict the escape of light rays from the fiber core, which are just within the critical angle, before bending or microbending occurs. By bending, we mean the actual bending of the whole fiber, but a microbend is a slight kink at the boundary between the core and the cladding. Notice that a considerable amount of light will

FIG. 13.8   Coupling of light with a LED.

FIG. 13.9   Attenuation caused by bending.

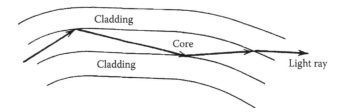

FIG. 13.10   Attenuation caused by microbending.

leak when the bend (or microbend) is sharp. In this context, a sharp bend is one having a radius of about 50 mm or less. Such bends can lead to a loss of a few decibels per kilometer.

Dispersion of light transmission in the optical fiber has two general causes. The first one is called material dispersion, and it results in the separation of a complex electromagnetic wave into its various frequency components. It is the result of the dependence of the velocity of light on the wavelength, or in other words, the dependence of the refractive index of the fiber material on the wavelength $\eta(\lambda)$, a circumstance that is similar to the chromatic aberration of a lens.

The second type of dispersion is called *intermodal dispersion*, which occurs even with purely monochromatic light waves, and is caused by different paths of light rays propagating in the fiber. This can be easily understood from the fact that two light rays are traveling along the fiber: one travels along the optical axis, and the other zigzags up and down at

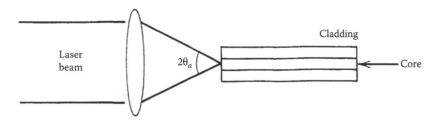

FIG. 13.7   Coupling of light into an optical fiber.

the critical angle. Since these two light rays travel different distances along the fiber core to arrive at the same destination, the relative path difference between these two light rays can be shown as

$$\frac{\Delta l}{l} = \frac{\eta_1 - \eta_2}{\eta_2} \qquad (13.8)$$

where l is the distance between the light ray. The intermodal dispersion within the optical fiber in terms of the difference in the periods of time that are taken by these two light rays can be shown as

$$\Delta T - \left(\frac{\eta_1 - \eta_2}{\eta_2}\right)\frac{L\eta_1}{c} = (\eta_1 - \eta_2)\left(\frac{\eta_1}{\eta_2}\right)\frac{L}{c} \qquad (13.9)$$

where L is the length of the optical fiber, and c is the velocity of light in vacuum.

In the preceding discussion, geometrical optics have been used to explain the propagation of light rays along optical fibers. However, the transmission of light along optical fibers can be better characterized by using electromagnetic wave theory. Similar to the laser oscillation modes, the propagation of light waves in an optical fiber is not arbitrary, but takes some discrete propagation modes, which are determined by the boundary conditions. For simplicity, we first assume that the optical fiber behaves as a waveguide with perfect conducting walls. Under such circumstances, the electric field at the waveguide boundary must be zero. An example of the propagation mode that meets such a requirement is illustrated in Fig. 13.11.

It should be noted that the electric field across the waveguide consists of interference patterns between the incident and reflected beams. To satisfy the boundary conditions (i.e., $E_b = 0$), the diameter of the waveguide must intersect an integral number of half-wavelengths of the incident light. This can be expressed as

$$\sin\theta = \frac{(n+1)\lambda}{2d\eta} \qquad n = 0, 1, 2, 3, \dots \qquad (13.10)$$

where θ is the incident angle, d is the separation of the conducting walls, η is the index of refraction, and n is the propagation mode. Thus, we can see that only certain values of the incident angles allow propagation along the waveguide; otherwise, the boundary condition would be violated.

An exact solution of the modal description for optical fibers may be obtained by solving Maxwell's equation subject to the boundary conditions. The detailed derivations are beyond the scope of this section, and interested readers can find them in a number of texts (e.g., [5–8]). Nevertheless, a similar solution for the conducting waveguide can be applied to optical fibers. The solution shows that only a number of specific angular directions of the incident beam can be allowed to propagate along the fiber. In fact, each of the allowed beam directions corresponds to a different mode of wave propagation in the fiber.

To further discuss the modal description in optical fibers, we can now equivalently examine the model field in a planar dielectric slab waveguide, as shown in Fig. 13.12. The dielectric slab corresponds to a core that is sandwiched by two dielectric cladding layers. This simple planar waveguide helps us to understand light-wave propagation in an optical fiber, which is difficult to deal with due to its cylindrical shape. A set of low-order transversal electric guided modes obtained from the solutions of Maxwell's equations are shown in Fig. 13.12. The order of the mode is equal to the number of transversal field minima (zero field points) across the waveguide, and is also related to the angle of the incident beam as given by Eq. (13.10). It can be seen that the steeper the incident angle, the higher the order of the guided wave mode. Unlike the perfect conducting waveguide, the electric

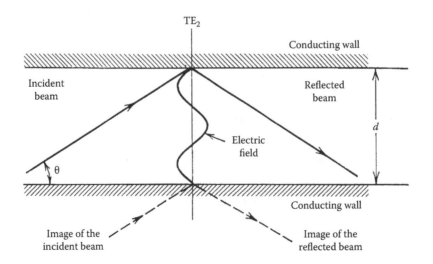

FIG. 13.11   Conducting waveguide. TE$_2$, transversal electric field of order 2.

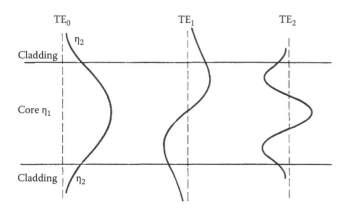

FIG. 13.12 Distribution of transversal electric field across the core–cladding region for different modes. TE, transversal electric field.

fields of the guided modes are not completely confined within the central core, but go beyond the core–cladding boundary and decay very rapidly in the cladding layer. For lower-order guided modes, the electric fields are more concentrated near the center of the core. However, for higher-order modes, the electric field distribution spreads the core–cladding boundary and penetrates further into the cladding layer. In addition to the guided modes, there also exist *leaky modes* in optical fibers. These leaky modes are only partially confined in the core region of the fiber. As they propagate along the fiber, their intensity attenuates continuously by radiating out from the fiber core. It can be shown that the upper and lower bounds of the propagation constant β for the existence of such leaky modes are given by

$$\eta_2 k < \beta < n_1 k \qquad (13.11)$$

where $k = 2\pi/\lambda$. Most of the leaky modes disappear a few centimeters along the fiber due to significant radiation losses. However, under certain circumstances a few leaky modes may persist in the fiber for more than 100 m.

13.4   TYPES OF OPTICAL FIBERS

In general, optical fibers can be divided into two types. They are *single-mode* and *multimode* fibers. For single-mode fibers, only one guided mode is permitted to propagate along the fiber core. This transmitting mode has the lowest order ($n = 0$) and corresponds to the transversal electric field profile that has no minima across the fiber cross section. All the other orders ($n \geq 1$) have an incident angle at the core–cladding boundary that is smaller than the critical angle and, consequently, quickly leaks into the cladding layer and is absorbed. By referring to Eq. (13.10), the condition for single-mode operation is

$$0 \approx \sin \theta < \frac{\lambda}{d} \qquad (13.12)$$

where d is the diameter of the fiber core. In terms of critical angle at the core–cladding boundary, the single-mode operation condition can be expressed as

$$0 = \frac{\pi}{2} - \phi_c < \frac{\lambda}{d} \qquad (13.13)$$

or

$$d < \frac{\lambda}{\pi/2 - \phi_c} \qquad (13.14)$$

This implies that in order to achieve single-mode operation, the fiber core can only have a very small diameter, ranging from 3 to 10 µm, depending on the wavelength of light to be transmitted.

The dispersion in a single-mode fiber is primarily due to absorption by the material of the core. Therefore, single-mode fibers are the most suitable media for *long-distance communication* and are capable of handling very broad band signals. Since the core diameter of a single-mode fiber is generally very small, launching light into the fiber is rather difficult, and keeping the insertion of light constant over a long period of time is even more challenging. Fortunately, these problems can be solved with the advances in optoelectronic technology, for which single-mode fibers have been deployed in various high-capacity communication networks.

In a multimode fiber, a number of guided modes (from two to a few thousand) are transmitted simultaneously. There are two types of multimode fibers: *step index fibers* and *graded index fibers*. A step index fiber is, in fact, the type of fiber that we have been discussing so far. In this type of fiber, the refractive index is constant within the core, but changes abruptly at the core–cladding interface, as shown in Fig. 13.13. Although it has a simple structure, this type of fiber suffers from severe intermodal dispersion. Ray diagrams for two different propagation modes (represented by an axial ray and a reflected ray) are shown in Fig. 13.13. One percent of optical path difference between these two light rays is common, which may lead to a time delay of 40 nsec between the two modes after they have propagated 1 km along the fiber. This makes it impossible to transmit information at a data rate of 20 Mb/sec over a 1 km communication link using step index fibers.

In contrast, a graded index fiber does not have a constant refractive index within its core, but has a decreasing index $\eta(r)$ as a function of radial distance. As shown in Fig. 13.14, the refractive index varies from a maximum value η_1 at the center of the core to a constant value η_2 at the core–cladding boundary. The profile of the refractive index can be generally described by the expression

$$\eta(r) = \begin{cases} \eta_1 \left[1 - 2\Delta \left(\dfrac{r}{a} \right)^{\alpha} \right]^{1/2} & r < a \\ \eta_2 & r \geq a \end{cases} \qquad (13.15)$$

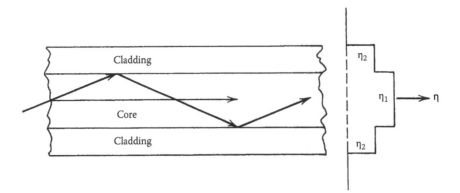

FIG. 13.13 Step index fiber.

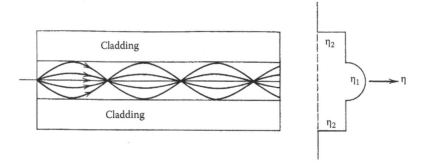

FIG. 13.14 Graded index fiber.

where a is the radius of the core, α is the profile parameter, r is the radial distance, and

$$\Delta = \frac{\eta_1^2 - \eta_2^2}{2\eta_1^2} \approx \frac{\eta_1 - \eta_2}{\eta_1} \qquad (13.16)$$

is known as the relative *refractive index difference*. The refractive index profile varies with the profile parameter α, as shown in Fig. 13.15. It can be seen that when $\alpha = \infty$, the profile actually represents that of the step index fiber; $\alpha = 2$

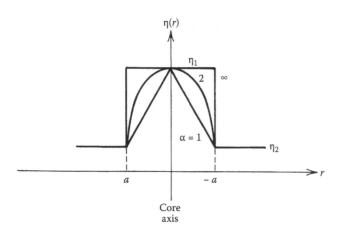

FIG. 13.15 Refractive index profiles for graded fiber.

corresponds to a parabolic profile, and $\alpha = 1$ illustrates a triangle profile.

The ray paths in a graded index fiber core are curved, as illustrated in Fig. 13.14. Since the inner rays travel through a higher refractive index and the outer rays travel through a lower index, the difference in the propagation time period for the outer and inner rays to be transmitted through the same fiber length can be made very small. In practice, such a time difference ΔT can be reduced to less than 1 nsec/km for a core diameter of about 50/µm. In other words, the graded index fibers have much less intermodal dispersions than the step index fibers. Since launching the light into a 50 µm core is fairly straightforward, it is possible to use butt–joint coupling with a semiconductor laser or a LED, as shown in Fig. 13.8.

13.5 FIBER-OPTIC COMMUNICATIONS

Owing to the successful reduction in optical fiber losses, more than 10 million km of fiber has been installed worldwide in various optical networks. Today bit rates for long-haul fiber links typically range between 400 and 600 Mb/sec and in some cases up to 4 Gb/sec, and are beyond the reach of copper cable communication networks. It is expected that fiber-optic communication will play a very important role in the development of the *information superhighway*, which will have a profound impact on many aspects of our society.

A typical fiber-optic communication system is presented in Fig. 13.16, in which light sources of dimensional compatibility

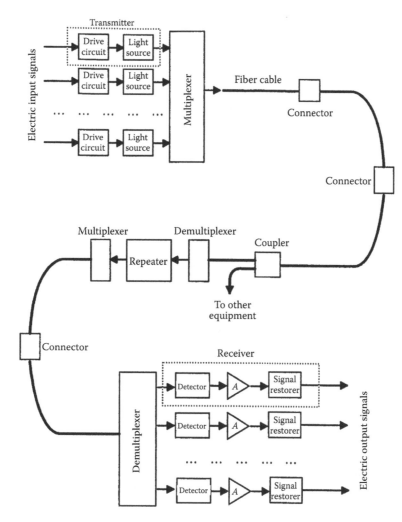

FIG. 13.16 Typical fiber-optic communication system.

with the fiber core can be used. In practice, LEDs and laser diodes can be used, since their output light can be easily modulated. The electrical input signal can be either analog or digital form, although digital signals are utilized in most telecommunication networks. The transmitter circuitry converts the electrical input signals into optical signals by simply varying the current flow through the light source. Output optical signals from the channels are then multiplexed with a multiplexer prior coupled to the optical cable. Connectors and couplers are utilized along the optical cable to extend the fiber link. This has the advantage of restoring the connection after repair and maintenance, and for switching between channels. At the end of the optical cable, the transmitted signals are first demultiplexed and then detected by an array of photocells, for which semiconductor PIN and avalanche photodiodes are commonly used. The array of detected signals is then amplified and reshaped before they are delivered to the user (e.g., telephone or computer terminals).

Optical signal propagation along the fiber will be progressively attenuated and distorted. This is primarily due to scattering, absorption, and dispersion along the fiber. Repeaters are used along the transmission line to amplify and reshape the signal before it becomes significantly deteriorated. An

optical repeater consists of a receiver and a transmitter and can be used to rejuvenate the signal. In other words, this repeater detects the signal, amplifies it, reshapes it, and then sends it down to the optical fiber link. Thus, we see that aside from light sources and photodetectors, long-distance fiber-optic communication requires transmitters, receivers, repeaters, connectors, couplers, multiplexers, and demultiplexers. These are the key components needed for a general optical communication system.

To take advantage of the transmission capacity of an optical fiber, signals from multichannels can be transmitted through a single fiber using multiplexing techniques. There are generally two types of multiplexing methods: TDM and WDM.

Although in practice time-division multiplexing and demultiplexing systems may be complicated, their operational principle is rather simple and straightforward, as can be seen by the two-channel system illustrated in Fig. 13.17. The signals from different channels are first encoded into series of short pulses and then launched into the fiber cable with a time delay δt. As can be seen, the multiplexer can be realized with a mirror and a beam splitter. It should be mentioned that the light waves launched into the fiber are linearly polarized. At the end of the fiber link, the transmitted signal

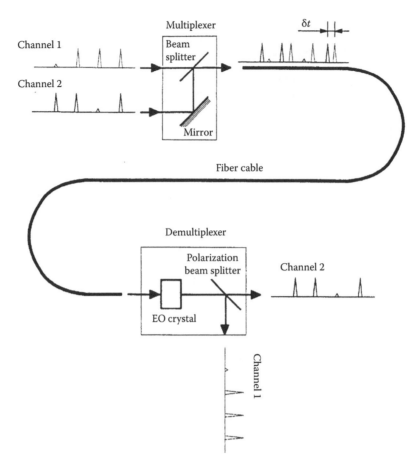

FIG. 13.17   TDM fiber-optic communication.

is demultiplexed by a demultiplexer composed of an electro-optic modulator and a polarization beam splitter. This can be accomplished by rotating the polarizer of channel 1 clockwise and the polarizer of channel 2 counterclockwise. The rotated polarized beams are then separated by a polarized beam splitter and then directed to two photodetectors.

Since the light source used for fiber-optic communication has a narrow spectral width, TDM utilizes only a very small transmission bandwidth. It is apparent that a drastic increase of the transmission capacity of a fiber link can be achieved by transmitting the signals using different wavelengths from different light sources. This is the basis of WDM. Conceptually, the WDM scheme is the same as that for *frequency-division multiplexing* used in microwave radio and satellite communication systems. A block diagram of the WDM system is given in Fig. 13.18, in which we see that the signal in each wavelength channel has been multiplexed via a TDM scheme.

Notice that angular dispersive elements, such as prisms and gratings, can be used as multiplexers in the WDM systems. In other words, light rays from different channels (and hence at different wavelengths) can be directed by a prism or grating at different incident angles. At the output of the multiplexer, these light rays become collinear and can be easily launched simultaneously into an optical fiber. Needless to

say, a wavelength-division demultiplexer works in exactly the reverse fashion as a WDM, by which it directs the light beams of different wavelengths (from the fiber) into their respective channels.

13.6   REMARKS

We emphasize again that the important aspects of using single-mode fiber for information transmission must be the wide bandwidth and low loss. Currently, fabrication of optical fibers has become a relatively mature technology, waiting for us to fully exploit its capability for various possible applications. Even with state-of-the-art optoelectronics-integrated circuit (OEIC) technologies, the potential capability of using optical fiber as applied to information transmission is still far below expectations. In other words, the future development of fiber-optic communication depends on the successful investigation of high-speed and high-density WDM technology, for which operation beyond 20 Gb/sec bit rates is feasible. Added to the recent development of OEIC chips, it is certain that OEIC technology will play a crucial role in developing high-density and high-speed optical communication systems. Nevertheless, coherent optical fiber communication will be the technology that will certainly be used for multichannel WDM applications. In other words, future

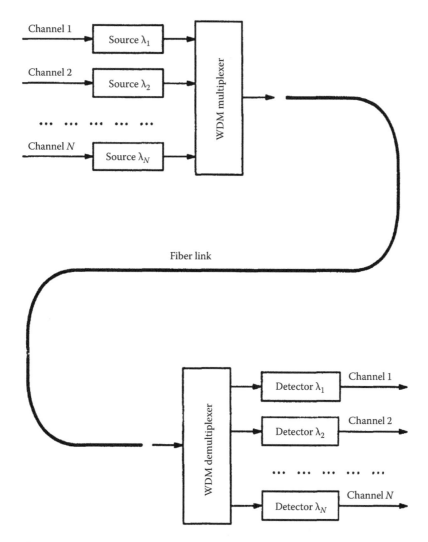

FIG. 13.18   WDM fiber-optic communication.

long-haul information transmission communication will rely on the development of OEIC coherent transmitters and receivers.

One of the new surge applications of fiber optics is the subscriber loop communicaton, which includes the loop-feeder and distributed communication links. The loop-feeder loop network is typically used for longer than a 10 km range, and the distributed subscriber loop is used in more localized communication between 1 and 3 km. Since absorption loss and dispersion are the main issues for long-haul communication, these unfavorable aspects are not a major concern for subscriber loop applications. Instead, the once-for-all installation, frequent upgrading, and maintenance are major issues for localized communication. We anticipate that activities of near-future fiber-optic communication will be more concentrated in localized transmission, for example, home subscriber deployment. Of course, the realization of this new surge of application will heavily depend on the dramatic development of WDM and OEIC technologies.

REFERENCES

1. A. Hasegawa and F. Tappert, Transmission of Stationary Nonlinear Optical Pulses in Dispersive Dielectric Fibers. I. Anomalous Dispersion, *Appl. Phys. Lett.*, vol. 23, 142 (1973).
2. L. F. Mollenauer and K. Smith, Demonstration of Soliton Transmission over More Than 4000 km in Fiber with Loss Periodically Compensated by Raman Gain, *Opt. Lett.*, vol. 13, 675–677 (1988).
3. E. Yamada, K. Suzuki, and M. Nakazawa, Gbit/s Single-Pass Soliton Transmission over 1000 km, *Electron. Lett.*, vol. 27, 1289 (1991).
4. L. F. Mollenauer, B. M. Nyman, M. J. Neubelt, G. Raybon, and S. G. Evangelides, Demonstration of Soliton Transmission at 2.4 Gbit/s over 12000 km, *Electron. Lett.*, vol. 27, 178 (1991).
5. G. Keiser, *Optical Fiber Communications*, 2nd ed., McGraw-Hill, New York, 1991.
6. W. Etten and J. Plaats, *Fundamentals of Optical Fiber Communications*, Prentice Hall, Englewood Cliffs, NJ, 1991.
7. G. P. Agrawal, *Fiber-Optic Communication Systems*, 2nd ed., Wiley-Interscience, New York, 1997.
8. F. T. S. Yu and X. Yang, *Introduction to Optical Engineering*, Cambridge University Press, Cambridge, 1997.

14 Synthetic Radar Imaging with Optics

The origin of radar systems [1] could have evolved from mimicking the acoustic detection of a bat, since bats do not actually see the target but are capable of determining its precise location using the returned acoustic pulse signal. On the other hand, radar pulse is an electromagnetic wave in nature, and it is capable of imaging normal visible light. However, an antenna of hundreds of feet is required to have a reasonable observable solution, and it is not practical to equip an aircraft with one. To overcome this impediment, antennas of a few feet long have been implemented on aircraft using the synthetic aperture radar (SAR) technique for imaging [2,3]. One of the major reasons for the success of SAR imaging is the narrow bandwidth of the microwave emitter, in which it has a long coherence length to ensure that the returned radar pulse can be coherently added with the local reference signal, for which the returned complex microwave front can be recorded on a square-law detector. Since one cannot get something from nothing, the major price paid must be the synthetic scanning (i.e., observation) time, as can be seen in this chapter.

14.1 SYNTHETIC APERTURE RADAR

One of the interesting applications of optical processing is SAR or side-looking antenna imaging [4,5]. Let us consider a side-looking radar system carried by an aircraft (Fig. 14.1), in which we assume that a sequence of pulsed radar signals is directed at the terrain and the return signals along the flight path are received. Let the cross-track coordinate of the radar image be the *ground-range coordinate*, and the along-track coordinate the *azimuth coordinate*. Then the coordinate joining the radar trajectory of the plane of any target under consideration can be defined as the *slant range* (Fig. 14.2). In other words, the *azimuth resolution* is in the order of $\lambda r_1/D$, where λ is the wavelength of the radar signals, r_1 is the slant range, and D is the along-track dimension of the antenna aperture. Since the radar wavelength is several orders of magnitude larger than the optical wavelength, a very large antenna aperture D is required to have an angular resolution comparable to that of a photoreconnaissance system. The required antenna length may be hundreds or even thousands of feet, which is impractical to realize on aircraft. Nevertheless, this difficulty can be resolved by means of the *synthetic aperture technique*, as will be shown in the following.

Let us assume that the aircraft carries a small side-looking antenna that is a few feet long. It emits a relatively broad beam radar signal to scan the terrain, by virtue of the aircraft motion. The radar pulses are emitted in a sequence of positions along the flight path, which can be treated as if they were the positions occupied by the elements of a linear antenna array. The return radar signal at each of the respective positions can be coherently recorded as a function of time with a local reference

signal, of which the amplitude and phase quantities (i.e., complex amplitude) are simultaneously encoded. Thus, we see that the returned complex waveforms can be synthesized as equivalent to a long-aperture-element antenna.

14.2 RADAR DATA FORMAT SYNTHESIS

For simplicity of analysis, we first consider a point target problem, which can be quickly extended to a multitarget situation owing to the spatial invariance condition. Since the radar signal is produced by periodic rectangular pulses modulated with a sinusoidal signal, the periodic pulses give the range information and the azimuth resolution is provided by the distance traveled by the aircraft between pulses, which is smaller than $\pi/\Delta p$, where Δp is the spatial bandwidth of the terrain reflections. Thus, the returned radar signal can be written as

$$S_1(t) = A_1(x_1, r_1) \exp\left\{ i\left[\omega t - 2kr_1 - \frac{k}{r_1}(x - x_1)^2 \right] \right\} \quad (14.1)$$

where A_1 is an appropriate complex constant.

Now if we consider that the terrain at range r_1 consists of a collection of N point targets, then by superposing the returned N target signals, we have

$$S(t) = \sum_{n=1}^{N} S_n(t)$$

$$= \sum_{n=1}^{N} A_n(x_n, r_1) \exp\left\{ i\left[\omega t - 2kr_1 - \frac{k}{r_1}(vt - x_n)^2 \right] \right\}$$

$$(14.2)$$

where v is the velocity of the aircraft.

By synchronously demodulating this returned radar signal, we have

$$S(t) = \sum_{n=1}^{N} \left| A_n(x_n, r_1) \right| \cos\left[\omega_c t - 2kr_1 - \frac{k}{r_1}(vt - x_n)^2 + \phi_n \right]$$

$$(14.3)$$

where ω_c is the arbitrary carrier frequency, and ϕ_n is an arbitrary phase angle.

In order to display the return radar signal of Eq. (14.3) on a cathode ray tube, the demodulated signal can be used to modulate the intensity of an electron beam, which is swept vertically across the cathode ray tube in synchronism with the return radar pulses (Fig. 14.3). If the modulated cathode ray display is imaged on a strip of photographic film

FIG. 14.1 Side-looking radar.

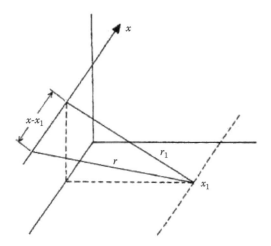

FIG. 14.2 Geometry of side-looking radar.

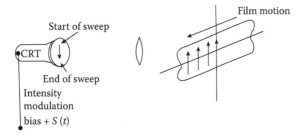

FIG. 14.3 Recording of the returned radar signals for subsequent optical processing.

(i.e., a square-law detector), which draws with a constant horizontal velocity, the successive radar range traces (or slides of the radar images) can be recorded side by side on a moving photographic film, as shown in Fig. 14.3. A recorded two-dimensional scanned radar format transparency can be obtained as illustrated in Fig. 14.4.

We further note that the vertical lines represent the successive range sweeps, and the horizontal dimension represents the azimuth position, which is the sample version of $s(t)$. This sampling is carried out in such a way that by the time the samples are recorded on the film, it is essentially indistinguishable from

FIG. 14.4 Radar signal recording format.

the unsampled version. In this recording, the time variable has been converted into a space variable, as defined in terms of distance along the recorded film. We emphasize that the minimum amount of time required for the azimuth resolution is in fact the amount of time paid for the azimuth resolution. Thus, we see that the recorded transparency represents the azimuth history of the returned radar signals.

By considering only the data recorded along the line y-y_1 on the film, the transmittance can be written as

$$T(x_1, y_1) = K_1 + K_2 \sum_{n=1}^{N} |A_n(x_n, r_1)|$$

$$\times \cos\left[\omega_x x - 2kr_1 - \frac{k}{r_1}\left(\frac{v}{v_f}x - x_n\right)^2 + \phi_n\right]$$

(14.4)

where K_S are bias constants, $x = v_f t$, v_f is the speed of film motion, and $\omega_f = \omega_c/v_f$.

For simplicity in illustration, we restrict ourselves to a *single target* problem (i.e., $n = 1$), for which the first exponential term of the cosine function in Eq. (14.4) can be written as

$$T_1(x_1, y_1) = C \exp(i\omega_x x) \exp\left[-i\frac{k}{r_1}\left(\frac{v}{v_f}\right)^2\left(x - \frac{v_f}{v}x_j\right)^2\right]$$

(14.5)

We see that it is essentially a one-dimensional *Fresnel lens* (or a one-dimensional hologram) equation. The first linear phase factor $(\omega_x x)$ represents a prism with an oblique angle of

$$\sin\theta = \frac{\omega_x}{k_1}$$

(14.6)

where $k_1 = 2\pi/\lambda_1$, with λ_1 a wavelength of the illuminating light source. The second term represents a cylindrical lens equation centered at

$$x = \frac{v_f}{v}x_j$$

(14.7)

which has a focal length of

$$f = \frac{1}{2}\frac{\lambda}{\lambda_1}\left(\frac{v_f}{v}\right)^2 r_1 \qquad (14.8)$$

Needless to say, for N targets, the corresponding cylindrical lens equations can be shown as

$$x = \frac{v_f}{v}x_n, \quad n = 1, 2, \cdots, N \qquad (14.9)$$

14.3   RADAR IMAGING WITH OPTICS

In order to reconstruct the recorded radar image format, the scanned radar transparency is illuminated by a monochromatic plane wave, as depicted in Fig. 14.5. It shows that the real and virtual images can be reconstructed relative to the positions of the point images along the foci of the lenslike structure of the recorded transparency, which were determined by the positions of the point targets.

However, the reconstructed images will be spread in the y direction since the transparency was realized in a one-dimensional function along $y = y_i$, which exerts no focal power in this direction.

Since it is our aim to reconstruct an image not only in the azimuth direction but also in the range direction, it is necessary to image the y coordinate directly onto the focal plane of the azimuth image. Thus, to construct the terrain map one must image the y coordinate of the transmitted signal onto a tilted plane determined by the focal distances of the azimuth direction. This imaging procedure can be accomplished by inserting a positive conical lens immediately behind the scanned radar image transparency, as shown in Fig. 14.6

The required transmittance function of the conical lens can be shown by

$$T_1(x_1, y_1) = \exp\left(-i\frac{k_1}{2f}x^2\right) \qquad (14.10)$$

Thus, we see that it is possible to remove the entire tilted plane of all the virtual diffraction to the point at infinity, while leaving the transmittance in the y direction unchanged.

If the cylindrical lens is placed at the focal distance from the scanned radar transparency, as shown in Fig. 14.6, it can create a virtual image of the y direction at infinity. In this setup, we see that the azimuth and range images (i.e., x and y directions) are all imaged at the point of infinity, and they can be brought back to a finite distance with a spherical lens. Thus, we see that radar imagery can be constructed at the output plane of the optical processor shown in Fig. 14.6, using a narrow slit aperture.

An example of radar imagery obtained with this optical processing setup is shown in Fig. 14.7, in which a variety of

FIG. 14.7   SAR radar image of Monroe County, Michigan, area. (From L. J. Cutrona. With permission.)

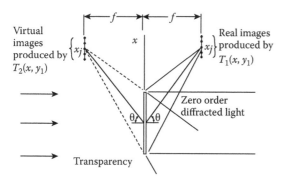

FIG. 14.5   Image reconstruction produced by the film transparency.

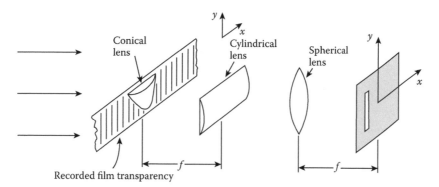

FIG. 14.6   Optical processing system for imaging synthetic antenna data.

scattered images in the Monroe County, Michigan, area can be visualized [4]. This includes the city streets, wooded areas, and farmlands. Lake Erie, with some ice floes, can also be seen on the right side in this side-looking radar imagery.

14.4 REMARKS

The major advantages of optical information processing are the unique space interconnectivity, high density, and parallel processing capability. These are the trademarks for using the optical information processing technique. We have shown that SAR imaging can be processed with optics, which was the first time that the naked human eye could actually see radar images with relatively good resolution. Since optical processing is generally a two-dimensional spatial processor, we have shown, by exploiting the two-dimensional processing capability, that a very large space-bandwidth signal, such as that of synthetic radar data, can actually be processed with a conventional coherent optical processor.

We further note that although optical data processing offers high density and parallel processing capabilities, its application to side-looking radar imaging is not an onboard processing technique. Nonetheless, with the advances of current digital processing technology, a synthetic aperture data imaging system has been implemented for onboard processing [6].

REFERENCES

1. M. I. Skolnik, *Introduction to Radar*, 3rd ed., McGraw-Hill, New York, 2001.
2. L. J. Cutrona, E. N. Leith, C. J. Palermo, and L. J. Porcello, Optical Data Processing and Filtering System, *IRE Trans.*, vol. IT-6, 386 (1960).
3. L. J. Cutrona, W. E. Vivian, E. N. Leith, and G. O. Hall, A High Resolution Radar Combat-Surveillance System, *IRE Trans.*, vol. MIL-5, 127 (1961).
4. L. J. Cutrona, E. N. Leith, L. J. Porcello, and W. E. Vivian, On the Application of Coherent Optical Processing Techniques to Synthetic-Aperture Radar, *Proc. IEEE*, 54, 1026 (1966).
5. F. T. S. Yu, *Introduction to Diffraction, Information Processing, and Holography*, MIT Press, Cambridge, MA, 1973.
6. I. G. Cumming and F. H. Wong, *Digital Processing of Synthetic Radar Data*, Artech House, Boston, 2005.

15 Wideband Signal Analysis with Optics

One of the major advantages of an optical processor is its inherent two-dimensional processing capability, which offers myriad high-spatial-resolution parallel processing operations. One of the applications is synthetic aperture radar processing, as described in Chapter 14. It is therefore reasonable to utilize the two-dimensional processing ability of optics for a large time-bandwidth signal application. Similar to the scanned radar image format, a broadband time signal can be obtained with a raster scan format for information processing. In other words, by converting a long string of large time-bandwidth signals into a two-dimensional raster scan format, a wideband time signal analysis can be continuously carried out with a conventional two-dimensional optical data processor.

15.1 TIME SIGNAL TO TWO-DIMENSIONAL SPATIAL CONVERSION

Another interesting application of optical processing is wideband time signal analysis [1,2]. One of the basic limitations of the multichannel optical analyzer is that the resolution is limited by the width of the input aperture, and this limitation can be alleviated by using a two-dimensional raster scan input transparency, which is similar to the synthetic aperture radar scanned format, shown in Fig. 15.1.

Thus, we see that if scanned format transparency is presented to the input plane of an optical Fourier transformer (Fig. 15.2), the complex light distributed at the output plane is the Fourier transform of the input scanned format. Thus, a wide space-bandwidth signal analysis can be continuously processed using a long string of scanned format transparency, similar to a strip of movie film.

Let us assume that a broadband signal is raster scanned on a spatial light modulator (or on a strip of transparency), in which we assume that the scan rate has been adjusted to accommodate the maximum frequency content of the broadband signal, as represented by

$$v \geq \frac{f_m}{R} \tag{15.1}$$

where v is the scan velocity, f_m is the maximum frequency content of the broadband signal, and R is the resolution limit of the optical system.

15.2 SPECTRUM ANALYSIS WITH OPTICS

If we present the scanned format transparency at the input plane of a coherent optical Fourier transformer as shown in

Fig. 15.2, then the complex light distributed at the output plane can be shown as

$$F(p,q) = C_1 \sum_{n=1}^{N} \mathrm{sinc}\left(\frac{qa}{2}\right)\left\{\mathrm{sinc}\left(\frac{pw}{2}\right) * F(p)\right.$$
$$\times \exp\left[i\frac{pw}{2}(2n-1)\right]\right\}$$
$$\times \exp\left\{-i\frac{q}{2}\left[h - 2(n-1)b - a\right]\right\} \tag{15.2}$$

where C_1 is a complex constant, and

$$F(p) = \int f(x')\exp(-ipx')\,dx'$$

$$x' = x + (2n-1)\frac{w}{2}, \text{ and } \mathrm{sinc}\,X \triangleq \frac{\sin X}{X} \tag{15.3}$$

For simplicity of analysis, we assume that the broadband format is a *single sinusoidal signal*:

$$f(x') = \sin p_0 x' \tag{15.4}$$

Then the output Fourier distribution can be shown as

$$F(p,q) = C_1 \mathrm{sinc}\left(\frac{qa}{2}\right) \sum_{n=1}^{N}\left\{\mathrm{sinc}\left(\frac{pw}{2}\right) * \frac{1}{2}\left[\delta(p-p_0)\right.\right.$$
$$\left.+ \delta(p+p_0)\right] \times \exp\left[i\frac{wp_0}{2}(2n-1)\right]\right\}$$
$$\times \exp\left\{-i\frac{q}{2}\left[w - 2(n-1)b\right]\right\} \tag{15.5}$$

To simplify the analysis, we consider only one of the components; thus, we have

$$F_1(p,q) = C_1 \mathrm{sinc}\left[\frac{w}{2}(p-p_0)\right]\mathrm{sinc}\left(\frac{qa}{2}\right)\exp\left(-i\frac{qw}{2}\right)$$
$$\times \sum_{n=1}^{N} \exp\left[i\frac{1}{2}(2n-1)(wp_0 + bq)\right] \tag{15.6}$$

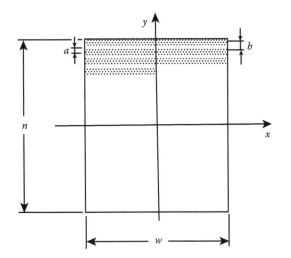

FIG. 15.1   Raster scan input signal transparency.

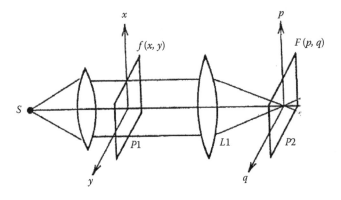

FIG. 15.2   Optical Fourier transformer. S, monochromatic point source; $P1$, input plane; $P2$, output Fourier plane.

The corresponding intensity distribution can be written as

$$I_1(p,q) = |F_1(p,q)|^2 = C_1^2 \operatorname{sinc}^2\left[\frac{w}{2}(p - p_0)\right]\operatorname{sinc}^2\left(\frac{qa}{2}\right)$$

$$\times \sum_{n=1}^{N} \exp[i2n\theta] \sum_{n=1}^{N} \exp[-i2n\theta] \qquad (15.7)$$

where

$$\theta = \frac{1}{2}(wp_0 + bq) \qquad (15.8)$$

By utilizing the following identity,

$$\sum_{n=1}^{N} e^{i2n\theta} \sum_{n=1}^{N} e^{-i2n\theta} = \left(\frac{\sin N\theta}{\sin \theta}\right)^2 \qquad (15.9)$$

Equation (15.7) can be deduced to the following form:

$$I_1(p,q) = C_1^2 \operatorname{sinc}^2\left[\frac{w}{2}(p - p_0)\right]\cdot\operatorname{sinc}^2\left(\frac{qa}{2}\right)\cdot\left(\frac{\sin N\theta}{\sin \theta}\right)^2$$

$$\qquad (15.10)$$

By carefully analyzing the preceding equation, we found that the first sinc factor represents a relatively narrow spectral line in the p direction, located at $p = p_0$, which is derived from the Fourier transform of a truncated sinusoidal signal within the width of the input transparency. The second sinc factor represents a relatively broad spectral band in the q direction, which is determined by the Fourier transform of a rectangular pulse width (i.e., the channel width). And the last factor deserves special mention for a large number of scanned lines N, by which the factor approaches a sequence of narrow purses located at

$$q = \frac{1}{b}(2\pi n - wp_0), \quad n = 1, 2, \ldots \qquad (15.11)$$

We note that this factor yields a very significant fine spectral resolution in the q direction.

Let us now confine ourselves within a relatively narrow region in the p direction, which is the half-width of the spectral spread:

$$\Delta p = \frac{2\pi}{w} \qquad (15.12)$$

This in fact equals the resolution limit of an ideal Fourier transform lens.

In the q direction, the irradiance is first confined within a relatively broad spectral band (which primarily depends on the channel width) centered at $q = 0$, and it is modulated by a sequence of narrow periodic pulses. The half-width of the broad spectral band is

$$\Delta q = \frac{2\pi}{a} \qquad (15.13)$$

We further notice that the separation of the narrow pulses is obtained from a similar equation,

$$\Delta q_1 = \frac{2\pi}{b} \qquad (15.14)$$

Therefore, as one investigates Eq. (15.14), there will be only a few pulses located within the spread of the broad spectral band for each given p_0, as depicted in Fig. 15.3. In it, we see that the actual location of any of the pulses is determined by the signal frequency. Thus, if the signal frequency changes,

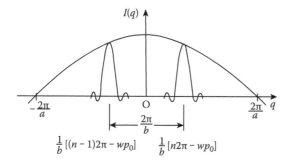

FIG. 15.3 Broad spectral band modulated by a sequence of narrow pulses.

the position of the pulses also changes, as shown in the following relationship:

$$dq = \frac{w}{b} dp_0 \tag{15.15}$$

We further note that the pulse displacement in the q direction is proportional to the displacement in the p direction. The output spectrum includes the required frequency discrimination and is equivalent to analysis of the entire signal in one continuous channel. In order to avoid the ambiguity in reading the output plane, all but one of the periodic pulses should be ignored. This may be accomplished by masking the entire output plane except the region, shown in Fig. 15.4. Thus, we see that as the input signal frequency advances, one pulse will eventually leave the open region and another pulse will emerge in the open region. This results in a single bright spot, as scanned out diagonally at the output plane. The frequency locus of the input signal can be determined in cycles per second. To remove the nonuniformity of the second sinc factor, a weighting density transparency may be implemented at the output plane of the processor.

Although our analysis was carried out on a one-frame basis, the operation can actually be performed in a continuously running manner, as the input scanned raster format is continuously moving in the q direction over the input plane of the optical processor. Thus, we see that a long string of broadband spectrum analysis can be continuously displayed at the output plane, as the strip of scanned format moves across the input aperture of the optical Fourier transformer.

15.3 PROCESSING WITH AREA MODULATION

Area modulation for movie sound track recordings has been used over the years, and the formats can also be used for spectrum analysis. Two formats have been used, namely, unilateral and bilateral area modulation [3,4], as illustrated in Fig. 15.5.

Although area modulation has the advantages of simple implementation and the ability to avoid nonlinearity of density modulation, there is a trade-off for input spatial-limited resolution. Let us now describe the area modulation transmittance functions for unilateral and bilateral, respectively, as follows:

$$T_1(x,y) = rect\left(\frac{x}{L}\right) rect\left\{\frac{y}{2[B + f(x)]}\right\} \tag{15.16}$$

and

$$T_2(x,y) = rect\left(\frac{x}{L}\right) rect\left\{\frac{y - [f(x) - B]/2}{[B + f(x)]}\right\} \tag{15.17}$$

where L is the width of the transmittance, B is the appropriate bias, $f(x)$ is the input function, and the rectangular function is described as

$$rect\left(\frac{x}{L}\right) \triangleq \begin{cases} 1, & |x| \leq L/2 \\ 0, & \text{otherwise} \end{cases} \tag{15.18}$$

If an area-modulated transmittance is provided at the input plane to an optical Fourier transformer (Fig. 15.2), the output light distributions will be, respectively,

$$
\begin{aligned}
G_1(p,0) &= 2C \int rect\left(\frac{x}{L}\right)[B + f(x)] \exp(-ipx)\,dx \\
&= 2C \int_{-L/2}^{L/2} B \exp(-ipx)\,dx \\
&\quad + 2C \int_{-L/2}^{L/2} f(x) \exp(-ipx)\,dx
\end{aligned} \tag{15.19}
$$

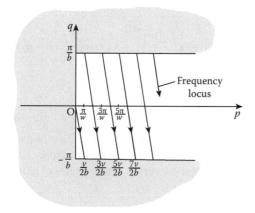

FIG. 15.4 Frequency loci at the output spectral plane.

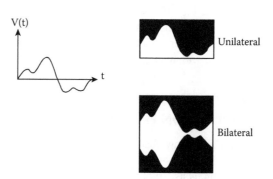

FIG. 15.5 Unilateral and bilateral area modulation formats.

FIG. 15.6 Speech spectrogram obtained by optical processing with area modulation. (Courtesy of C. C. Aleksoff.)

and

$$
\begin{aligned}
G_2(p,0) &= \int T_2(x,y)\exp(-ipx)\,dx \\
&= C\left[\int_{-L/2}^{L/2} B\exp(-ipx)\,dx + \int_{-L/2}^{L/2} f(x)\exp(-ipx)\,dx\right] \\
&= \frac{1}{2}G_1(p,0)
\end{aligned}
\tag{15.20}
$$

It is evident that if is a bilateral sinusoidal area function is implemented as

$$
f(x) = A\sin(p_0 x + \theta)
\tag{15.21}
$$

then the output light distribution can be shown as

$$
G_1(p,0) = K1\delta(p) + K_2\delta(p - p_0) + K_3\delta(p + p_0)
\tag{15.22}
$$

Thus, we see that two spectral points located at $p = p_0$ and $p = -p_0$ can be respectively detected.

One of the apparent advantages of using area modulation as applied to optical information processing is that a larger class of recording material can be used, since the input transmittance does not need to be restricted within the linearity region of the encoding device.

For illustration, Fig. 15.6 shows a section of sound spectrogram as obtained with the area modulation technique. In it, we see that optical generation of a speech spectrogram can easily be obtained with a simple optical Fourier transformer.

15.4 BROADBAND AREA MODULATION PROCESSING

In this section, we show that area modulation can also be extended, similar to density modulation as applied to broadband spectrum analysis. In other words, either a bilateral or unilateral area modulation scanned raster format can be easily

FIG. 15.7 Geometry of area-modulated input for wideband processing.

applied by spatial light modulator or film, as illustrated in Fig. 15.7.

For example, the transmittance of a bilateral area modulation can be written as

$$
f(x,y) = \sum_{n=1}^{N} rect\left(\frac{x}{L}\right) rect\left\{\frac{y - nD}{2\left[B + f_n(x)\right]}\right\}
\tag{15.23}
$$

where N is the number of scanned lines.

If this area-modulated transmittance of Eq. (15.23) is presented at the input plane of an optical Fourier transformer, then the output light distribution can be shown as

$$
\begin{aligned}
F(p,q) = \sum_{n=-N}^{n=N} C\int_{-L/2}^{L/2} &\left[B + f_n(x)\right]sinc\left\{q\left[B + f_n(x)\right]\right\} \\
&\times \exp\left[-i(px - nDq)\right]dx
\end{aligned}
\tag{15.24}
$$

For the sake of simple analysis, we restrict ourselves to a single sinusoidal signal problem. Then the corresponding output intensity distribution can be shown as

$$
\begin{aligned}
I_1(p,q) = &\left[K_1 sinc\left(\frac{LP}{2}\right)\right]^2 + K_2\left\{sinc\left[\frac{L}{2}(p - p_0)\right]\right\}^2 \\
&\times \left(\frac{\sin N\theta}{\sin\theta}\right)^2
\end{aligned}
\tag{15.25}
$$

We see that the result is similar to the density-modulating case, as we have shown before. For a large number of scanned lines N, the last factor converges to a sequence of narrow periodic pulses located at

$$
q = (1/D)(2n\pi - Lp_0)\quad n = 1,2\ldots
\tag{15.26}
$$

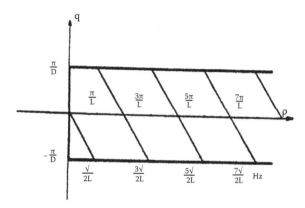

FIG. 15.8 Output spectrum frequency locus.

FIG. 15.9 Output spectrum of an area-modulated sinusoidal input. The pair of inserted white lines represents the slit boundaries.

Thus, we see that as the input frequency changes, the location of the pulses also changes:

$$dq = (L/D)dp_0 \qquad (15.27)$$

in which we see that the displacement in the q direction is proportional to the displacement in the p direction. Since the pulse width decreases as the number of scan lines N increases, the output spectrum would yield a frequency discrimination equivalent to that obtained with a continuous signal of NL.

We further note that as the input frequency advances, one of the pulses leaves the open aperture at $q = -\pi/D$ and a new pulse enters at $q = \pi/D$, which results in a diagonal frequency locus, as shown in Fig. 15.8.

We emphasize again that area modulation offers the advantage of simple implementation, in which the nonlinearity of density modulation can be avoided. However, the available space-bandwidth product for the processor is smaller. Aside from this drawback, compared with density modulation, by efficiently utilizing the two-dimensional area modulation format, its large space-bandwidth product may still warrant some specific applications, for example, application to digital implantation of a scanned format.

For illustration, Fig. 15.9 shows an example of the output spectral distribution as obtained from a sinusoidal input using area modulation. The inserted parallel lines represent the slit aperture of the spectrum locus limits. In this figure, we can see that there was no apparent spurious effect, which is due to

the nonlinearity of density modulation. However, one of the major drawbacks of using area modulation is that it requires a larger input spatial domain.

15.5 REMARKS

One of the major advantages of a conventional optical processor is its two-dimensional spatial data processing. In this chapter, we have shown that broadband time signals can be easily implemented using a raster scan technique, by which a continuous spectrogram analysis can be continuously carried out with a strip of encoded format or spatial light modulator. With the availability of real-time electro-optic devices (e.g., spatial modulators and detectors), a near-real-time optical information processor can be realized in practice, as has been illustrated in Chapter 11.

REFERENCES

1. C. E. Thomas, Optical Spectrum Analysis of Large Space-Bandwidth Signals, *Appl. Opt.*, vol. 5, 1782 (1966).
2. F. T. S. Yu, *Introduction to Diffraction, Information Processing, and Holography*, MIT Press, Cambridge, MA, 1973.
3. A. Tai and F. T. S. Yu, Wide-Band Spectrum Analysis with Area Modulation, *Appl. Opt.*, vol. 18, 460 (1979).
4. F. T. S. Yu, S. Jutamulia, and S. Yin, *Introduction to Information Optics*, Academic Press, San Diego, 2001.

16 Information in Myth of Science and Creation

Since every substance has a price tag of information that includes all the building blocks in our universe, one cannot simply ignore information when dealing with science. We have shown that there is a profound connection between information and entropy, a quantity that has been well accepted in science. Without this connection, information would be more difficult to apply in science. Two of the most important pillars in modern physics must be Einstein's relativity theory and Schrödinger's quantum mechanics. However, there exists a profound connection between them, by means of Heisenberg's uncertainty principle. Due to the uncertainty relation, we see that every bit of information takes time and energy to transfer, create, and observe. Since one cannot create something from nothing, anything to be created needs a huge amount of energy and a great deal of entropy to make it happen. My question is, can we afford it?

16.1 SPACE AND INFORMATION

Let us start with a bit of information provided by a unit space. Notice that a physical space we are dealing with is in fact not empty; otherwise, it would be a null space or absolute empty, which does not exist in practice [1]. In reality, any physical space should fill with a substance or substances, such as all the subspaces within our universe. We further stress that a virtual reality is an imaginary reality without the support of the current laws of science and rule of time, while the real reality is supported by the laws of science and rule of time.

Let us look now at a unit information cell or a unit space, as illustrated in Fig. 16.1a.

In order to make this unit cell a physically reality, we should first look at the time–space reality. Time is one of the most intriguing variables in our physical science that governs all the laws of physics. Without time, there would be no space, no substance, and no life [1]. Thus, time must be the supreme law of science, by which all the laws of science cannot be dissociated with time.

Let us look at a unit cell as becoming a physical real space. To do so, we would take the ultimate limit in the current laws of physics: the speed of light. With the imposition by speed of light C, each dimension can be described by $d = C\Delta t$, where Δt is time duration. Nonetheless, this unit space is still not a physical space yet, since it is absolutely empty or null. As we know, an absolute empty space cannot be embedded in a real space [1], and we have to fill up the space with a substance or substances. By filling up the space with a substance or substances, it will become a real unit space as supported by the laws of science.

If we further assume that this unit cell provides a dual-level signal at a time (i.e., a white or black color), as shown in Fig. 16.1b, then the information content provided by this unit cell can be written as [2,3]

$$I = \log_2 2 = 1 \text{ bit} \tag{16.1}$$

There is a question that remains to be asked: What would be the lowest dimension of a unit information cell? The answer is that the lowest limit must be imposed by Heisenberg's uncertainty principle, as described in Section 2.4. In other words, it is the time duration Δt and the light speed C that determine the size of a unit information cell, under the constraint of current laws of science.

Let us further extend this concept to a three-dimensional $N \times M \times H$ cubic cell structure, as shown in Fig. 16.2.

Then the information provided at a given time of these cubic cells can be shown as

$$I = (N \times M \times H) \text{ bits} \tag{16.2}$$

in which we have again assumed that the occurrence of a black or white cell at a time is equally probable, which is the maximum information contents of the cubic cell provided.

Let us now extend the cubic cell example slightly, for which we assume each unit cell is equally capable of giving one of W distinguishable gray-level signals at any given time; then the overall information provided by this three-dimensional cubic structure is given as

$$I = (N \times M \times H)(\log_2 W) \text{ bits} \tag{16.3}$$

This is in fact the amount of information needed to create this three-dimensional cubic information structure, with the assumption of 100% certainty.

16.2 TIME AND INFORMATION

Vocal voice is a typical example of a temporal information or time signal, and the telegram (Morse code) signal is another example. A page of an article or a picture is a typical example of a space information or spatial signal. Needless to say, spatial and temporal signals can be interfaced or interweaved for temporal transmission. For instances, a TV display is an example of exploiting temporal information transmission for spatial information display, an old-fashioned movie sound track is an example of exploiting spatial information for

FIG. 16.1 Information cell: (a) virtual unit cell and (b) binary information cell.

FIG. 16.2 Cubic $N \times M \times H$ information cells.

temporal information transmission, and a continuously running TV video is a typical illustration for spatial and temporal information transmission.

Let us now look at the information content of a time signal as depicted in Fig. 16.3a.

As we know, any temporal signal can be digitized in binary form for temporal transmission or time signal transmission, as

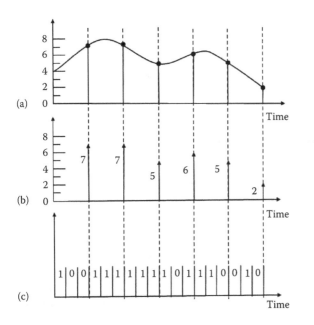

FIG. 16.3 Analog-to-digital signal conversion. (a) Analog signal. (b) Sampled signal. (c) Digital signal.

illustrated (i.e., 0 and 1) in Fig. 16.3b and c. This is precisely the binary or digital format we use in current communication and computer systems. Most people, including some engineers and scientists, know how the digital system works, yet some of them may not know why we developed it.

Now, let us start with the major differences between the digital and analog systems, as follows. The digital system operates in binary form (i.e., 0, 1), while the analog system operates in analog form (i.e., multilevel). The digital system provides lower (temporal) information content (e.g., 1 bit per level), while the analog system provides higher (temporal) information content (e.g., more bits per level). Since the information content of a digital system is lower than that of an analog system, why do we go through all the trouble to transform the analog to digital, and then back to analog for the receiver? The answer is that by exploiting the transmission in light speed, it can carry a lot of temporal information at that velocity, which is precisely the price paid for the digital time transmission.

The major purpose of using digital transmission is for noise immunity; otherwise, we would not pay for a longer data rate transmission. The reason is that in digital transmission, the signal can be *repeated* (or refreshed), in contrast with analog signal transmission. One may see that after a few cycles of amplifications, an analog signal will be severely contaminated by noise, while in digital signal transmission, the transmitted signal can be easily refreshed by means of a repeater. Thus, a digital signal can be transmitted over thousands of miles by consecutively repeating with digital repeaters, and the transmitted digital signal is just as good as the original.

As an example, if one consecutively copies a compact disc or DVD many times, one will discover that the latest copy is just as good as the original one. Although the digital transmitted signals, strictly speaking, are not real-time ones, they appear to be very close to real-time signals because they were transmitted at light speed. And this is precisely the price we pay for digital transmission at light speed.

Time is one of the most esoteric quantities in the myth of science; it has no weight or volume, no beginning or end. It cannot stand still or even go backwards. Besides that, every physical substance temporally exists, constantly transforming or changing with time. In fact, every piece of information can only convey with time. Without the existence of time, all the information would not be able to convey and observe. In other words, every piece of information takes time and energy to be transmitted, and it is not free. A trivial example is that if one could travel at the speed of light, as illustrated in Fig. 16.4, one could not see himself with a mirror in front of him since there is no expenditure of time. Thus, we see that it takes time for a piece of information to convey.

FIG. 16.4 Travel at speed of light.

16.3   ENTROPY AND INFORMATION

In reality, our world, including the present universe, is fully filled with information. Then there is a question, is there a connection between information and science? In other words, without a physical quantity to represent information, information would be very difficult to apply and connect with science.

Nevertheless, one of the most intriguing laws in the second law of thermodynamics must be the law of entropy, as developed by Boltzmann [4]:

$$S = k \ln N \qquad (16.4)$$

where k is Boltzmann's constant, ln is the natural log, and N is the total number of possible distinguishable states. In view of this, the information measure (e.g., in bits) is

$$I = -\log_2(p) \qquad (16.5)$$

where p is the probability of a possible state. There exists a profound connection between information and the law of entropy. Both equations are statistically in nature, and both are written in logarithmic forms. It is apparent that every bit of information is closely associated to an amount of entropy, a quantity that has been well accepted in science [5]. In other words, every piece of information is equivalent to an amount of entropy, and it can be simply traded as shown in Section 4.3,

$$\Delta I \leftrightharpoons \Delta S \qquad (16.6)$$

Furthermore, this relationship is one of the most significant connections of information with science; otherwise, information would not have any physical meaning as applied to science.

16.4   SUBSTANCE AND INFORMATION

Every substance has a piece of information that includes all the elementary particles, the basic building blocks of all elements, atoms, paper, our planet, our solar system, our galaxy, and even our universe. In other words, the universe is flooded with information (i.e., spatial and temporal), or the information fills up the whole universe. Strictly speaking, when one is investigating the origin of the universe, which includes time–space as well as the existence of life, the aspect of information cannot be absent. Then, one would ask, what would be the amount of information, aside from the needed energy, that is required to create a specific substance? Or equivalently, what would be the cost of entropy to create it?

There are several kinds of spatial information, or simply information, that we need to define before our discussion. From a physical standpoint, there are two types of spatial information that I will define: static (or pro-passive type) and dynamic (or proactive type) information. Static information is primarily created by the consequences of Mother Nature's law

of physics. For instance, an amount of information is needed to create an atom or an elementary particle. In other words, static information is not created by a bioevolutional process or man-made. Static information was mostly formulated by complicated nuclei and chemical reactions or interactions. It is not created due to intelligent or human intervention.

Samples of static information, such as the information for an elementary particle, an electron, a proton, and various atomic particles (e.g., hydrogen atom, oxygen atom, uranium, and gold), are typical examples. In fact, to actually produce those subatomic particles, aside from the extreme nuclei interaction (as we will discuss later), an amount of information (or entropy) is required for a successful creation. This process cannot be done by simple human intervention or by an evolutional process, but must be done by means of statistical consequences from nuclei and chemical interactions, similar to how our universe, solar system, and planets were formed.

On the other hand, dynamic information is mostly man-made or by biological (i.e., evolutional) processes. In fact, the dynamic information that I intend to mention is not created by Mother Nature, but is either generated by evolutional processes or man-made. Nevertheless, dynamic information has two subcategories: artificial and self-organized information. Artificial information is man-made, while self-organized information is self-produced or by means of adaptation, such as the information contents of DNA, genes, the immune system, and biological cells.

Practically all the things made by human are considered artificial, which includes tools and devices such as watches, computers, battleships, and automobiles. Nonetheless, each of the devices has its own signature associated with a specific amount of information. And this is the amount of information or artificial information needed, aside from time and energy, for the creation to happen. Again, the information needed for creation is not free, and it will cost a lot of entropy.

16.5   UNCERTAINTY AND INFORMATION

One of the most fascinating principles in quantum mechanics [6] is Heisenberg's uncertainty principle [7]:

$$\Delta x \cdot \Delta p \geq h \qquad (16.7)$$

where Δx and Δp are the position and momentum errors, respectively, and h is Planck's constant. The principle implies that one cannot detect or observe the position and momentum (e.g., of a particle) simultaneously, within an error of Planck's constant or the h region. Needless to say, Heisenberg's uncertainty relation can also be written in various forms [7], as follows:

$$\Delta E \cdot \Delta t \geq h \qquad (16.8)$$

$$\Delta v \cdot \Delta t \geq 1 \qquad (16.9)$$

where ΔE, Δt, and Δv are energy, time, and spectral resolution, respectively.

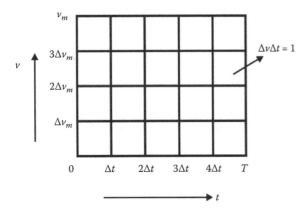

FIG. 16.5   Unit information cell.

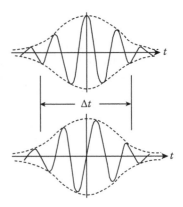

FIG. 16.6   Set of Gaussian orthogonal wavelets: cosine and sine wavelets.

We emphasize that the lower bound of $\Delta v \cdot \Delta t \geq 1$ offers a very significant relationship for information or, equivalently, equals an information cell, as shown by Gabor [8] and illustrated in Fig. 16.5, where v_m and T denote the frequency and time limits of a time signal. As we can see, every bit of information can be efficiently conveyed or transmitted if and only if it is operating under the constraint of Heisenberg's uncertainty principle (i.e., $\Delta v \cdot \Delta t \geq 1$). This relationship implies that the signal bandwidth should be either equal to or smaller than the system bandwidth, that is, $1/\Delta t \leq \Delta v$, for which we see that Δt and Δv can be traded. It is, however, the *unit region*, and not the shape of the information cell, that determines the limit.

We further stress that the boundary of the information cells (i.e., Heisenberg's uncertainty limit) was determined by the mean square error criterion, for which a signal conveyed under a Gaussian noisy environment can be transmitted with high efficiency. In this case, a Gaussian wavelet (i.e., signal) is the optimum choice.

Similarly, with reference to the uncertainty relation $\Delta E \cdot \Delta t \geq h$, we see that it is Planck's region, or the h region, that defines the boundary, and not the shape. Since Planck's region is related to the unit information cell as we have described, it can be easily discerned that every bit of information needs an incremental quantity of energy ΔE and an amount of time Δt to efficiently transfer, create, and observe, and again, it is not free.

By referring to Fig. 16.5, the total number of information cells can be written as

$$N' = (v_m/\Delta v)(T/\Delta) \tag{16.10}$$

We note that the band-limited signal must be a very special type, for which the function has to be well behaved. It contains no discontinuity or no sharp angles, and has only rounded-off features. This type of signal must be an analytic function (i.e., signal). Let us stress again that the shapes of the information cells are not particularly critical; it is the unit area $\Delta v \cdot \Delta t = 1$ or the h region for $\Delta E \cdot \Delta t = h$ that determines the limit. Nevertheless, in principle the region can be further settled to

an even lower limit using a set of orthogonal Gaussian wavelets, as shown in Fig. 16.6.

Then the total number of information cells would be twice the amount, that is, $N = 2N'$. Then Heisenberg's uncertainty limits can be respectively written in the following forms:

$$\Delta x \cdot \Delta p \geq (1/2)h$$
$$\Delta E \cdot \Delta t \geq (1/2)h \tag{16.11}$$
$$\Delta v \cdot \Delta t \geq 1/2$$

Thus, we see that the boundary of the original limits (i.e., h region or 1) can be reduced to half, which coincides with Pauli's half-spin exclusive principle in quantum mechanics [9].

To wrap up this section, we provide a set of sound spectrograms that show that the uncertainty principle holds in practice, as depicted in Fig. 16.7a and b. In these figures, we see that time resolution (i.e., time striation) Δt can be observed for the wideband analysis of Fig. 16.7a, compared with the narrowband analysis of Fig. 16.7b. In viewing the narrowband analysis in Fig. 16.7b, we found that finer spectral resolution Δv can be resolved at the expense of time resolution Δt. And these results are quite consistent with the uncertainty principle of Heisenberg; one cannot observe fine spectral resolution Δv and time resolution Δt simultaneously [10].

16.6   CERTAINTY AND INFORMATION

Heisenberg's uncertainty principle was derived under the assumption of invasive observation, in which the specimen under investigation was using an external illuminator. Although the principle appears not to be violated, this does not mean that the certainty limit cannot be exploited. Thus, by reversing the inequality, we have

$$\Delta t \Delta v \leq 1 \tag{16.12}$$

It is, however, reasonable to name this reversal relationship the *certainty principle*, in contrast to the uncertainty principle. This means that when the observation signal (e.g., object light beam) propagates within a time window Δt, the complex light

(a)   Time

(b)   Time

FIG. 16.7 (a) Wideband sound spectrogram. (b) Narrowband sound spectrogram.

field preserves a high degree of certainty. Thus, as the bandwidth Δv of the light beam (or signal) is decreasing, the signal property is self-preserving (i.e., unchanged) within a wider time window Δt, or vice versa. This is precisely the temporal coherence limit of an electromagnetic signal, or light beam. If one multiplies the preceding certainty inequality with the velocity of light c, we have

$$c\Delta t \leq c/\Delta v \qquad (16.13)$$

This is the *certainty distance* (or coherence length) of a signal beam (or light source), as written by

$$\Delta d \geq c/\Delta v \qquad (16.14)$$

This means that within the coherence length Δd, the signal $u(t)$ would be highly correlated with preceding time signal $u(t + \Delta t)$, as written by

$$\Gamma_{12}(\Delta t) = \lim_{T \to \infty} \frac{1}{T} \int_0^T u_1(t)u_2^*(t + \Delta t)\, dt \qquad (16.15)$$

where $\Delta t \leq 1/\Delta v$, * denotes the complex conjugate, u_1 is the received signal, and u_2^* is the signal before Δt. This

expression is named the *certainty function*, which is similar to the mutual coherence function in coherence optics [11]. The *degree of certainty* (i.e., mutual coherence) can be determined by the following equation:

$$\gamma_{12}(\Delta t) = \frac{\Gamma_{12}(\Delta t)}{[\Gamma_{11}(0)\,\Gamma_{22}(0)]^{1/2}} \qquad (16.16)$$

As mentioned earlier, the shape of an information cell (i.e., $\Delta v \cdot \Delta t = 1$) is not a major issue, as long it is within a unit cell. We stress that it is this unit region that has not been fully exploited, such as applied to signal transmission, observation, processing, measurement, imaging, and possibly others.

Let us show a couple of applications that fall within the certainty region. One of the successful applications within the certainty region (i.e., within the unit region) is the wave front reconstruction of Gabor [12]. It is well known that successful holographic construction is dependent on the coherence length (i.e., temporal coherence) of the light source, in which the complex light beam distribution of the object beam can be encoded on a square-law detector (e.g., photographic film). In other words, the bandwidth of the light source controls the certainty relationship such that the object beam and the reference beam are highly correlated (i.e., mutually coherent). Otherwise, the complex wave front of the object beam cannot be properly recorded or detected.

Another example is applied to synthetic aperture radar (SAR) imaging [13]. It is required that the returned radar signal be combined with a highly coherent local reference signal, so that the complex distribution of the returned radar wave front can be synthesized on a square-law medium (e.g., photograph film). Let us now provide the reconstructed imaging results as obtained from a holographic technique and from SAR imaging, shown in Fig. 16.8a and b, respectively.

Figure 16.8a shows an image reconstructed from a hologram, which was encoded within the certainty distance (Δd) of a laser (i.e., light source). This holographic construction was achieved with the object, and the reference beams are mutually coherent such that the complex wave front of the object beam was properly encoded on a photographic plate. Figure 16.8b shows a SAR image that was obtained from a synthetic aperture format, which was synthesized by a series of reflected radar wave fronts with a mutually coherent local reference signal. We further stress that a microwave radar source has a very narrow bandwidth. In practice, its certainty distance (or coherence length) Δd can be made over tens of thousands of feet.

16.7 RELATIVITY AND INFORMATION

By referring to Heisenberg's uncertainty relation,

$$\Delta v \cdot \Delta t \geq 1 \qquad (16.17)$$

we see that every bit of observable information is limited by this fundamental constraint, in which the spectral resolution

(a)

(b)

FIG. 16.8 Imaging within the certainty limit: (a) hologram image and (b) SAR image.

and time resolution of the observable information contents can be traded. In other words, it is the unit information region, and not the shape of the region, that determines the fundamental constraint. For example, the shape can be either elongated or compressed, as long as it is restricted by a unit region.

Since time is one of the most intriguing variables in science, it can only move forward and not backward. Nonetheless, referring to Einstein's special theory of relativity, time may slow down somewhat if one travels closer to the speed of light.

Let us now take a look at the time dilation expression from Einstein's special theory of relativity [14]:

$$\Delta t' = \frac{\Delta t}{\sqrt{1 - v^2/c^2}}$$ (16.18)

where $\Delta t'$ is the dilated time window, Δt is the time window, v is the velocity of the observer, and c is the speed of light. If we assume that the observer is traveling at a velocity v and observing an object that is standing still (i.e., $v = 0$), then the observer would have a wider time window $\Delta t'$, instead of Δt.

Then as applied to Heisenberg's uncertainty relationship of Eq. (16.17), we have

$$\Delta v \cdot \Delta t' \geq 1$$ (16.19)

in which we see that a finer spectral resolution Δv can be obtained, since the observer has a wider time window $\Delta t' \geq \Delta t$.

We further note that as the velocity of the observer v approaches the speed of light (i.e., $v \to c$), the observer's time window $\Delta t'$ becomes infinitely large (i.e., $\Delta t' \to \infty$). This means that the observer in principle can observe the object as long as he or she wishes when the observer travels at the speed of light. In this case, the observer would have an infinitesimally small (or finer) spectral resolution (i.e., $\Delta v \to 0$).

On the other hand, if an observer who is standing still (i.e., $v = 0$) is observing a moving object traveling at velocity v, then the observation time window of the observer can be written as

$$\Delta t = \Delta t' \sqrt{1 - v^2/c^2}$$ (16.20)

in which we see that the observation time window decreases as the object velocity increases. By implementing Δt in Heisenberg's uncertainty relation, that is, $\Delta v \cdot \Delta t \geq 1$, a broader (i.e., poorer) spectral resolution of Δv is observed, since $\Delta t \leq \Delta t'$.

Again, if the velocity of the object approaches the speed of light (i.e., $v \to c$), the observer will have no time to observe since $\Delta t \to 0$. The spectral resolution is the lowest (i.e., poorest), since $\Delta v \to \infty$.

Two of the most important pillars in modern physics must be Einstein's relativity theory and Schrödinger's quantum mechanics. As we know, Einstein's relativity deals mostly with very large objects, while Schrödinger's mechanics deals with very small particles. Yet there exists a profound connection between these two pillars, by means of Heisenberg's uncertainty principle, in which we have shown that the observation time window can be altered with respect to the time dilation of Einstein's special theory of relativity.

Let us look again at the significance of Heisenberg's uncertainty principle as written in following forms:

$$\Delta E \cdot \Delta t \geq h$$
$$\Delta p \cdot \Delta x \geq h$$ (16.21)

Again, we stress that the constraints of the preceding relations are not by the shape of the information cell but by Planck's region of h. Since each h region is limited by ΔE and Δt (or by Δp and Δx), they can simply be *traded*. In other words, it is Planck's region that limits the constraint of the trading limit.

However, the time window Δt is more difficulty to vary, since time can only move forward and cannot slow down or even stand still. Nonetheless, we have shown in the preceding that the observation time window can be enlarged as the observer travels closer to the speed of light.

16.8 CREATION AND INFORMATION

We have noted that every substance has a distinctive signature or a piece of information. And this piece of information is equivalent to the cost of entropy, as we have shown that entropy and information can be traded. In other words, the cost of entropy is needed to produce the equivalent amount of information. But in practice, the cost of entropy is usually very excessive.

Now let us go back to the example of the three-dimensional cubic cells shown in Fig. 16.2; we had shown that the amount of information provided is

$$I = (N \times M \times H)(\log_2 W) \text{ bites} \qquad (16.22)$$

The equivalent amount of entropy can be shown as

$$S = (N \times M \times H)(\log_2 W)k \ln 2 \qquad (16.23)$$

This is precisely the price paid or the *minimum cost of entropy* to create this cubic cell structure.

For simplicity, we let this cubic structure have a mass of m grams; what would be the *minimum amount of energy* required to create it?

With reference to Einstein's total energy equation, $E \approx mc^2$ (Appendix D), we see that this must be the *minimum cost of energy* needed, if one wishes to reverse energy into mass, which we have assumed that we can do.

Let us provide a simple example: First, what will be the cost of entropy to create the three-unit binary information cell shown in Fig. 16.9? Second, what would be the cost of energy to build one if we assume that the overall mass of this structure is 1 kg?

The answer to the second question is that it requires a minimum cost of energy of about 1000 atomic bombs dropped on Nagasaki at the end of World War II; this is under the assumption that we can reverse the total energy back to mass. Even though we assume one could reverse it, it is still not sufficient to create the three-unit information cell shown in Fig. 16.9. To do so, we still need 3 bits of information (or minimum cost of entropy) to make it happen. And the cost of entropy needed can be shown as $S = 3 k \ln (2)$.

We further note that the minimum cost of energy to create the 3-bit cubic structure is a necessary condition but not sufficient to create the whole structure. However, the 3-bit information, or the minimum cost of entropy, to make it happen is

$S = 3 k \ln (2)$. In fact, the 3 bits of information (or equivalent amount of entropy) provided the sufficient condition to create the structure. One question may be asked: Do the 3 bits of information come from Mother Nature or are they man-made? The apparent answer is that they are very likely man-made.

16.9 PRICE TAG AND INFORMATION

Let me iterate again that every bit of information has a price tag. For instance, every bit of information is attached to a cost of entropy, and it is limited by Δt and ΔE of Heisenberg's uncertainty relationship. This means that every bit of information takes time and energy to transmit, process, record, retrieve, learn, and create. And it is not free. Then the question is, can one afford to pay for it?

For example, we show a mechanical clock and its dissembled parts as depicted in Fig. 16.10a and b, respectively. If we provide all the needed parts to a layman to reassemble it, what would be the amount of information (or the cost of entropy) he needs to reassemble it? And as a follow-up question, how long would it take for him to reassemble it?

If we further assume that the layman does not know how a clocks work, then he needs to first learn how it works. This

(a)

(b)

FIG. 16.9 Three-unit information cell.

FIG. 16.10 (a) Mechanical clock. (b) Its dissembled parts.

represents an amount of information (or cost of entropy) that the layman needs to learn. Second, after the layman has learned how a clock works, he has to figure out how to assemble it. This will take another set of information (e.g., provided by an instructional manual) to assemble the clock.

In the preceding, we have so far shown two major costs of entropy that the layman needs to pay for reassembling the clock. However, in practice these costs are excessive. We further note that the information needed for the layman to assemble a clock is the *dynamic-artificial type* made by humans, and not derived from the static type provided by Mother Nature's law of physics.

As for the follow-up question, "How long will it take for the layman to actually reassemble the clock?" the answer is that it will take the layman a lengthy time and effort (i.e., additional energy) to reassemble. My guess is that it will take the layman a couple of days to a couple weeks to reassemble it.

Now, let us extend the example further: What would be the cost of energy to actually create a mechanical clock as shown in Fig. 16.10a, for which we assume the mass of the clock is about 1 kg?

By referring to Einstein's total energy equation (Eq. (16.1)), the required minimum energy is equivalent to about 1000 atomic bombs dropped on Nagasaki at the end of World War II, by which we had assumed that we can reverse it. But even though we assume that we are able to convert energy to mass, the newly created mass still will not give us a clock.

If we even assume that the purpose of this energy-to-mass creation by Mother Nature is to build a clock, then first we have to convert this newly created mass into all the necessary *raw materials* (e.g., iron ore) for building the clock. Let me point out that the amount of information needed for mass for the necessary raw materials conversion comes from the statistical consequences of Mother Nature's law of physics (i.e., nuclei and chemical reactions). Although Mother Nature may have the ability to create the needed materials, she is still unable to create man-made objects (e.g., clocks, computers, and airplanes). This must be one of the major reasons why Mother Nature left the rest of the tasks for developing a clock to us.

Even though we have all the needed raw materials for constructing a clock, we still have to process them before we can use them to build the necessary parts. And this is not the end of the story yet; we still need an additional amount of information to assemble it. If we add all the needed information, it would sum up to a great amount. Thus, the overall information needed (i.e., entropies contributed by Mother Nature plus man-made ones) requires a huge amount of entropy to create such a clock, as shown in Fig. 16.10a.

In the preceding example, we see that even though Mother Nature has all the available energy, she still cannot create man-made objects, not even a simple piece of paper. Although humans are a part of Mother Nature, there are things that they can make that Mother Nature cannot.

Another question that may be asked is, what will be the costs of energy and entropy to create an adult chimpanzee? And what are the chances of this creature eventually surviving in this hostile planet Earth?

This example is somewhat more complicated than the mechanical clock problem. Again, if we assume that the weight of an adult chimpanzee is about 40 kg, then the minimum energy required for conversion into mass would be about 40,000 atomic bombs dropped on Nagasaki. And the cost of entropy required for creating this chimpanzee (e.g., flesh, blood vessels, bones, nerves, and DNA) would be huge. We note that all the required information is primarily derived from the *dynamic bioevolutional processes*, which cannot be accidentally created by Mother Nature's laws of physics. Thus, it would cost an extremely large amount of entropy to instantly create this chimpanzee. Even though we assume that Mother Nature may instantly create this creature, without basic living skills in this hostile planet Earth, the chance of this chimpanzee eventually surviving is extremely small.

16.10   REMARKS

We have shown the mythical relationship between science and information. Since every substance has a price tag for information, which includes all the building blocks in our universe, one cannot simply ignore information when dealing with science. We have shown that there is a profound connection between information and entropy, in which information and entropy can be traded. Since entropy is a quantity well accepted in physics, it makes information a bit easier to connect with science.

Two of the most important pillars in modern physics must be Einstein's relativity theory and Schrödinger's quantum mechanics; one deals with very large objects, and the other deals with very small particles. Yet there exists a profound relationship between them, by means of Heisenberg's uncertainty principle, in which we have shown that an observation time window can be elongated if the observer can keep up with light speed. From the uncertainty relationship, we see that every bit of information takes time and energy to transfer, to create, and to observe. In other words, every bit cannot be conveyed without a cost.

We have further shown that one of the most important aspects in digital communication is that a huge amount of information can be conveyed by light speed. Although the uncertainty principle was derived under the assumption of invasive observation, and it appears not to be violated, this does not mean that one cannot exploit it within the certainty limit. We have shown that successful holographic imaging and SAR imaging is in fact the result obtained within the certainty region observation.

Since one cannot create something from nothing, we have shown that every substance in principle can be created with a huge amount of energy and a great deal of entropy. The question is, can we afford it?

REFERENCES

1. F. T. S. Yu, Time: The Enigma of Space, *Asian J. Phys.*, vol. 26, no. 3, 149–158 (2017).
2. C. E. Shannon and W. Weaver, *The Mathematical Theory of Communication*, University of Illinois Press, Urbana, 1949.

3. F. T. S. Yu, *Optics and Information Theory*, Wiley-Interscience, New York, 1976.

4. F. W. Sears, *Thermodynamics, the Kinetic Theory of Gases, and Statistical Mechanics*, Addison-Wesley, Reading, MA, 1962.

5. L. Brillouin, *Science and Information Theory*, 2nd ed., Academic Press, New York, 1962.

6. E. Schrödinger, An Undulatory Theory of the Mechanics of Atoms and Molecules, *Phys. Rev.*, vol. 28, no. 6, 1049 (1926).

7. W. Heisenberg, Über den anschaulichen Inhalt der quantentheoretischen Kinematik und Mechanik, *Z. Phys.*, vol. 43, 172 (1927).

8. D. Gabor, Communication Theory and Physics, Phil. Mag, vol. 41, no. 7, 1161 (1950).

9. W. Pauli, Über den Zusammenhang des Abschlusses der Elektronengruppen im Atom mit der Komplexstruktur der Spektren, *Z. Phys.*, vol. 31, 765 (1925).

10. F. T. S. Yu, Information Content of a Sound Spectrogram, *J. Audio Eng. Soc.*, vol. 15, 407–413 (1967).

11. F. T. S. Yu, *Introduction to Diffraction, Information Processing and Holography*, MIT Press, Cambridge, MA, 1973.

12. D. Gabor, A New Microscope Principle, *Nature*, 161, 777 (1948).

13. L. J. Cultrona, E. N. Leith, L. J. Porcello, and W. E. Vivian, On the Application of Coherent Optical Processing Techniques to Synthetic-Aperture Radar, *Proc. IEEE*, 54, 1026 (1966).

14. A. Einstein, *Relativity, the Special and General Theory*, Crown Publishers, New York, 1961.

17 Time: The Enigma of Space

One of the important aspects of science is the substantiated physical realities, which were built by the fundamental laws of physics and cannot be simply substituted by unsubstantiated virtual reality. In writing this chapter, we have mostly based it on the constraints of the current laws of physics to illustrate enigmatic time as the origin for creating our physical space (i.e., the universe). The differences between physical reality and virtual reality are that physical reality exists within the rule of time and is supported by the laws of science, while virtual reality is created without the constraints of time and is mostly not substantiated by the laws of physics.

17.1 TIME AND ENERGY

One of the most enigmatic variables in the laws of science is time. So what is time? Time is a variable and not a substance. It has no mass, no weight, no coordinate, and no origin, and it cannot be detected or even seen. Yet time is an everlasting existing variable within our known universe. Without time, there would be no physical matter, no physical space, and no life. The fact is that all physical matter is associated with time, including our universe. Therefore, when one is dealing with science, time is one of the most enigmatic variables, is ever present, and cannot be simply ignored. Strictly speaking, all the laws of science, as well every physical substance, cannot exist without the existence of time.

On the other hand, energy is a physical quantity that governs the existence of every substance, which includes the entire universe. In other words, without the existence of energy, there would be no substance and no universe. Nonetheless, based on our current laws of science, all the substances were created by energy and every substance can also be converted back to energy. Thus, energy and substance are exchangeable, but this requires some physical conditions (e.g., nuclei and chemical interactions) to start the conversion. Since energy can be derived from mass, mass is equivalent to energy. Hence, all mass can be treated as an *energy reservoir*. In fact, our universe is compactly filled with mass and energy. Without the existence of time, the trading (or conversion) between mass and energy could not happen.

17.2 EINSTEIN'S ENERGY EQUATION

Let us now start with Einstein's energy equation, which was derived by his special theory of relativity [1] (see Appendix D):

$$\varepsilon \approx mc^2 \tag{17.1}$$

where m is the rest mass, and c is the velocity of light. Notice that all the laws in science are approximations, for which we have intentionally used the approximated sign. Strictly speaking, the energy equation should be more appropriately presented with an inequality sign, as described by

$$\varepsilon < mc^2 \tag{17.2}$$

This means that in practice, the total energy should be smaller than or at most approach the rest mass m times the square of the speed of light (i.e., c^2).

17.3 ENERGY CONVERSION AND RESERVOIR

In view of Einstein's energy equation of Eq. (17.1), we see that it takes time to convert mass into energy. Thus, without the inclusion of time variable, the conversion would not take place. Nonetheless, Einstein's energy equation represents the *total amount of energy* that can be converted from a rest mass m. In other words, every mass can be viewed as an energy reservoir. Thus, by incorporating with the time variable, Einstein's energy equation can be represented by a partial differential equation [2]:

$$\frac{\partial \varepsilon(t)}{\partial t} = c^2 \frac{\partial m(t)}{\partial t} \tag{17.3}$$

where $\partial \varepsilon(t)/\partial(t)$ is the rate of increasing energy conversion, $\partial m(t)/\partial(t)$ is the corresponding rate of mass reduction, and c is the speed of light.

17.4 TRADING MASS AND ENERGY

One of the important aspects in the preceding equation is that energy and mass can be traded, for which the rate of energy conversion from a mass can be written in terms of electromagnetic (EM) radiation or radian energy (see Appendix E) [2]:

$$\frac{\partial \varepsilon}{\partial t} = c^2 \frac{\partial m}{\partial t} = [\nabla \cdot S(v)] = -\frac{\partial}{\partial t}\left[\left(\frac{1}{2} \in E^2(v) + \frac{1}{2}\mu H^2(v)\right)\right] \tag{17.4}$$

where $\in$ and μ are the permittivity and permeability of the physical space, respectively; v is the radian frequency variable; E^2 and H^2 are the respective electric and magnetic field intensities; the negative sign represents the *outflow* energy per unit time from a unit volume; $(\nabla \cdot)$ is the divergent operator; and S is the *Poynting vector* or *energy vector* of an EM radiator [3], as can be shown by $S = E \times H$. In view of Eq. (17.4), we see that radian energy (i.e., radiation) diverges from the mass, as mass reduces with time.

Similarly, the conversion from energy to mass can also be presented as (see Appendix E)

$$\frac{\partial m}{\partial t} = \frac{1}{c^2}\frac{\partial \varepsilon}{\partial t} = -\frac{1}{c^2}[\nabla \cdot S(v)] = \frac{1}{c^2}\frac{\partial}{\partial t}\left[\frac{1}{2}\in E^2(v) + \frac{1}{2}\mu H^2(v)\right]$$

(17.5)

The major difference between this equation and Eq. (17.4) is the energy convergent operator −∇, where we see that the rate of energy, in the form of EM radiation, *converges* into a small volume for the mass creation, instead of diverging from the mass. Since mass creation is inversely proportional to c^2, a huge amount of energy is required to produce a small quantity of mass. Nevertheless, in view of the cosmological environment, the availability of a huge amount of energy has never been a problem.

Incidentally, a *black hole* [4,5] can be considered one of the *energy convergent operators*. Instead, the convergent force is relies more on the black hole's intense gravitational field. A black hole remains an intriguing physical substance about which we can learn more. Its gravitational field is so intense that even light cannot be escaped.

By the constraints of the current laws of science, observation is limited by the speed of light. If light is totally absorbed by a black hole, this does not mean that the black hole is an infinite energy sink [5]. Nonetheless, every black hole can actually be treated as an energy convergent operator, which is responsible for the energy to mass conversion as it will be exploded.

17.5 PHYSICAL SUBSTANCE AND SPACE

In our physical world, all matter are substances that include all the elemental particles; the electric, magnetic, and gravitation fields; and energy. The reason is that they were all created by means of energy or mass. Our physical space (e.g., our universe) is fully compacted with substances (i.e., mass and energy) and has no *absolute empty subspace* left within it. In fact, all physical substances exist with time, and no physical substance can exist forever or without time, which includes our universe. Thus, without time there would be no substance and no universe. Since every physical substance describes itself as a *physical space*, it is constantly changing with respect to time. In fact, every physical substance is itself a *temporal space* (or a *physical subspace*), as will be discussed in the subsequence sections.

In view of physical reality, no physical substance can exist without time; thus, if there is no time, no substances, which includes all the building blocks in our universe and the universe itself, can exist. On the other hand, time cannot exist without the existence of substances. Therefore, time and substance must mutually coexist or inclusively exist. In other words, substance and time have to simultaneously exist (i.e., one cannot exist without the other). Nonetheless, if our universe has to exist with time, then our universe will eventually get old and die. So the aspects of time are not as simple as we

have known them to be. For example, as a species living in a far distant galaxy moves closer to the speed of light, their time moves somewhat slower than ours [1]. Thus, we see that the relativistic aspects of time may not be the same at different subspaces in our universe (e.g., at the edge of our universe).

Since substances (i.e., mass) were created by energy, energy and time have to simultaneously exist. As we know, no conversion, either from mass to energy or from energy to mass, can start without the inclusion of time. Therefore, time and substance (i.e., energy and mass) have to simultaneously exist. Thus, we see that all physical substances, including our universe and us, coexist with time (or a function of time).

17.6 ABSOLUTE AND PHYSICAL SUBSPACES

Let us define various subspaces in the following, as they will be used in the subsequence sections.

An *absolute empty space* has no time, no substance, no coordinate, and no bounded or unbounded event. It is a virtual space, and it does not exist in practice.

A *physical space* is described by dimensional coordinates; it exists in practice, compactly filled with substances, supported by the current laws of science and the rule of time (i.e., time can only move forward and cannot move backward). Physical space and absolute empty space are mutually exclusive. In other words, a physical space cannot be embedded in an absolute empty space, and it cannot have an absolute empty subspace in it.

A *temporal space* is a time variable of physical space that is supported by the laws of science and rule of time. In fact, *all physical spaces are temporal spaces*.

A *spatial space* is described by dimensional coordinates and may not be supported by the laws of science and the rule of time.

A *virtual space* is imaginary and is generally not supported by the laws of science and the rule of time.

As we have noted, absolute empty space cannot exist in physical reality. Since every physical space needs to be completely filled with substances and no absolute empty subspace can be left within it, *every physical space is created by substances*. For example, our universe is a gigantic physical space created by mass and energy (i.e., substances) and has no empty subspaces in it. Yet, in physical reality all the mass (and energy) exist with time. Without the existence of time, there would be no mass, no energy, and no universe. Thus, we see that every physical substance coexists with time. In fact, every physical space is a temporal space, which includes us and our universe.

Since a physical space cannot be embedded within an absolute empty space and it cannot have any absolute empty subspace in it [6], our universe must be embedded in a more complex physical space. If we accept that our universe is embedded in a more complex space, then our universe must be a *bounded subspace*.

How about time? Since our universe is embedded in a more complex space, that space may share the same rule of time. However, the complex space that embeds our universe may not have the same laws of science as we do. Whether our universe is bounded or not is not the major issue of our current interest, since it takes a deeper understanding of our current universe before we can move on to the next level of complex space abstraction. It is, however, our aim, abiding within our current laws of science, to investigate the essence of time as the *enigma origin of our universe.*

17.7 TIME AND PHYSICAL SPACE

One of the most intriguing questions in our life must be the existence of time. So far, we know that time comes from nowhere, and it can only move forward, not backward. Although time may slow down somewhat, based on Einstein's special theory of relativity [1], so far time cannot move backward or even stand still. So if time exists, how do we know there is a physical space?

One answer is that there be a profound connection between time and physical space. In other words, if there is no time, then there would be no physical space.

A physical space is in fact a temporal space, in contrast to a virtual space. In other words, temporal space can be described by time, while virtual space is an imaginary space without the constraint of time. In other words, temporal space is supported by the laws of science, while virtual space is not.

A television video image is a typical example of trading time for space. For instance, each TV displayed image of (dx, dy) takes an amount of time to be displayed. Since time is a forward-moving variable, it cannot be traded back at the expense of a displayed image (dx, dy). In other words, it is time that determines the physical space, and it is not the physical space that can bring back the time that has been expended. And it is the size (or dimension) of space that determines the amount of time required to create the space (dx, dy).

Based on our current constraints of science, the *speed of light is the limit.* Since every physical space is created by substances, a physical space must be described by the speed of light. In other words, the dimension of a physical space is determined by the velocity of light, where the space is filled with substances (i.e., mass and energy).

Another issue is why the speed of light is limited. It is limited because our universe is a gigantic physical space that is filled with substances that cause a time delay on an EM wave's propagation. Nevertheless, if there were physical substances travel beyond the speed of light (which remains to be found), their velocities would also be limited, since our physical space is fully compacted with physical substances. Let me further note that a substance can travel in space without a time delay if and only if the space is absolutely empty. However, absolute empty space cannot exist in practice, since every physical space (including our universe) has to be fully filled with substances (i.e., energy and mass), with no empty subspaces left within it.

17.8 ELECTROMAGNETICS AND LAWS OF PHYSICS

Strictly speaking, all our laws of physics are evolved in the regime of EM science. Besides, all physical substances are part of EM-based science, and combined, all the living species on Earth are primarily dependent on the source of energy provided by the sun. About 78% of the sunlight that reaches the surface of our planet is well concentrated within a narrow band of visible spectrum. In response to our species' existence, which includes all living species on Earth, a pair of visible eyes (i.e., antennas) evolved in us humans, which help us for our survival. And this narrow band of visible light led us to the discovery of an even wider band of EM spectral distribution in nature. It is also the major impetus allowing us to discover all the physical substances that are part of EM-based physics. In principle, all physical substances can be observed or detected with EM interaction, and the speed of light is the current limit.

A picture that is worth a thousand words [7] is a trivial example to shown that EM observation is one of the most efficient aspects in information science. Yet, the ultimate physical limitation is also imposed by limitation of the EM regime, unless new laws of science emerge. The essence of Einstein's energy equation shows that mass and energy are exchangeable. It shows that energy and mass are equivalent, and energy is a form of EM radiation in view of Einstein's equation. We further note that all physical substances within our universe were created from energy and mass, which includes the dark energies [8] and dark matter [9]. Although the dark substances may not be observed directly using EM interaction, we may indirectly detect their existence, since they are basically energy base substances (i.e., EM-based science). It may be interesting to note that our current universe is composed of 72% dark energy, 23% dark matter, and 5% other physical substances.

Although dark matter contributes about 23% of our universe, it represents a total of 23% of gravitational fields. With reference to Einstein's differential energy equation (Eq. (17.3)), dark energies and dark matter dominate the entire universal energy reservation, well over 95%. Furthermore, if we accept the big bang theory for our universe creation [10], then creation could have been started with Einstein's energy differential form of Eq. (17.3) (see Appendix E):

$$\frac{\partial \epsilon}{\partial t} = c^2 \frac{\partial m}{\partial t} = [\nabla \cdot S(v)] = -\frac{\partial}{\partial t}\left[\frac{1}{2}\in E^2(v) + \frac{1}{2}\mu H^2(v)\right]$$

(17.6)

where $[\nabla \cdot S(v)]$ represents a divergent energy operation. In this equation, we see that a broad-spectral-band intense radian energy diverges (i.e., explodes) at the speed of light from a compacted matter M, where M represents a gigantic mass of energy reservoir. It is apparent that the creation is ignited by time and the exploded debris (i.e., matter and energy) starts to spread out in all directions, similar to an expanding air balloon. The boundary (i.e., radius of the sphere) of the universe expands at the speed of light, as the created debris is dispersed. It took about 16 billion chaotic light years [11,12] to come up with the present state of constellation, in which the boundary is still expanding at the speed of light beyond the current observation. With reference to a recent report using the Hubble Space Telescope, we can see galaxies about 16 billion light years away from us. This means that the creation process is not stopping yet, and at the same, time the universe might have started to de-create itself, since the big bang started, due to intense convergent gravitational forces from all the newly created debris of matter (e.g., galaxies and dark matter). To wrap up this section, we would stress that one of the viable aspects of Eq. (17.6) is the transformation from a spatially dimensionless equation to a Space–Time function (i.e., $\nabla \cdot S$); it describes how our universe was created with a huge explosion.

17.9   TRADING TIME FOR SPACE

Let us now take one of the simplest connections between physical space and time [13]:

$$d = vt \qquad (17.7)$$

where d is the distance, v is the velocity, and t is the time variable. Notice that this equation may be one of the most profound connections between time and physical space (or temporal space). Therefore, a three-dimensional (Euclidean) physical (or temporal) space can be described by

$$(dx, dy, dz) = (vx, vy, vz)t \qquad (17.8)$$

where (vx, vy, vz) are the velocities vectors, and t is the time variable. Under the current laws of science, the speed of light is the limit. Then, by replacing the velocity vectors equal to the speed of light c, a temporal space can be written as

$$(dx, dy, dz) = (ct, ct, ct) \qquad (17.9)$$

Thus, we see that time can be traded for space and space cannot be traded for time, since time is a forward variable. In other words, once time is expended, we cannot get it back.

Needless to say, a *spherical temporal space* can be described by

$$r = ct \qquad (17.10)$$

where radius r increases at the speed of light. Thus, we see that the boundary (i.e., edge) of our universe is determined by radius r, which is limited by the light speed, as illustrated in the composite diagram of Fig. 17.1. In view of this figure, we see that our universe is expanding at the speed of light well beyond the current observable galaxies. Figure 17.2 shows a discrete temporal diagram, in which we see that the size of our universe is continuously expanding as time moves forward.

Assuming that we have already accepted the *big bang creation*, sometime in the future (i.e., billions of light years later) our universe will eventually stop expanding and then start to shrink back, preparing for the next cycle of big bang explosion. The forces for the collapsing universe are mainly due to the intense gravitational field, mostly from giant black holes

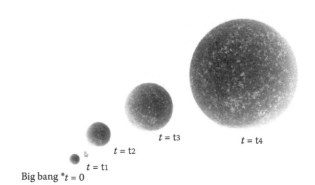

FIG. 17.2   Discrete temporal universe diagram. t is time.

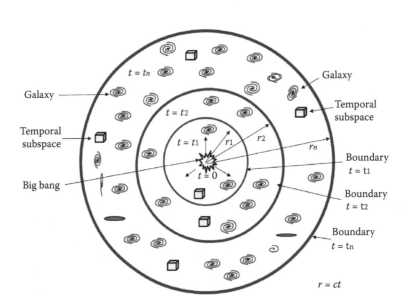

FIG. 17.1   Composite temporal space universe diagram. $r = ct$, r is the radius of our universe, t is time, and c is the velocity of light.

and matter that were derived from merging (or swallowing) with smaller black holes and other debris (i.e., physical substances). Since a black hole's gravitational field is so intense, even light cannot be escaped; however, a black hole is by no means an infinite energy reservoir. Eventually, the storage capacity of a black hole will reach a limit for explosion, as started for the mass-to-energy and debris creation. In other words, there will be one dominant giant black hole within the shrinking universe, to initiate the next cycle of universe creation. Therefore, every black hole can be treated as a *convergent energy sink*, which relies on its intense gravitation field to collect all the debris of matter and energies.

Referring to big bang creation, a gigantic energy explosion was the major reason for the universe's creation. In fact, it can be easily discerned that the creating process has never slowed down since the birth of our universe, as we see that our universe is still continuingly expanding even today. This is by no means an indication that all the debris created came from the big bang's energy (e.g., Mc^2); there might have been some leftover debris from a preceding universe. Therefore, the overall energy within our universe cannot be restricted to just the amount that came from the big bang creation.

In fact, the conversion processes between mass and energy have never been totally absent since the birth of our universe, but they are on a much smaller scale now. In fact, right after birth, our universe started to slow down the divergent process due to the gravitational forces produced by the created matter. In other words, the universe will eventually reach a point when overall divergent forces will be weaker than the convergent forces, which are mostly due to gravitational fields coming from the newly created matter, including black holes.

As we had mentioned earlier, our universe currently has about 23% dark matter, which represents about 23% of the gravitational fields within the current universe. The intense localized gravitational field could have been produced from a group or a giant black hole, derived from merging with (or swallowing up) some smaller black holes, nearby dark matter, and debris. Since a giant black hole is not an infinite energy sink, eventually it will explode for the next cycle of universal creation. And it is almost certain that the next big bang creation will not occur at the same center of our present universe. One can easily discern that our universe will never shrink to a few inches in size, as commonly speculated. It will, however, shrink to a smaller size until one of the giant black holes (e.g., swallowed-up sufficient physical debris) reaches the big bang explosive condition to release its gigantic energy for the next cycle of universal creation.

The speculation of a possible collapsing universe remains to be observed. Nonetheless, we have found that our universe is still expanding, as observed by the Doppler shifts of the distant galaxies at the edge of our universe, about 16 billion light years away [11,12]. This tells us that our universe has not reached its half-life yet. In fact, the expansion has never stopped since the birth of our universe, and our universe has also been started to de-create since the big bang started, which is primarily due to convergent gravitational forces from the newly created debris (e.g., galaxies, black holes, and dark matter).

17.10   RELATIVISTIC TIME AND TEMPORAL SPACE

Relativistic time at a different subspace in the vast universal space may *not* be the same as that based on Einstein's special theory of relativity [1]. Let us start with the relativistic time dilation:

$$\Delta t' = \frac{\Delta t}{\sqrt{1 - v^2/c^2}} \qquad (17.11)$$

where $\Delta t'$ is the relativistic time window, compared with a standstill subspace; Δt is the time window of the standstill subspace; v is the velocity of the moving subspace; and c is the velocity of light. We see that the time window $\Delta t'$ of the moving subspace, relative to the time window of the standstill subspace, appears to get *wider as velocity increases*. Therefore, for the species living in an environment that travels closer to the speed of light (e.g., at the edge of the universe), their time appears to be slower than ours, as illustrated in Fig. 17.3. In this figure, we see an old man traveling at a speed closer to the velocity of light; his relative observation time appears to be wider as he is looking at us, and the laws of science within his subspace may not be the same as ours.

17.11   TIME AND PHYSICAL SPACE

Every physical (or temporal) subspace is created by substances (i.e., energy and mass), and substances coexist with time. In this context, we see that our universe was essentially created by time and energy, and the universe is continuously evolving (i.e., changing) with time. Although relativistic time may not be the same at the different subspaces in our universe, the rule of time may remain the same. As for the species living closer to the speed of light, relativistic time may not be noticeable to them, but their laws of science within their subspace may be different from ours. Nonetheless, our universe was simultaneously created by time and a gigantic energy explosion. Since our universe cannot be embedded in an empty space, it must be embedded in a more complex space that remains to be found.

From an inclusive point of view, mass is energy or energy is mass, which was discovered by Einstein almost a century ago [1]. And it is this basic fundamental law of physics that we

FIG. 17.3   Effects on relativistic time.

have used for investigating the origin of time. Together with a huge energy explosion (i.e., big bang theory [10]), time is the igniter for the creation of our universe. As we know, without the existence of time, the creation of our universe would not have happened.

As we have shown, time can be traded for space, but space cannot be traded for time. Our universe is in fact a temporal physical subspace, and it is continuously evolving or changing with time. Although every temporal subspace is created by time (and substances), it is not possible for us to trade any temporal subspace for time. Since every physical substance has a life, our universe (a gigantic substance) cannot be excluded. With reference to the report from a recent Hubble Space Telescope observation [11,12], we are capable of viewing galaxies about 16 billion light years away and have also learned that our universe is still by no means slowing down in expansion. In other words, our universe has still not reached its half-life, based on our estimation. As we have shown, time ignited the creation of our universe, yet the created physical substances presented to us the existence of time.

17.12 WE ARE NOT ALONE

Within our universe, we can easily estimate there were billions and billions of civilizations that had been emerged and faded away in the past 16 billion light years. Our civilization is one of the billions and billions of current consequences within our universe, and it will eventually disappear. We are here, and will be here, for just a very short moment. Hopefully, we will be able to discover substances that travel well beyond the limit of light before the end of our existence, so that a better observational instrument can be built. If we point the new instrument at the right place, we may see the edge of our universe beyond the limit of light. We are not alone with almost absolute certainty. By using the new observational equipment, we may find a planet that once upon a time had harbored a civilization for a period of twinkle thousands of (Earth) years.

17.13 REMARKS

We have shown that time is one of the most intriguing variables in the universe. Without time, there would be no physical substances, no space, and no life. With reference to Einstein's energy equation, we have shown that energy and mass can be traded. In other words, mass is equivalent to energy, and energy is equivalent to mass, for which all mass can be treated as an energy reservoir. We have also shown that a physical space cannot be embedded in an absolute empty space, and cannot even have any absolute empty subspace in it. In reality,

every physical space has to be fully packed with physical substances (i.e., energy and mass). Since no physical space can be embedded in an absolute empty space, it is reasonable to assume that our universe is a subspace within a more complex space, which remains to be found. We have also shown that it takes time to create a physical space, and it cannot bring back the time that has been used for the creation. Since all physical substances exist with time, all physical spaces are created by time and substances (i.e., energy and mass). This means that our cosmos was created by time and a gigantic energy explosion, in which we see that every substance coexists with time. That is, without time, the creation of physical substances would not have happened. We have further noted that our universe is in a temporal space, and it is still expanding based on current observation. This shows us that our universe has not reached its half-life yet, as we have accepted the big bang creation. And finally, we are not alone with almost absolute certainty. Someday, we may find a planet that once upon a time had harbored a civilization for a period of light years.

REFERENCES

1. A. Einstein, *Relativity, the Special and General Theory*, Crown Publishers, New York, 1961.
2. F. T. S. Yu, Gravitation and Radiation, *Asian J. Phys.*, vol. 25, no. 6, 789–795 (2016).
3. J. D. Kraus, *Electro-Magnetics*, McGraw-Hill Book Company, New York, 1953, p. 370.
4. M. Bartrusiok, *Black Hole*, Yale University Press, New Haven, CT, 2015.
5. G. O. Abell, D. Morrison, and S. C. Wolff, *Exploration of the Universe*, 5th ed., Saunders College Publishing, New York, 1987, pp. 47–88.
6. F. T. S. Yu, Science and the Myth of Information, *Asian J. Phys.*, vol. 24, no. 24, 1823–1836 (2015).
7. F. T. S. Yu, *Optics and Information Theory*, Wiley-Interscience, New York, 1976.
8. L. Amendola and S. Tsujikawa, *Dark Energy: Theory and Observation*, Cambridge University Press, Cambridge, 2010.
9. G. Bertone, ed., *Particle Dark Matter: Observation, Model and Search*, Cambridge University Press, Cambridge, 2010.
10. M. Bartrusiok and V. A. Rubakov, *Introduction to the Theory of the Early Universe: Hot Big Bang Theory*, World Scientific Publishing, Princeton, NJ, 2011.
11. J. O. Bennett, M. O. Donahue, M. Voit, and N. Schneider, *The Cosmic Perspective Fundamentals*, Addison Wesley Publishing, Cambridge, MA, 2015.
12. R. Zimmerman, *The Universe in a Mirror: The Saga of the Hubble Space Telescope*, Princeton Press, Princeton, NJ, 2016.
13. F. T. S. Yu, Time, Space, Information and Life, *Asian J. Phys.*, vol. 24, no. 2, 217–223 (2015).

18 Time–Space Quantum Entanglement

In writing this chapter, we started with the fundamental differences between science and mathematics. One is physical reality, and the other is abstract reality. By using exact math to analyze approximated science, it is not a guarantee to obtain reliable and accurate solutions. Since we live in a temporal subspace, every substance within the universe obeys the laws of science and the rule of time. We show that instant quantum entanglement at a large distance only exists in timeless space. But timeless space is not temporal space, and it cannot exist within temporal space. Particle entanglement has to be at least two to tangle (it takes two to tangle). Since every entanglement involves time and space, temporal and spatial entanglement can be defined. We have also shown that quantum entanglement operates within the certainty limit of Heisenberg. In view of Einstein's relativity, quantum entanglement can be extended to the relativistic regime, namely, relativistic quantum entanglement.

18.1 SCIENCE AND MATHEMATICS

Mathematics is an axiom of absolute certainty, while science is a law of approximation. One conflicting issue is that using exact math to analyze inexact science, although rigorous, by no means guarantees the obtainment of accurate and reliable solutions. In practice, an accurate solution has to be proven by experiments, and reliability is determined by repeated outcomes. Thus, we see that there is a fundamental conflict between theoretical science using precise mathematical analyses. The conflict could have been mounted by the inexact results, for which theoretical science has to constantly update the limitation of physical reality, in the course of a tedious, detailed evaluation. In other words, just depending on rigorous mathematics in solving physical problems may not produce a reliable solution. In some cases, the result deviates from physical reality. In order for mathematicians and theoretical scientists to reconcile, it is recommended that mathematicians be consciously aware of the physical limitation in science, and scientists have a deeper knowledge of mathematical abstraction. In short, science is an inexact physical reality, and math is an exact virtual reality.

18.2 TIME AND TEMPORAL SPACE

In Chapter 17, we showed that our universe is a bounded temporal space [1], and every subspace within the universe is a temporal subspace. Note that the current physical limit is imposed by the speed of light. We also showed that every physical space is a temporal subspace, and it is compacted with substances (e.g., energy and mass). Also, every physical space is created by time, and the created subspace cannot bring back the time that had been used for space creation. In fact, every physical substance (or particle) can be described by a function of space and time. Since every physical particle is itself a temporal particle (or subspace), it can be described by a function of space and time:

$$f(x;t), \text{ where } t > 0 \qquad (18.1)$$

where x denotes the space variable, and t is time.

As we showed in Chapter 15, one of the most fascinating principles in quantum mechanics is Heisenberg's uncertainty principle [2]:

$$\Delta x \cdot \Delta p \geq h$$
$$\Delta E \cdot \Delta t \geq h \qquad (18.2)$$
$$\Delta v \cdot \Delta t \geq 1$$

where h is Planck's constant, and Δx, Δp, ΔE, Δt, and Δv are the incremental position, momentum, energy, time, and bandwidth, respectively. As we impose Pauli's exclusive principle [3], the uncertainty relations can be deduced to the following forms:

$$\Delta x \cdot \Delta p \geq h/2$$
$$\Delta E \cdot \Delta t \geq h/2 \qquad (18.3)$$
$$\Delta v \cdot \Delta t \geq 1/2$$

We further note that these incremental variables can be bilaterally traded [4], as long as they are restricted within Heisenberg's limit, the $h/2$ or $1/2$ region. In other words, it is the region, not the shape of the region, that limits the trading.

Similarly, a set certainty principle [4] can also be written, respectively, in the following:

$$\Delta x \cdot \Delta p \leq h/2$$
$$\Delta E \cdot \Delta t \leq h/2 \qquad (18.4)$$
$$\Delta v \cdot \Delta t \leq 1/2$$

The essence of the certainty principle is that position and momentum, energy and time, and time and bandwidth can be seen as preserving Heisenberg's limit, for instance, as waves and particles that travel or entangle within a certainty region. For example, side-looking radar imaging [5], holography [6], quantum particle entanglement [7], and possibly others, are actually operating within the certainty region of Heisenberg.

18.3   QUANTUM ENTANGLEMENT

Quantum entanglement is a physical phenomenon that occurs when pairs (or a group of particles) interact in such a way that the quantum state of each particle cannot be independently described, even when particles are separated by a large distance. Instead, a quantum state must be described [8–10]. Some scientists believe that, in principle, instant (i.e., with no time delay) quantum entanglement between particles is possible, even for a large-distance separation. This is one of the analytical implications that has puzzled quantum scientists for decades.

Theoretically speaking, instant entanglement between particles at a large distance is possible if and only if the distance is a "timeless distance" (i.e., timeless space). However, our universe is a temporal space, and it cannot have any timeless space (or absolute empty space) in it [1]. It is therefore apparent that any quantum entanglement between distant particles cannot get started without the expenditure of time (i.e., $t > 0$).

In practice, every physical substance can be written as a function of time and spatial variables, for which entanglement between particles is included. In principle, entanglement can also be presented in the form of a spatial and time variable function:

$$q(X;t), \text{ where } t > 0 \tag{18.5}$$

It is therefore apparent that temporal entanglement between particles can be written similar to the mutual coherence function [11]:

$$Q_{12}(\Delta t) = \lim_{T \to \infty} \frac{1}{T} \int_0^T q_1(t) q_2^*(t + \Delta t)\, dt \tag{18.6}$$

where $q_1(x; t)$ and $q_2(x; t)$ are the entanglement functions from Q_1 and Q_2 particles, respectively. The corresponding degree of temporal entanglement can be determined by

$$\xi_{12}(\Delta t) = \frac{Q_{12}(\Delta t)}{\left[Q_{11}(0) Q_{22}(0)\right]^{1/2}} \tag{18.7}$$

On the other hand, the spatial entanglement function can also be written as

$$Q_{12}(\Delta x) = \lim_{T \to \infty} \frac{1}{T} \int_0^T q_1(x;t) q_2^*(x + \Delta x;t)\, dt \tag{18.8}$$

and the degree of spatial entanglement can be shown as

$$\xi_{12}(\Delta x) = \frac{Q_{12}(\Delta x)}{\left[Q_{11}(0) Q_{22}(0)\right]^{1/2}} \tag{18.9}$$

where Δx represents the spatial separation variable between Q_1 and Q_2 particles. In view of the mutual entanglement, it can

be understood that quantum entanglement operates within the certainty regime of Heisenberg, as shown by

$$\Delta E \cdot \Delta t \le h/2 \tag{18.10}$$

in which we see that every change in quantum state takes ΔE energy and Δt time to tangle, and is not free.

18.4   RELATIVISTIC QUANTUM ENTANGLEMENT

Let us take the time dilation equation of Einstein,

$$\Delta t' = \frac{\Delta t}{\sqrt{1 - v^2/c^2}} \tag{18.11}$$

in which we see that the relativity time window widens as the speed (or relative speed) of the subspace increases.

Let us now present a fictitious example in Fig. 18.1, in which we see two entangled particles Q_1 and Q_2 that are respectively situated in two distant subspaces; one travels at a velocity v, and the other is assumed to stand still. We further assume that the two subspaces are located within a timeless space (although this is not possible in practice) so that entanglement between Q_1 and Q_2 could be instant (i.e., without any time delay). We note once again that, in practice, it is impossible for any temporal subspace to be embedded in a timeless space, or vice versa. Only theoretician can make it happen.

We further note that, as based on the current laws of science and rule of time, our universe was created as a bounded temporal space. The speed of all the known physical induced fields (such as gravitation and the electric and magnetic fields) cannot go beyond the speed of light. And the speed of quantum entanglement between distant particles should not be excluded. Beyond the speed of light is still a myth, unless physical (i.e., not mathematically proven) evidence has been confirmed. The reason is that the changes in a particle's quantum state are still within the physics of an electromagnetic regime. And the induced entanglement (i.e., entangle speed) between distant particles beyond the speed of light still cannot be officially confirmed physically.

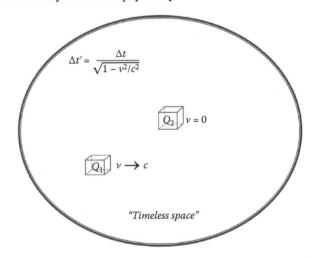

FIG. 18.1 Relativistic quantum entanglements in a timeless environment.

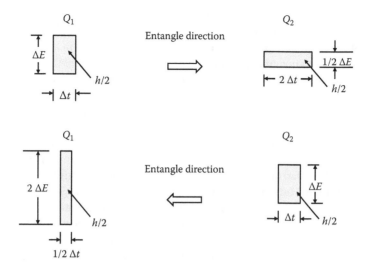

FIG. 18.2 Quantum particle entanglements, relativistic time $\Delta t' = 2\Delta t$.

Now, if we let Q_1 switch its quantum state with an incremental expenditure of an incremental time Δt, then Q_2 is assumed to instantly respond without any time delay (i.e., in timeless space). And we see that the incremental time Δt used in the course of switching Q_2's quantum state would be longer, because of the relativistic time effect. For instance, if we assume that the time window for Q_1, relative to Q_2, is two times wider, that is, $\Delta t' = 2\Delta t$, then it will take two times longer for Q_2 to complete the changes of its quantum state, as can be seen in Fig. 18.2. We further note that, as accepted by Heisenberg's certainty limit, the incremental amount of energy ΔE used for the entanglement is conserved, as illustrated in the figure.

In our physical world, one cannot get something from nothing, and there is always a price to pay. As for distant quantum particle entanglement, the minimum price paid must be $1/2\,h$ for each induced quantum entangle, as limited by Heisenberg's region, where h is Planck's constant. Thus, we see that it takes Δt and ΔE to tangle.

We further note that the relativistic time at different subspaces in our universe may not be the same, as based on Einstein's special theory of relativity. For subspaces at the edge of our universe, their laws of science may not be the same, but the time within their subspace is the same. The relative speeds of time, however, are different.

18.5 THE ART OF QUANTUM LIMITED-SUBSPACE

As we have known in a temporal subspace, it is the time that ignited the creation of a physical subspace and the created subspace cannot bring back the amount of time that had been used for the creation. We had also shown that our temporal universe was created by a huge divergent explosion of energy as given by [1]

$$\frac{\partial \varepsilon}{\partial t} = c^2 \frac{\partial m}{\partial t} = [\nabla \cdot S(v)] = -\frac{\partial}{\partial t}\left[\frac{1}{2}\in E^2(v) + \frac{1}{2}\mu H^2(v)\right]$$

(18.12)

where m and ε are the mass and energy, c is the light speed, t is the time variable, $(\nabla \cdot S)$ is the divergent energy operator, $S = E \times H$ is the Poynting (or energy) vector, E and H are the electric and magnetic field vectors, and ($\cdot$ and x) represent dot and cross products, respectively. Let me stress again that the essence of this equation must be the transformation from a spatially dimensionless equation to a **Space–Time** function (i.e., $\nabla \cdot S$), which precisely shows how our universe was created with a huge explosion, and our universe is still expanding!

Since every temporal subspace within the universe can be described by time, where speed of light is the current limit, as given by $d = c\,\Delta t$, then a subspace as limited by the Heisenberg's Certainty Principle represents a **Quantum Limited-Subspace (or a Unit Quantum Space)** as illustrated in Fig. 18.3. Sets of Rectangular and Spherical Quantum Limited-Spaces for (ΔE, Δt) and (Δv, Δt) are shown in (a) and (b), respectively.

One of the interesting aspects of the Quantum Limited-Subspace is that it can be used in describing the **Coherence Theory of Light**. It can be shown within the Quantum Limited-Subspace that a high **Degree of Mutual Coherence** can be obtained [11]. For instance, the narrower the spectral bandwidth Δv of an electromagnetic radiator (e.g., light source), a larger volume of Quantum Limited-Subspace (i.e., $r = c\,\Delta t \leq c/\Delta v$) can be obtained. In other words, longer **Mutual Coherence Length** can be achieved using narrower source bandwidth Δv.

$$\Delta E \cdot \Delta t = h \qquad \Delta v \cdot \Delta t = 1$$

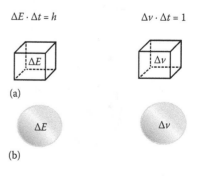

FIG. 18.3 Quantum Limited-Subspace: (a) Cartesian Representation, side $= c\Delta t$; (b) Spherical Representation, radius $r = c\Delta t$.

Now, let us point out that two well-known experiments as applied within the Quantum Limited-Subspace are the **Wave Front Reconstruction** of Gabor [6] and the **Synthetic Aperture Radar** (or side-looking Radar) imaging [5]. These are the typical examples that the solution was exploiting within Quantum Limited-Subspace as illustrated in Fig. 18.4.

Since wave propagation is **directional**, we have assumed that three complex wave fields $u_1(x;t)$, $u_2(x;t)$, and $u_3(x;t)$ located at u_1, u_2, and u_3 within the path of wave propagation can be obtained, where the EM wave is originated from a common electromagnetic source. By referring to the complex wave field u_1 with respect to u_2, their **Mutual Coherence Function** between point u_1 and point u_2 can be written as

$$\Gamma_{12}(\Delta x; \Delta t) = \lim_{T \to \infty} \frac{1}{T} \int_0^T u_1(x;t) u_2^*(x;t+\Delta x)\, dt \qquad (18.13)$$

where $(x;t)$ represent the space and time variables, and * denotes the complex conjugate. The Degree of Coherence between them can be determined by

$$\gamma_{12}(\Delta t) = \frac{\Gamma_{12}(\Delta t)}{\left[\Gamma_{11}(0)\,\Gamma_{22}(0)\right]^{1/2}} \qquad (18.14)$$

On the other hand, as referring to location u_1 with respect to u_3, their Mutual Coherence Function between them is given by

$$\Gamma_{13}(\Delta x; \Delta t) = \lim_{T \to \infty} \frac{1}{T} \int_0^T u_1(x;t) u_{13}^*(x+\Delta x;t+\Delta t)\, dt \quad (18.15)$$

As we can see, the wave field at point u_3 is located outside the coherence domain of u_1's Quantum Limited Sphere and the Degree of Mutual Coherence between u_1 and u_3 is anticipated to be much **lower**.

Before we move on to the Mutual Quantum Entanglement application, I would like to stress that Quantum Entanglement was one of the intriguing solutions obtained by Quantum Mechanics of Schrödinger [9,10]. Although the evaluation was based on the laws of physics with very rigorous mathematical

manipulation, it does not guarantee that the evaluated solution falls within the limit of physical reality. We note that every physical subspace is created under the laws of science and rule of time, where speed of light is our current limit. Unless physical experiments have been proven that the speed of quantum entanglement is beyond the speed of light, we would assume that the entanglement is still within the limit light speed for our justifiable illustration. Thus, we shall use Heisenberg Certainty Limit in applying to our discussion (i.e., $\Delta E \cdot \Delta t \le h$ or $\Delta v \cdot \Delta t \le 1$) in the following:

In our temporal universe, one cannot get something from nothing; there is always a price to pay, namely, ΔE and t. Then, a couple of questions may be asked; if the entanglement is **"instant,"** then there will be a conflict of physical reality with our temporal universe, since a timeless subspace cannot be existed within a temporal space!

If the entanglement, on the other hand, is beyond the speed of light, then the entangled substance has to be found or experimentally proven. This newly discovered substance would be one of the greatest discoveries in our modern history in science. This new substance eventually would be able to change most the laws of Science. This includes Einstein's relativity theory and his famous energy equation, which have been used to pave the way showing how our temporal universe was created [1]!

Since Quantum Entangles is bilateral, we assume that there are four measurable entangled particles Q_1, Q_2, Q_3, and Q_4 located within and outside their Entanglement Quantum Limited-Subspaces as shown in Fig. 18.5.

Let us denote that a measurable entangle strength between Q_1 and Q_2 is given by $q_{12}(x;t)$; then the **Mutual Entangle Function** between Q_1 and Q_2 can be written as

$$Q_{12}(\Delta x; \Delta t) = \lim_{T \to \infty} \frac{1}{T} \int_0^T q_{12}(x;t) q_{12}^*(x+\Delta x;t+\Delta t)\, dt \qquad (18.16)$$

And their **Degree of Mutual Entanglement** can also be determined as given by

$$\xi_{12}(\Delta t) = \frac{Q_{12}(\Delta t)}{\left[Q_{11}(0)\,Q_{22}(0)\right]^{1/2}} \qquad (18.17)$$

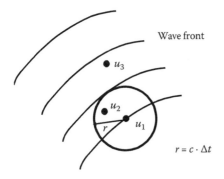

FIG. 18.4 Coherence Quantum Limited-Subspace. r is the radius of coherence subspace with reference to u_1, and c is speed of light.

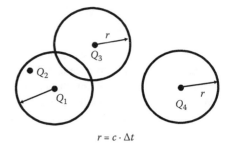

FIG. 18.5 Entanglement Quantum Limited-Subspaces. r is the radius of the Entanglement Quantum Limited-Subspace and c is the velocity of light.

With reference to Fig. 18.5, we see that the Degree of Mutual Entanglement between Q_1 and Q_3 is expected to be much **weaker**, since they are located beyond their Entanglement Quantum Limited-Subspace.

18.6 REMARKS

In concluding this chapter, we note the fundamental differences between science and mathematics: one is approximation, and the other is absolute certainty. Using exact math to analyze approximated science does not guarantee the obtainment of reliable and accurate solutions. Since we live in a temporal subspace, every substance within our universe obeys (at least approximately) the laws of science and the rule of time. The aspect of instant quantum entanglement only exists in timeless space, but timeless space is not temporal space, and it cannot be a subspace within a temporal space. Quantum entanglement has to be at least two or multiple particles to entangle; thus, we see that it takes at least two to tangle. Since every entanglement involves time and space, entanglement is defined in terms of temporal and spatial entanglement. In practice, mutual entanglement measurements can be obtained. We have also shown that quantum entanglement operates within the certainty limit of Heisenberg. As anticipated, the degree of mutual entanglement decreases as time and space separation increases. And finally, we have shown that quantum entanglement can be extended to the relativistic time regime, in which we have shown that entangled time changes with respect to the relative velocity of subspaces. We have also added a section on the art of Quantum Limited-Subspace. Examples such as Coherence and Entangled Quantum Unit-Space have been given.

In concluding this text, I would remark once again that: Since we are living in the temporal universe, one cannot get something from nothing and there is always a price to pay, namely, $(\Delta t, \Delta E)$ or a unit physical subspace. Any science that existed within the temporal universe is physically real and the one beyond the temporal universe is still fictitious, unless it is physically proven by repeated experiments.

REFERENCES

1. F. T. S. Yu, Time: The Enigma of Space, *Asian J. Phys.*, vol. 26, no. 3, 149–158 (2017).
2. W. Heisenberg, Über den anschaulichen Inhalt der quantentheoretischen Kinematik und Mechanik, *Z. Phys.*, vol. 43, 172 (1927).
3. W. Pauli, Über den Zusammenhang des Abschlusses der Elektronengruppen im Atom mit der Komplexstruktur der Spektren, *Z. Phys.*, vol. 31, 765 (1925).
4. F. T. S. Yu, Science and the Myth of Information, *Asian J. Phys.*, vol. 24, no. 12, 1823–1836 (2015).
5. L. J. Cultrona, E. N. Leith, L. J. Porcello, and W. E. Vivian, On the Application of Coherent Optical Processing Techniques to Synthetic-Aperture Radar, *Proc. IEEE*, 54, 1026 (1966).
6. D. Gabor, A New Microscope Principle, *Nature*, vol. 161, 777 (1948).
7. J. A. Wheeler and W. H. Zurek, *Quantum Theory and Measurement*, Princeton University Press, Princeton, NJ, 2014.
8. A. Einstein, B. Podolsky, N. Rosen, and N. Podolsky, Can Quantum-Mechanical Description of Physical Reality Be Considered Complete? *Phys. Rev.*, vol. 47, no. 10, 777–780 (1935).
9. E. Schrödinger, Discussion of Probability Relations between Separated Systems, *Math. Proc. Cambr. Philos. Soc.*, vol. 31, no. 4, 555–563 (1935).
10. E. Schrödinger, Probability Relations between Separated Systems, *Math. Proc. Cambr. Philos. Soc.*, vol. 32, no. 3, 446–452 (1936).
11. F. T. S. Yu, *Introduction to Diffraction, Information Processing and Holography*, MIT Press, Cambridge, MA, 1973.

Appendix A: Linear Difference Equation with Constant Coefficients

An equation of the form

$$U(x+n) + P_{n-1}U(x+n-1) + \cdots + P_1 U(x+1) + P_0 U(x) = f(x)$$
(A.1)

where P_i takes on the values $i = 0, 1, \ldots, n-1$ and $f(x)$ is a function of x, or constant, is called a *linear* difference equation. If $f(x) = 0$, then the equation is called a *homogeneous linear* difference equation. If $f(x) = 0$ and all the coefficients are constants, then the equation is known as a homogeneous linear difference equation with constant coefficients.

It is noted that linear difference equations with variable coefficients are extremely difficult to solve. However, if the variable coefficients are under very severe restrictions, a linear difference equation may be solvable. Linear difference equations with constant coefficients are generally solvable.

Now let us consider a linear homogeneous difference equation with constant coefficients:

$$U(x+n) + A_{n-1}U(x+n-1) + \ldots + A_1 U(x+1) + A_0 U(x) = 0$$
(A.2)

where A_i, $i = 0, 1, \ldots, n-1$, are constants. It is noted that Eq. (A.2) has *n particular* solutions. These particular solutions can be found by letting

$$U(x) = a^x$$
(A.3)

where $a \neq 0$.

By substituting Eq. (A.3) into Eq. (A.2), we have

$$a^{x+n} + A_{n-1}a^{x+n-1} + \cdots + A_1 a^{x+1} A_0 a^x = 0$$
(A.4)

or

$$a^x(a^n + A_{n-1}a^{n-1} + \cdots + A_0) = 0$$
(A.5)

Therefore, if a is a solution of Eq. (A.5), then it is necessary that

$$a^n + A_{n-1}a^{n-1} + \cdots + A_0 = 0$$
(A.6)

where Eq. (A.6) is also known as a *characteristic* equation. Conversely, if $a_1, a_2, \ldots, a_n$ are n distinct solutions of Eq. (A.6), then $a_1^x, a_2^x, \ldots, a_n^x$ are independent solutions of the homogeneous Equation (A.2). Furthermore, the linear combination

$$U(x) = \sum_{k=1}^{n} C_k a_k^x$$
(A.7)

is also a solution, where C_k are arbitrary constants. We can prove that Eq. (A.7) is also the general solution of Eq. (A.2). It is noted that for the case of multiple roots of a_i of degree m,

$$U(x) = a_i^x, xa_i^x, \ldots, x^{m-1}a_i^x$$
(A.8)

is also an independent solution.

Appendix B: Solution of the *A Priori* Probabilities of Eqs. (5.37) and (5.38)

Let $p_n(t)$ be the probability of n molecules arriving at a trapdoor at time t, and let λ be the average rate of molecules arriving at the trapdoor. Then for $n = 0$, the well-known Kolmogorov differential equation reduces to

$$\frac{\partial p(t)}{\partial t} = -\lambda p(t) \tag{B.1}$$

with the initial condition

$$p(0) = 0 \tag{B.2}$$

The solution is therefore

$$p(t) = e^{-\lambda t} \tag{B.3}$$

where $p(t)$ is the probability that no molecules will arrive at the trapdoor.

It is noted that $\bar{n} = \lambda \, \Delta t$, the average number of molecules arriving at the trapdoor per Δt. Thus, the *a priori* probability of p_2 is

$$p_2 = p(t) \big|_{t=\Delta t} = e^{-\bar{n}} \tag{B.4}$$

Hence,

$$p_1 = 1 - p_2 = 1 - e^{-\bar{n}} \tag{B.5}$$

Appendix C: Probability Energy Distribution

Let a physical measuring device or apparatus be maintained at a constant temperature of T degrees Kelvin. This temperature remains practically constant as long as a very small quantity of heat (compared with the total energy E_0 of the device) is involved. Let the entropy equation of the device be

$$S = k \ln N_0 \qquad (C.1)$$

where k is Boltzmann's constant, and N_0 is the amount of complexity of the device. Then we have

$$\frac{d \ln N_0(E_0)}{dE_0} = \frac{1}{kT} \qquad (C.2)$$

It is apparent that the solution of Eq. (C.2) is

$$N_0(E_0) = Ce^{E_0/kt} \qquad (C.3)$$

We now introduce a small quantity of heat (energy ΔE) into this device. It is assumed that the excess energy is very small by comparison with E_0, so that the system temperature T is not altered appreciably. However, this excess energy, although very small, causes certain small energy fluctuations within the device, such as

$$E_0 = E - \Delta E \qquad (C.4)$$

Thus, from Eq. (C.3) we have

$$N_0(E_0) = Ke^{-\Delta E/kt} \qquad (C.5)$$

where $K = Ce^{E_0/kt}$.

Since the most probable energy distribution is the one for which the complexity N of the device is maximum, we have

$$N = N_0(E_0)\Delta N(\Delta E) = \Delta N(\Delta E)Ke^{-\Delta E/kT} \qquad (C.6)$$

This equation shows that in the physical device, each of the small changes in complexity ΔN corresponds to a certain energy ΔE involved. Thus, the probability distribution of energy can be written as

$$p(\Delta E) = K \exp\left(-\frac{\Delta E}{kT} \right) \qquad (C.7)$$

where K is an arbitrary constant. It is also noted that Eq. (C.7) is essentially the well-known Gibbs formula in statistical thermodynamics.

Appendix D: Einstein's Energy Equation

Einstein's energy equation can be derived by his general theory of relativity, by which the mass of a particle can be written as

$$m = \frac{m_o}{\sqrt{(1 - v^2/c^2)}} = m_o(1 - v^2/c^2)^{-1/2} \qquad \text{(D.1)}$$

where m_o is the rest mass of the particle, v is its velocity, and c is the speed of light.

With reference to the binomial expansion, the above equation can be written as

$$m = m_o\left(1 + 1/2\frac{v^2}{c^2} + \text{terms of order } \frac{v^4}{c^4}\right) \qquad \text{(D.2)}$$

By multiplying the preceding equation with the velocity of light c^2 and noting that the terms with the orders of v^4/c^2 are negligibly small, the above equation can be approximated by

$$m \approx m_o + 1/2\,m_o v^2 \frac{1}{c^2} \qquad \text{(D.3)}$$

which can be written as

$$(m - m_o)c^2 \approx 1/2\,m_o v^2 \qquad \text{(D.4)}$$

The significance of the preceding equation is that $m - m_o$ represents an increase in mass due to motion, which is the kinetic energy, and $(m - m_o)c^2$ is the extra energy gain due to motion. What Einstein postulated is that there must be energy associated with the mass, even at rest. And this was exactly what he proposed:

$$\varepsilon \approx mc^2 \qquad \text{(D.5)}$$

where ε represents the total energy of the mass and

$$\varepsilon_o \approx m_o c^2 \qquad \text{(D.6)}$$

is the energy of the mass at rest, where $v = 0$ and $m = m_o$.

Appendix E: Gravitation and EM Field Interaction

With reference to Einstein's energy equation as given by

$$\frac{\partial \varepsilon}{\partial t} = c^2 \frac{\partial m}{\partial t} \qquad (E.1)$$

the rate of the gravitational field can be written as

$$\frac{\partial G}{\partial t} = K \frac{c^2}{r^2} \frac{\partial m}{\partial t} = \frac{K}{r^2} \frac{\partial \varepsilon}{\partial t} \qquad (E.2)$$

where K is a proportionality constant, and r is the distance from the mass. Thus, we see that there exists a profound relationship between a reduction in the gravitational field with respect to mass reduction.

The converted energy in the form of the electromagnetic (EM) field can be shown as

$$\varepsilon_{em} = \varepsilon_e + \varepsilon_m = \frac{1}{2}\epsilon E^2 + \frac{1}{2}\mu H^2 \qquad (E.3)$$

by which the *outflow* energy can be written as

$$-\frac{\partial}{\partial t}\left(\frac{1}{2}\epsilon E^2 + \frac{1}{2}\mu H^2\right)\Delta V = \int_\sigma S \cdot d\sigma \qquad (E.4)$$

where σ is the surface vector of the unit volume ΔV, S is the energy vector per unit area per unit time passing through the surface σ (i.e., surface of the unit volume ΔV), and the dot represents dot product operation.

By dividing the preceding equation by ΔV and taking the limit of ΔV approaching zero, we have

$$\frac{\partial \varepsilon_{em}}{\partial t} = -\frac{\partial}{\partial t}\left(\frac{1}{2}\epsilon E^2 + \frac{1}{2}\mu H^2\right) = \nabla \cdot S \qquad (E.5)$$

where $(\nabla \cdot)$ denotes the divergent operation, $S = E \times H$ is the energy *Poynting vector* from an EM energy radiator, and x denotes the cross product operation.

Thus, we see that there exists a relationship between the mass reduction rate and the rate of outflow energy as given by

$$\frac{\partial m}{\partial t} = \frac{1}{c^2}\frac{\partial \varepsilon}{\partial t} = \frac{1}{c^2}(\nabla \cdot S) = -\frac{1}{c^2}\frac{\partial}{\partial t}\left(\frac{1}{2}\epsilon E^2 + \frac{1}{2}\mu H^2\right) \qquad (E.6)$$

And the corresponding gravitational field reduction can be written as

$$\frac{\partial g}{\partial t} = \frac{1}{4\pi\gamma r^2 c^2}\nabla \cdot (E \times H) = \frac{1}{4\pi\gamma r^2 c^2}(\nabla \cdot S) \qquad (E.7)$$

Although gravitational field strength is small by the inverse square-law constraint, in the cosmic environment, the *diverging energy flow* can be very large due to galaxy collision or neutron star explosion.

The general solution to include all possible converted energy densities can be written as

$$\varepsilon_{em}(v;t) = \varepsilon_e(v;t) + \varepsilon_m(v;t) = \frac{1}{2}\epsilon E^2(v;t) + \frac{1}{2}\mu H^2(v;t) \qquad (E.8)$$

And the total energy converted from the mass is given by

$$\varepsilon_t = \int dv \int \varepsilon_{em}(v;t)\,dt = mc^2 \qquad (E.9)$$

Similarly, we have

$$\frac{\partial m(v;t)}{\partial t} = \frac{1}{c^2}\frac{\partial \varepsilon(v;t)}{\partial t} = -\frac{1}{c^2}\frac{\partial}{\partial t}\left(\frac{1}{2}\epsilon E^2(v;t) + \frac{1}{2}\mu H^2(v;t)\right)$$
$$= \frac{1}{c^2}\nabla \cdot S(v;t) \qquad (E.10)$$

And the corresponding gravitational field change is

$$\frac{\partial g(v;t)}{\partial t} = \frac{1}{4\pi\gamma r^2 c^2}[\nabla \cdot S(v;t)] \qquad (E.11)$$

The overall gravitational field reduction by the entire radiation bandwidth Δv can be evaluated as follows:

$$\frac{\partial}{\partial t}\int_{\Delta v} g(v;t)\,dv = \frac{1}{4\pi\gamma r^2 c^2}\int_{\Delta v} \nabla \cdot S(v;t)\,dv \qquad (E.12)$$

As for reversing energy back into the mass case, we have

$$\frac{\partial m}{\partial t} = -\frac{1}{c^2}\frac{\partial \varepsilon}{\partial t} = \frac{1}{c^2}\frac{\partial}{\partial t}\left(\frac{1}{2}\epsilon E^2 + \frac{1}{2}\mu H^2\right) = -\frac{1}{c^2}(\nabla \cdot S) \qquad (E.13)$$

The major difference is that energy is *inflow* into the unit volume ΔV, instead of outflow, as can be seen by the following equation:

$$\frac{\partial \varepsilon}{\partial t} = -c^2 \frac{\partial m}{\partial t} = \frac{\partial}{\partial t}\left(\frac{1}{2}\epsilon E^2 + \frac{1}{2}\mu H^2\right) = -(\nabla \cdot S) \quad (E.14)$$

where $(-\nabla\cdot)$ represents a *convergent operation* in which the negative sign represents the inflow energy into a unit volume ΔV.

As for a general solution, we have

$$\frac{\partial m(v;t)}{\partial t} = \frac{1}{c^2}\frac{\partial \varepsilon(v;t)}{\partial t} = \frac{1}{c^2}\frac{\partial}{\partial t}\left(\frac{1}{2}\epsilon E^2(v;t) + \frac{1}{2}\mu H^2(v;t)\right)$$
$$= -\frac{1}{c^2}\nabla \cdot S(v;t) \quad (E.15)$$

And the amount of energy needed for the energy-to-mass conversion is

$$\varepsilon = \int dt \int [-\nabla \cdot S(v;t)]\, dv = mc^2 \quad (E.16)$$

which equals the famous energy equation of Einstein. In practice, however, the amount of energy required would be very excessive.

Index